Electronic Amplifiers for Automatic Compensators

Electronic Amplifiers for Automatic Compensators

Contributors

Anamarija Juhas, Ladislav A. Novak et al.

AURIS
Reference

www.aurisreference.com

Electronic Amplifiers for Automatic Compensators

Contributors: Anamarija Juhas, Ladislav A. Novak et al.

Published by Auris Reference Limited
www.aurisreference.com

United Kingdom

Electronic Amplifiers for Automatic Compensators

ISBN: 978-1-78154-919-3

British Library Cataloguing in Publication Data
A CIP record for this book is available from the British Library

Printed in the United Kingdom

Exclusively distributed by CBS Publishers & Distributors Pvt. Ltd.

Sales & Distribution Rights only for India, Pakistan, Bangladesh, Sri Lanka, Nepal and Bhutan.This book is not to be sold outside these territories.

Contents

List of Abbreviations

ASK	Amplitude Shift Keying
AGC	Automatic gain control
BER	Bit error rate
CMOS	Complementary metal oxide semiconductor
CCTA	Current conveyor transconductance amplifier
CDBAs	Current differencing buffered amplifiers
CDTAs	Current differencing transconductance amplifiers
CFAs	Current feedback amplifiers
DWDM	Dense wavelength division multiplexed
DDCCTA	Differential difference current conveyor transconductance amplifier
DVCC	Differential voltage current conveyor
DPA	Doherty Power Amplifier
DRS	Double Rayleigh scattering
DO-CIBAs	Dual output current inverter buffered amplifiers
ECG	Electrocardiogram
EDF	Erbium doped fiber
EDFA	Erbium Doped Fiber Amplifier
ESA	Excited state absorption
FRA	Fiber Raman Amplifier
FWM	Four Wave Mixing processes
FSK	Frequency Shift Keying
GCFTA	Generalized current follower transconductance amplifier
GSA	Ground state absorption
IIN	Impedance Inverting Network
IBO	Input back-off
LSK	Load Shift Keying
LNA	Low noise amplifier
MRI	Magnetic Resonance Imaging
MISO	Multiple-input single-output
OOK	On-Off Keying
OTA	Operational transconductance amplifier
OTRAs	Operational transresistance amplifiers
OFDM	Orthogonal frequency division multiplexing
OBO	Output Back-Off
PAPR	Peak-to-average power ratio (PAPR)
PCN	Phase Compensation Network
PSK	Phase Shift Keying
PUF	Power utilization factor
PWM	Pulse width modulated
SOA	Semiconductor Optical Amplifier
SST	Shifted second tire
SOI	Silicon-on-insulator
SRCO	Single resistance controllable oscillators
SBS	Stimulated Brillouin Scattering
TDF	Thulium doped fiber
TDFA	Thulium Doped Fiber Amplifier

THD	Total harmonic distortion
TIA	Transimpedance amplifier
VDBA	Voltage differencing buffered amplifier
VDIBA	Voltage differencing inverted buffered amplifier
WDM	Wavelength multiplexing
ZDS	Zero derivative switching
ZVS	Zero voltage switching

List of Contributors

Anamarija Juhas
Department of Power, Electronics and Communication Engineering, Faculty of Technical Sciences, University of Novi Sad, 21000 Novi Sad, Serbia

Ladislav A. Novak
Department of Power, Electronics and Communication Engineering, Faculty of Technical Sciences, University of Novi Sad, 21000 Novi Sad, Serbia

Anthony N. Laskovski
The University of Newcastle Australia

Mehmet R. Yuce
The University of Newcastle Australia

Inderpreet Kaur
Rayat and Bahra Institute of Engineering, Mohali, India

Neena Gupta
PEC University of Technology (Formally Punjab Engineering College), Chandigarh, India

Paolo Colantonio
University of Roma Tor Vergata Italy

Franco Giannini
University of Roma Tor Vergata Italy

Rocco Giofrè
University of Roma Tor Vergata Italy

Luca Piazzon
University of Roma Tor Vergata Italy

Essra E. Al-Bayati
Department of Electronic and Communications Engineering, Al-Nahrain University, Baghdad, Iraq

R. S. Fyath
Department of Computer Engineering, Al-Nahrain University, Baghdad, Iraq

Roman SOTNER
Dept. of Radio Electronics, Brno University of Technology, Technicka 3082/12, 616 00 Brno, Czech Republic

Jan JERABEK
Dept. of Telecommunications, Brno University of Technology, Technicka 3082/12, 616 00 Brno, Czech Republic

Norbert HERENCSAR
Dept. of Telecommunications, Brno University of Technology, Technicka 3082/12, 616 00 Brno, Czech Republic

Neeta Pandey
Department of Electronics and Communication Engineering, Delhi Technological University, Delhi 110042, India

Sajal K. Paul
Department of Electronics Engineering, Indian School of Mines, Jharkhand 826004, India

Kevin Tom
Centre for Telecommunications and Microelectronics Victoria University, Melbourne-3011, Australia

Vandana Basoo
Centre for Telecommunications and Microelectronics Victoria University, Melbourne-3011, Australia

Mike Faulkner
Centre for Telecommunications and Microelectronics Victoria University, Melbourne-3011, Australia

Thomas Lejon
Ericsson AB, Business Unit Access, SE 16480, Sweden

Preface

An amplifier is an electronic device that increases the voltage, current, or power of a signal. Amplifiers are used in wireless communications and broadcasting, and in audio equipment of all kinds. The text *Electronic Amplifiers for Automatic Compensators* presents the design and operation of electronic amplifiers for use in automatic control and measuring systems. Various classes of nonnegative waveforms containing dc component, fundamental and kth harmonic (k ≥ 2), which proved to be of interest in waveform modelling for power amplifier (PA) design, are considered in first chapter. Second chapter provides a broad background in the development of biomedical engineering, and the recent contribution of electronics to this field. Third chapter discusses about hybrid fiber amplifier. Fundamentals of the Doherty power amplifier have been presented in fourth chapter. Fifth chapter presents a novel design scheme for distributed amplifiers (Das) suitable for frontend amplification in 40 and 100 Gb/s optical receivers. Sixth chapter deals with some interesting new applications in the field of analog signal processing focused on signal generation. The objective of seventh chapter is to propose a new active building block, namely, differential difference current conveyor transconductance amplifier (DDCCTA). In last chapter, we discuss a suitable scheme to characterize the linear amplification using nonlinear components (LINC) performance of class-E amplifier.

Chapter 1

CONFLICT SET AND WAVEFORM MODELLING FOR POWER AMPLIFIER DESIGN

Anamarija Juhas and Ladislav A. Novak

Department of Power, Electronics and Communication Engineering, Faculty of Technical Sciences, University of Novi Sad, 21000 Novi Sad, Serbia

ABSTRACT

Various classes of nonnegative waveforms containing dc component, fundamental and kth harmonic ($k \geq 2$), which proved to be of interest in waveform modelling for power amplifier (PA) design, are considered in this paper. In optimization of PA efficiency, nonnegative waveforms with maximal amplitude of fundamental harmonic and those with maximal coefficient of cosine term of fundamental harmonic (optimal waveforms) play an important role. Optimal waveforms have multiple global minima and this fact closely relates the problem of optimization of PA efficiency to the concept of conflict set. There is also keen interest in finding descriptions for various classes of suboptimal waveforms, such as nonnegative waveforms with at least one zero, nonnegative waveforms with maximal amplitude of fundamental harmonic for prescribed amplitude of kth harmonic, nonnegative waveforms with maximal coefficient of cosine part of fundamental harmonic for prescribed coefficients of kth harmonic, and nonnegative cosine waveforms with at least one zero. Closed form descriptions for all these suboptimal types of waveforms are provided in this paper. Suboptimal waveforms may also have multiple global minima and therefore be related to the concept of conflict set. Four case studies of usage of closed form descriptions of nonnegative waveforms in PA modelling are also provided.

INTRODUCTION

The origin of the concept of conflict set goes back to J. C. Maxwell (Maxwell 1831–1879), who informally introduced most of features of what today is called conflict set [1]. From this reason Maxwell set or Maxwell stratum is also used as synonyms for conflict set. Roughly speaking, conflict set associated with a smooth function f with m parameters is the set of m-tuples in parameter space for which f has multiple global minima. Conflict set is also intimately related to singularity theory and catastrophe theory [1].

Although without explicit reference, many max-min/ min-max engineering design problems related to nonsmooth optimizations in parameter spaces (e.g., see [2]), including problems related to the optimization of efficiency of power amplifiers (PAs) (e.g., see [3–12]), are connected to the concept of conflict set. The concept of conflict set has been also used in mathematics (e.g., see [13–16]) and physics (e.g., see [17, 18]), including subjects like black holes [19].

Nonnegative waveforms with maximal amplitude of fundamental harmonic and those with maximal coefficient of cosine term of fundamental harmonic (optimal waveforms) have multiple global minima and therefore are closely related to the concept of conflict set. The suboptimal waveforms such as

1. nonnegative waveforms with at least one zero,
2. nonnegative waveforms with maximal amplitude of fundamental harmonic for prescribed amplitude of kth harmonic,
3. nonnegative waveforms with maximal coefficient of cosine part of fundamental harmonic for prescribed coefficients of kth harmonic,
4. nonnegative cosine waveforms with at least one zero

may also have multiple global minima [9, 11, 12] and therefore be related to the concept of conflict set, as well. These suboptimal waveforms are clearly of interest in shaping/modellingdrain (collector/plate) waveforms in PA design (e.g., see [3– 12, 20, 21]). Fejer in his seminal paper [22] provided general description of all nonnegative trigonometric polynomials with n consecutive harmonics in terms of $2n + 2$ parameters satisfying one nonlinear constraint. He also derived closed form solution to the problem of finding maximum possible amplitude of the first harmonic of nonnegative cosine polynomials with consecutive harmonics. Fuzik [3] (see also [10]) considered cosine polynomials with dc, fundamental and kth harmonic, for arbitrary $k \geq 2$ and provided closed form solution for coefficients of optimal waveform. Rhodes in [7] provided closed form expression for maximum possible amplitude of fundamental harmonic of nonnegative waveforms containing consecutive odd harmonics. A subclass of nonnegative cosine waveforms with dc, fundamental and third harmonic, having factorized form description has been considered in [23].

High efficiency PA with arbitrary output harmonic terminations has been analysed in [9], along with maximal efficiency, fundamental output power, and load impedance.

Factorized form of nonnegative waveforms up to second harmonic with at least one zero has been suggested in [11] in the context of continuous class B/J mode of PA operation.

General description of all nonnegative waveforms up to second harmonic in terms of four independent parameters has been provided in [12]. This includes nonnegative waveforms with at least one zero, as a special case.

End point of conflict set normally corresponds to socalled maximally flat waveform, which also belongs to class of suboptimal waveforms. First comprehensive usage of maximally flat waveforms, in the context of analysis of PA, goes to Raab [20]. General description of maximally flat waveforms with arbitrary number of harmonics has been presented in [21], along with closed form expressions for efficiency of class-F and inverse class-F PA with maximally flat waveforms. Description of maximally flat cosine waveforms with consecutive harmonics has been presented in [8] in the context of finite harmonic class-C PA.

In this paper we provide general descriptions of a number of optimal and suboptimal nonnegative waveforms containing dc component, fundamental and an arbitrary kth harmonic, $k \geq 2$, and show how they are related to the concept of conflict set. According to our best knowledge, this paper provides the very first usage of conflict set in the course of solving problems related to optimization of PA efficiency. Main results are stated in six propositions (Propositions 1, 6, 9, 18, 22, and 26), four corollaries (Corollaries 2–5), twenty remarks, and three algorithms. Four case studies of usage of closed form descriptions of nonnegative waveforms in PA efficiency analysis are considered in detail in Section 7.

This paper is organized in the following way. In Section 2 we introduce concepts of minimum function and gain function (Section 2.1), conflict set (Section 2.2), and parameter space (Section 2.3). In Sections 3–6 we provide general descriptions of various classes of nonnegative waveforms containing dc component, fundamental and kth harmonic with at least one zero. General case of nonnegative waveforms with at least one zero is presented in Section 3.1.The case withexactly two zeros is considered in Section 3.2. An algorithm for calculation of coefficients of fundamental harmonic of nonnegative waveforms with two zeros, for prescribed coefficients of kth harmonic, is presented in Section 3.3. Description of nonnegative waveforms with maximal amplitude of fundamental harmonic for prescribed amplitude of kth harmonic is provided in Section 4. Nonnegative waveforms with maximal coefficient of cosine part of fundamental harmonic for prescribed coefficients of kth harmonic are considered in Section 5.1. An illustration of results of Section 5.1 for particular case $k=3$ is given in Section 5.2. Section 6.1 is devoted to nonnegative cosine waveforms with at least one zero and arbitrary $k \geq 2$, whereas Section 6.2 considers cosine waveforms with at least one zero for $k=3$. In Section 7 four case studies of application of descriptions of nonnegative waveforms with fundamental and kth harmonic in PA modelling are presented. In the Appendices, list of some finite sums of trigonometric functions, widely used throughout the paper, and brief account of the Chebyshev polynomials are provided.

Minimum Function, Gain Function, and Conflict Set

In this section we consider minimum function and gain function (Section 2.1), conflict set (Section 2.2), and parameter space (Section 2.3) in the context of nonnegative waveforms with fundamental and kth harmonic.

We start with provision of a brief account of the facts related to the concepts of minimum function and conflict set. For this purpose let us denote by $(x; u)$ a family of smooth functions of n variables depending on m parameters, where $x \in R^n$ is n-tuple of variables and $u \in R^m$ is m-tuple of parameters. The minimum function $F: R^m \to R$, associated with the function f, is defined as (u) $= \min x(x; u)$. Therefore, the domain of the minimum function is parameter space of the function f. The minimum function (u) is continuous, but not necessarily smooth function of parameters [13, 24]. It is a smooth function if $(x; u)$ possesses unique global minimum at nondegenerate critical point [13] (critical point is degenerate if at least first two consecutive derivatives are equal to zero). In this context, the conflict set can be defined as the set of the parameters for which function f has global minimum at a degenerate critical point or/and multiple global minima [13].

For a wide class of minimum functions, when the number of parameters is not greater than four, the behaviour of minimum function in a neighbourhood of any point can be described by one of "normal forms" from a finite list as stated in [24]. For example, for smooth function $f: R \times R^2 \to R$, the minimum function $(u_1, u_2) = \min_x(x, u_1, u_2)$ near the origin can be locally reduced to one of the following three normal forms [25]: $-|u_1|$, $\min(u_1, u_2, u_1 + u_2)$, or $\min x(x^4 + u_1 x^2 + u_2 x)$. In this example, the conflict set is the set of all points (u_1, u_2) for which minimum function (u_1, u_2) is not differentiable because function (x, u_1, u_2) possesses at least two global minima [25].

Minimum Function and Gain Function. In what follows We Consider Family of Waveforms of Type

$$w(\tau; \gamma, A, \alpha) = 1 - \gamma(\cos \tau + A \cos(k\tau + \alpha)),$$

(1)

where τ stands for ωt, $\gamma > 0$, $k \geq 2$, $A > 0$, and $\alpha \in [0, 2\pi)$. Waveforms of type (1) include all possible shapes which can occur, but not all possible waveforms containing fundamental and kth harmonic. However, shifting of waveforms of type (1) along τ-axis could recover all possible waveforms with fundamental and kth harmonic.

The problem of finding nonnegative waveform of type (1) having maximum amplitude of fundamental harmonic plays an important role in optimization of PA efficiency. This extremal problem can be reformulated as problem of finding nonnegative waveform from family (1) having maximum possible value of coefficient γ. Nonnegative waveform of family (1) with maximum possible value of coefficient γ is called "optimal" or "extremal" waveform.

Furthermore, let us introduce an auxiliary waveform

$$f(\tau; A, \alpha) = -\cos \tau - A \cos(k\tau + \alpha),$$

(2)

which is smooth function of one variable τ and two parameters A and α. In terms of $(\tau; A, \alpha)$, the above extremal problem reduces to the problem of finding maximum possible value of coefficient γ that satisfies

$$1 + \gamma f(\tau; A, \alpha) \geq 0.$$

(3)

Clearly, for any prescribed pair (A, α), there is a unique maximal value of coefficient γ for which inequality (3) holds for all τ. This maximal value of γ associated with the pair (A, α) we denote it by (A, α) and call it "gain function."

Let

$$F_{\min}(A, \alpha) = \min_{\tau} f(\tau; A, \alpha)$$

(4)

be the minimum function associated with $f(\tau; A, \alpha)$. According to (3), (A, α) and $F_{\min}(A, \alpha)$ satisfy the following relation: $1 + G(A, \alpha)F_{\min}(A, \alpha) = 0$. Since $F_{\min}(A, \alpha)$ is obviously nonzero it follows immediately that

$$G(A, \alpha) = -\frac{1}{F_{\min}(A, \alpha)}.$$

(5)

A relation analogue to (5), for $k=2$ (fundamental and second harmonic), has been derived in [4]. According to our best knowledge, it was the first appearance of gain function expressed via associated minimum function. The consideration presented in [4] has been restricted to the particular case when $\alpha = \pi$. The same problem for $\alpha = \pi$ and arbitrary $k \geq 2$ has been investigated in [3] (see also [10]).

According to above consideration, the problem of finding 3-tuple (γ, A, α) with maximum possible value of γ for which (3) holds is equivalent to the problem of finding maximum value of gain function

$$\gamma_{\max} = \max_{A,\alpha} G(A,\alpha) = -\frac{1}{\max_{A,\alpha} F_{\min}(A,\alpha)}$$

(6)

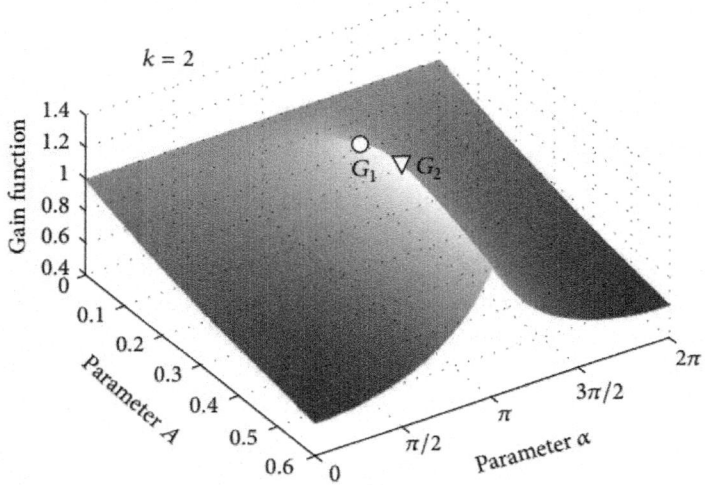

Figure 1: Graph of (A, α) for $k=2$. Points G_1 and G_2 denote beginning of the ridge and maximum of gain function, respectively.

and corresponding pair (A^*, α^*) that satisfies

$$\gamma_{\max} = \max_{A,\alpha} G(A,\alpha) = G(A^*,\alpha^*),$$

$$\max_{A,\alpha} F_{\min}(A,\alpha) = F_{\min}(A^*,\alpha^*).$$

(7)

Thus the optimal waveform $w^*(\tau)$ is determined by parameters $\gamma_{\max}$, A^*, and α^*; that is,

$$w^*(\tau) = w(\tau; \gamma_{\max}, A^*, \alpha^*).$$

(8)

Optimal waveform has two global minima (this claim will be justified in Section 4, Remark 21). Consequently, the pair (A^*, α^*), which corresponds to maximum of gain function (A, α), belongs to conflict set in (A, α) parameter space.

Figure 1 shows graph of gain function (A, α) for $k=2$. Notice that it has sharp ridge and that maximum of gain function (point G_2) lies on the ridge. This maximum corresponds to the optimal waveform (solution of the considered extremal problem). The beginning of the ridge (point G_1) corresponds to the waveform which possesses global minimum at degenerate critical point, that is, corresponds to maximally flat waveform (e.g., see [21]). Gain function (A, α) is not differentiable on the ridge and consequently is not differentiable at the point where it has global maximum. This explains why the approach based on critical points does not work and why conflict set is so important in the considered problem.

Positions of global minima of $(\tau; A, \alpha)$ for $k=2$ are presented in Figure 2. According to Proposition 1, conflict set is the ray defined by $A > 1/4$ and $\alpha=\pi$. Waveforms$(\tau; A, \alpha)$ with parameters that belong to the conflict set have two global minima. The waveform corresponding to the end point of the ray $(A = 1/4$ and $\alpha=\pi)$ has global minimum at degenerate critical point (so-called maximally flat waveform [21]).

Nonnegative waveforms of type (1) with $\gamma = (A, \alpha) = -1/F_{min}(A, \alpha)$ have at least one zero. To show that, it is sufficient to see that $(\tau; \gamma, A, \alpha) = 0$ for τ satisfying $(\tau; A, \alpha) = F_{min}(A, \alpha)$.

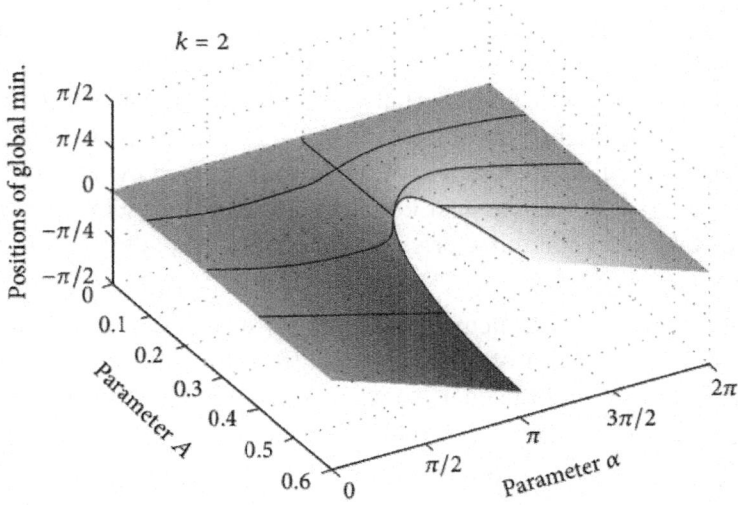

Figure 2: Positions of global minima of $(\tau; A, \alpha)$ for $k=2$.

The problem of finding maximum value of fundamental harmonic cosine part of nonnegative waveform of the form

$$w_a(\tau; \gamma_a, b, A, \alpha) = 1 - \gamma_a(\cos\tau + b\sin\tau + A\cos(k\tau + \alpha)), \tag{9}$$

where $\gamma_a > 0$, is also related to the problem of finding maximum of the minimum function. Optimal waveform of family (9) has two global minima (this claim will be justified in Section 5, Remark 25), and therefore corresponding 3-tuple of parameters belongs to the conflict set in parameter space of family (9).

Let us introduce an auxiliary waveform

$$f_a(\tau; b, A, \alpha) = -\cos\tau - b\sin\tau - A\cos(k\tau + \alpha),$$
(10)

and corresponding minimum function $F_{a,\min}(b, A, \alpha) = \min_\tau f_a(\tau; b, A, \alpha)$. Inequality $w_a(\tau; \gamma_a, b, A, \alpha) \geq 0$ can be rewritten as $1 + \gamma_a F_{a,\min}(b, A, \alpha) \geq 0$ and therefore the highest value of γ_a is attained for $\gamma_a = -1/F_{a,\min}(b, A, \alpha)$. It immediately follows that nonnegative waveform of type (9) with $\gamma_a = -1/F_{a,\min}(b, A, \alpha)$ has zero for τ satisfying $f_a(\tau; b, A, \alpha) = F_{a,\min}(b, A, \alpha)$.

Conflict Set

Historically, conflict set came into being from the problems in which families of smooth functions (such as potentials, distances, and waveforms) with two competing minima occur. The situation when competing minima become equal refers to the presence of conflict set (Maxwell set, Maxwell strata) in the associated parameter space.

There are many facets of conflict set. For example, in the problem involving distances between two sets of points, the conflict set is the intersections between iso-distance lines [14]. Conflict set also arises in the situation when two wave fronts coming from different objects meet [15, 25]. In the study of black holes, conflict set is the line of crossover of the horizon formed by the merger of two black holes [19]. In the classical Euler problem, conflict set is a set of points where distinct extremal trajectories with the same value of the cost functional meet one another [18].

Conflict set is very difficult to calculate, both analytically and numerically (e.g., see [15]), because of apparent nondifferentiability in some directions. In optimization of PA efficiency, some authors already reported difficulties in finding optimum via standard analytical tools [4, 5].

In this section, we consider conflict set in the context of family of waveforms of type (2) for arbitrary $k \geq 2$. In this context, for prescribed integer $k \geq 2$, conflict set is said to be a set of all pairs (A, α) for which $f(\tau; A, \alpha)$ possesses multiple global minima.

Suppose that τ' and τ'' are the positions of global minima of $(\tau; A, \alpha)$. Then, the conflict set is specified by the following set of relations:

$$f\left(\tau'; A, \alpha\right) = f\left(\tau''; A, \alpha\right),$$

(11)

$$f'\left(\tau'; A, \alpha\right) = 0, \qquad f'\left(\tau''; A, \alpha\right) = 0,$$

(12)

$$f''\left(\tau'; A, \alpha\right) > 0, \qquad f''\left(\tau''; A, \alpha\right) > 0,$$

(13)

$$(\forall \tau)\, f\left(\tau; A, \alpha\right) \geq f\left(\tau'; A, \alpha\right).$$

(14)

Relations (12) and (13) say that $(\tau; A, \alpha)$ has minima at τ' and τ'', while relations (11) and (14) imply that these minima are equal and global.

The following proposition describes the conflict set of family of waveforms of type (2).

Proposition 1: Conflict set of family of waveforms of type (2) is the set of all pairs (A, α) such that $A > 1/k2$ and $\alpha = \pi$.

The proof of Proposition 1, which is provided at the end of this section, also implies that the following four corollaries hold.

Corollary 2: The conflict set has end point at $(A, \alpha) = (1/k^2, \pi)$. This end point corresponds to the maximally flat waveform [21].

Corollary 3: Waveforms of type (2) with parameters that belong to conflict set have two global minima at $\pm\tau_\Delta$, where $0 < \tau_\Delta \leq \pi/k$.

Corollary 4: Every waveform with fundamental and kth harmonic has either one or two global minima.

Corollary 5: Conflict set can be parameterised in terms of τ_Δ as follows:

$$\alpha = \pi, \quad A\left(\tau_\Delta\right) = \frac{\sin \tau_\Delta}{k \sin k\tau_\Delta}, \quad 0 < \tau_\Delta \leq \frac{\pi}{k}.$$

(15)

Notice that (τ_Δ) is monotonically increasing function on interval $0 < \tau_\Delta \leq \pi/k$.

Proof of Proposition 1: Without loss of generality, we can restrict our consideration to the interval $-\pi < \tau \leq \pi$. This is an immediate consequence of the fact that $(\tau; A, \alpha)$ is a periodic function.

Suppose that τ' and τ'', where $\tau' < \tau''$, are points at which $(\tau; A, \alpha)$ has two equal global minima. Then conflict set is specified by relations (11)–(14). From (11)–(13) it follows that relations

$$f\left(\tau'; A, \alpha\right) - f\left(\tau''; A, \alpha\right) = 0,$$

$$f'\left(\tau'; A, \alpha\right) + f'\left(\tau''; A, \alpha\right) = 0,$$

$$f'\left(\tau'; A, \alpha\right) - f'\left(\tau''; A, \alpha\right) = 0,$$

$$f''\left(\tau'; A, \alpha\right) + f''\left(\tau''; A, \alpha\right) > 0 \tag{16}$$

also hold. Let

$$\tau_{sr} = \frac{\left(\tau' + \tau''\right)}{2}, \qquad \tau_\Delta = \frac{\left(\tau'' - \tau'\right)}{2} \tag{17}$$

be a pair of points associated with (τ', τ''). Clearly

$$-\pi < \tau_{sr} < \pi, \tag{18}$$

$$0 < \tau_\Delta < \pi, \tag{19}$$

$$\tau'' = \tau_{sr} + \tau_\Delta, \qquad \tau' = \tau_{sr} - \tau_\Delta. \tag{20}$$

The first and second derivatives of $(\tau; A, \alpha)$ are equal to

$$f'(\tau; A, \alpha) = \sin \tau + kA \sin(k\tau + \alpha),$$

$$f''(\tau; A, \alpha) = \cos \tau + k^2 A \cos(k\tau + \alpha). \tag{21}$$

By using (20)-(21), system (16) can be rewritten as

$$\sin \tau_{sr} \sin \tau_\Delta + A \sin(k\tau_{sr} + \alpha) \sin k\tau_\Delta = 0, \tag{22}$$

$$\sin \tau_{sr} \cos \tau_\Delta + kA \sin(k\tau_{sr} + \alpha) \cos k\tau_\Delta = 0, \tag{23}$$

$$\cos \tau_{sr} \sin \tau_\Delta + kA \cos(k\tau_{sr} + \alpha) \sin k\tau_\Delta = 0, \tag{24}$$

$$\cos \tau_{sr} \cos \tau_\Delta + k^2 A \cos(k\tau_{sr} + \alpha) \cos k\tau_\Delta > 0. \tag{25}$$

From (19) it follows that $\sin \tau_\Delta > 0$. Multiplying (24) and (25) with $-\cos \tau_\Delta$ and $\sin \tau_\Delta > 0$, respectively, and summing the resulting relations, we obtain $kA \cos(k\tau_{sr} + \alpha)[k \sin \tau_\Delta \cos k\tau_\Delta - \sin k\tau_\Delta \cos \tau_\Delta] > 0$. The latest relation immediately implies that

$$k \sin \tau_\Delta \cos k\tau_\Delta - \sin k\tau_\Delta \cos \tau_\Delta \neq 0.$$

$$(26)$$

Equations (22) and (23) can be considered as a system of two linear equations in terms of $\sin \tau_{sr}$ and $A \sin(k\tau_{sr} + \alpha)$. According to (26), the determinant of this system is nonzero and therefore it has only trivial solution:

$$\sin \tau_{sr} = 0, \qquad \sin \left(k\tau_{sr} + \alpha \right) = 0.$$

$$(27)$$

According to (18), $\sin \tau_{sr} = 0$ implies

$$\tau_{sr} = 0.$$

$$(28)$$

According to (20), $\tau_{sr} = 0$ implies

$$\tau'' = \tau_\Delta, \qquad \tau' = -\tau_\Delta.$$

$$(29)$$

Furthermore, $\tau_{sr} = 0$ and $\sin(k\tau_{sr} + \alpha) = 0$ imply that $\sin \alpha = 0$. From (29) it follows that τ_Δ is position of global minimum of $(\tau; A, \alpha)$. Clearly $(\tau_\Delta; A, \alpha) \leq (0; A, \alpha)$, which together with $\sin \alpha = 0$ leads to

$$1 - \cos \tau_\Delta + A \cos \alpha \left(1 - \cos k\tau_\Delta \right) \leq 0.$$

$$(30)$$

From $\tau\Delta = 0$ (see (19)), $A>0$ and (30) it follows that $\cos \alpha < 0$, which together with $\sin \alpha = 0$ yields

$$\alpha = \pi.$$

$$(31)$$

Since τ_Δ is position of global minimum, it follows that $(\tau_\Delta; A, \pi) \leq (\pi/k; A, \pi)$. Accordingly $(1 + \cos k\tau_\Delta) \leq \cos \tau_\Delta - \cos \pi/k$, which together with $A>0$ implies that $\cos \tau_\Delta - \cos \pi/k \geq 0$. This relation along with (19) yields

$$0 < \tau_\Delta \leq \frac{\pi}{k}.$$

$$(32)$$

Substitution of (31) and (28) in (24) leads to

$$A = \frac{\sin \tau_\Delta}{k \sin k\tau_\Delta}.$$

$$(33)$$

Notice that $\sin k\tau_\Delta / \sin \tau_\Delta$ is monotonically decreasing function on interval (32). Therefore parameter A is monotonically increasing function on the same interval with $\lim_{\tau\Delta \to 0+} A = 1/k^2$. Consequently $A > 1/k^2$, which completes the proof.

Parameter Space

In parameter space of family of waveforms (2) there are two subsets playing important role in the classification of the family instances.These are conflict set and catastrophe set.

Catastrophe set is subset of parameter space of waveform $(\tau; A, \alpha)$. It consists of those pairs (A, α) for which the corresponding waveforms $(\tau; A, \alpha)$ have degenerate critical points at which first and second derivatives are equal to zero. Thus, for finding catastrophe set we have to consider the following system of equations:

$$f'\left(\tau_d; A, \alpha\right) = 0,$$

$$f''\left(\tau_d; A, \alpha\right) = 0,$$

(34)

where τ_d is a degenerate critical point of waveform $f(\tau; A, \alpha)$. Conflict set in parameter space of waveform $(\tau; A, \alpha)$, as shown in Proposition 1, is the ray described by $A > 1/k^2$ and $\alpha=\pi$. It is intimately connected to catastrophe set.

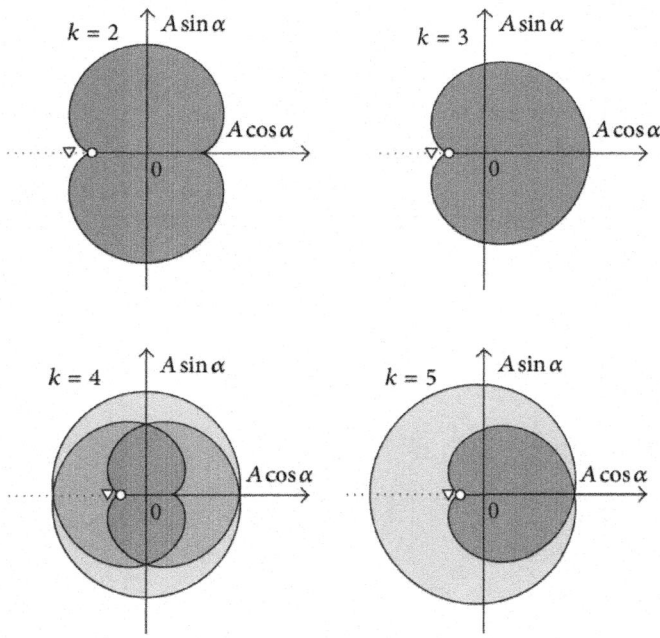

Figure 3: Catastrophe set (solid line) and corresponding conflict set (dotted line) for $k \leq 5$. In each plot, white triangle dot corresponds to optimal waveform and white circle dot corresponds to maximally flat waveform.

In what follows in this subsection we use polar coordinate system (A cos α, A sin α) instead of Cartesian coordinate system (A, α). Examples of catastrophe set and conflict set for $k \leq 5$ plotted in parameter space (A cos α, A sin α) are presented in Figure 3. Solid line represents the catastrophe set while dotted line describes conflict set. The isolated pick points (usually called cusp) which appear in catastrophe curves correspond to maximally flat waveforms, with maximally flat minimum and/or maximally flat maximum. There are two such picks in the catastrophe curves for k=2 and k=4 and one in the catastrophe curves for k=3 and k=5. Notice that the end point of conflict set is the cusp point.

Catastrophe set divides the parameter space (A cos α, A sin α) into disjoint subsets. In the cases k=2 and k = 3 catastrophe curve defines inner and outer part. For $k > 3$ catastrophe curve makes partition of parameter space in several inner subsets and one outer subset (see Figure 3).

Notice also that multiplying $f(\tau; A, \alpha)$ with a positive constant and adding in turn another constant, which leads to waveform of type $w(\tau; \gamma, A, \alpha)$ (see (1) and (2)), do not make impact on the character of catastrophe and conflict sets. This is because in the course of finding catastrophe set first and second derivatives of $(\tau; A, \alpha)$ are set to zero. Clearly (34) in terms of $(\tau; A, \alpha)$ are equivalent to the analogous equations in terms of $(\tau; \gamma, A, \alpha)$. Analogously, in the course of finding conflict set we consider only the positions of global minima (these positions for waveforms $(\tau; A, \alpha)$ and $w(\tau; \gamma, A, \alpha)$ are the same).

NONNEGATIVE WAVEFORMS WITH AT LEAST ONE ZERO

In what follows let us consider a waveform containing dc component, fundamental and kth ($k \geq 2$) harmonic of the form

$$T_k(\tau) = 1 + a_1 \cos \tau + b_1 \sin \tau + a_k \cos k\tau + b_k \sin k\tau. \tag{35}$$

The amplitudes of fundamental and kth harmonic of waveform of type (35), respectively, are

$$\lambda_1 = \sqrt{a_1^2 + b_1^2}, \tag{36}$$

$$\lambda_k = \sqrt{a_k^2 + b_k^2}. \tag{37}$$

As it is shown in Section 2.1, nonnegative waveforms with maximal amplitude of fundamental harmonic or maximal coefficient of fundamental harmonic cosine part have at least one zero. It is also shown in Section 2.2 (Corollary 4) that waveforms of type (35) with nonzero amplitude of fundamental harmonic have either one or two global minima. Consequently, if nonnegative waveform of type (35) with nonzero amplitude of fundamental harmonic has at least one zero, then it has at most two zeros.

In Section 3.1 we provide general description of nonnegative waveforms of type (35) with at least one zero. In Sections 3.2 and 3.3 we consider nonnegative waveforms of type (35) with two zeros.

General Description of Nonnegative Waveforms with at Least One Zero

The main result of this section is presented in the following proposition.

Proposition 6: Every nonnegative waveform of type (35) with at least one zero can be expressed in the following form:

$$T_k(\tau) = \left[1 - \cos\left(\tau - \tau_0\right)\right]\left[1 - \lambda_k r_k(\tau)\right],$$

(38)

where

$$r_k(\tau) = (k-1)\cos\xi$$

$$+ 2\sum_{n=1}^{k-1}(k-n)\cos\left(n\left(\tau - \tau_0\right) + \xi\right),$$

(39)

providing that

$$\lambda_k \leq \left[(k-1)\cos\xi + k\frac{\sin(\xi - \xi/k)}{\sin(\xi/k)}\right]^{-1},$$

(40)

$$|\xi| \leq \pi.$$

(41)

Remark 7: Function on the right hand side of (40) is monotonically increasing function of $|\xi|$ on interval $|\xi| \leq \pi$ (for more details about this function see Remark 15). From (57) and (65) it follows that relation

$$0 \leq \lambda_k \leq 1$$

(42)

holds for every nonnegative waveform of type (35). Notice that, according to (40), $\lambda_k = 1$ implies $|\xi| = \pi$. Substitution of $\lambda_k = 1$ and $|\xi| = \pi$ into (55) yields $(\tau) = 1 - \cos k(\tau - \tau_0)$. Consequently, $\lambda_k = 1$ implies that amplitude λ_1 of fundamental harmonic is equal to zero.

Remark 8: Conversion of (38) into additive form leads to the following expressions for coefficients of nonnegative waveforms of type (35) with at least one zero:

$$a_1 = -\left(1 + \lambda_k \cos \xi\right) \cos \tau_0 - k\lambda_k \sin \xi \sin \tau_0, \tag{43}$$

$$b_1 = -\left(1 + \lambda_k \cos \xi\right) \sin \tau_0 + k\lambda_k \sin \xi \cos \tau_0, \tag{44}$$

$$a_k = \lambda_k \cos \left(k\tau_0 - \xi\right), \tag{45}$$

$$b_k = \lambda_k \sin \left(k\tau_0 - \xi\right), \tag{46}$$

providing that λ_k satisfy (40) and $|\xi| \leq \pi$.

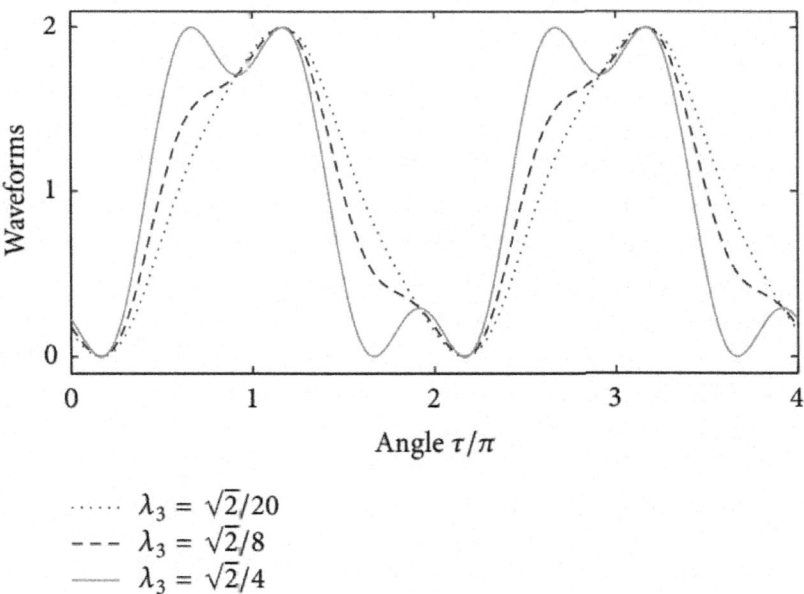

$$\cdots\cdots \quad \lambda_3 = \sqrt{2}/20$$
$$--- \quad \lambda_3 = \sqrt{2}/8$$
$$\longrightarrow \quad \lambda_3 = \sqrt{2}/4$$

Figure 4: Nonnegative waveforms with at least one zero for $k=3$, $\tau_0 = \pi/6$, and $\xi = 3\pi/4$.

Three examples of nonnegative waveforms with at least one zero for $k=3$ are presented in Figure 4 (examples of nonnegative waveforms with at least one zero for $k=2$ can be found in [12]). For all three waveforms presented in Figure 4, we assume that $\tau_0 = \pi/6$ and $\xi = 3\pi/4$. From (40) it follows that $\lambda_3 \leq \sqrt{2}/4$. Coefficients of waveform with $\lambda_3 = \sqrt{2}/20$ (dotted line) are $a_1 = -0.8977$, $b_1 = -0.3451$, $a3 = 0.05$, and $b3 = -0.05$. Coefficients of waveform with $\lambda_3 = \sqrt{2}/8$ (dashed line) are $a_1 = -0.9453$, $b_1 = -0.1127$, $a_3 = 0.125$, and $b_3 = -0.125$.

Coefficients of waveform with $\lambda_3 = \sqrt{2}/4$ (solid line) are $a_1 = -1.0245$, $b_1 = 0.2745$, $a_3 = 0.25$, and $b_3 = -0.25$. First two waveforms have one zero, while third waveform (presented with solid line) has two zeros.

Proof of Proposition 6: Waveform of type (35), containing dc component, fundamental and kth harmonic, can be also expressed in the form

$$T_k(\tau) = 1 + \lambda_1 \cos(\tau + \varphi_1) + \lambda_k \cos(k\tau + \varphi_k),$$

$$(47)$$

where $\lambda_1 \geq 0$, $\lambda_k \geq 0$, $\varphi_1 \in (-\pi, \pi]$, and $\varphi_k \in (-\pi, \pi]$. It is easy to see that relations between coefficient of (35) and parameters of (47) read as follows:

$$a_1 = \lambda_1 \cos\varphi_1, \qquad b_1 = -\lambda_1 \sin\varphi_1,$$

$$(48)$$

$$a_k = \lambda_k \cos\varphi_k, \qquad b_k = -\lambda_k \sin\varphi_k.$$

$$(49)$$

Let us introduce ξ such that

$$|\xi| \leq \pi, \quad \xi = (k\tau_0 + \varphi_k) \bmod 2\pi.$$

$$(50)$$

Using (50), coefficients (49) can be expressed as (45)-(46).

Let us assume that (τ) is nonnegative waveform of type (35) with at least one zero; that is, $T_k(\tau) \geq 0$ and $T_k(\tau_0) = 0$ for some τ_0. Notice that conditions $(\tau) \geq 0$ and $T_k(\tau_0) = 0$ imply that $T_k'(\tau_0)(\tau_0) = 0$. From $(\tau_0) = 0$ and $T_k'(\tau_0)(\tau_0) = 0$, by using (50), it follows that

$$\lambda_1 \cos(\tau_0 + \varphi_1) = -(1 + \lambda_k \cos\xi),$$

$$\lambda_1 \sin(\tau_0 + \varphi_1) = -k\lambda_k \sin\xi,$$

$$(51)$$

respectively. On the other hand, $\lambda_1 \cos(\tau + \varphi_1)$ can be rewritten as

$$\lambda_1 \cos(\tau + \varphi_1) = \lambda_1 \cos(\tau_0 + \varphi_1) \cos(\tau - \tau_0)$$

$$- \lambda_1 \sin(\tau_0 + \varphi_1) \sin(\tau - \tau_0).$$

$$(52)$$

Substitution of (51) into (52) yields

$$\lambda_1 \cos(\tau + \varphi_1) = -(1 + \lambda_k \cos\xi) \cos(\tau - \tau_0)$$

$$+ k\lambda_k \sin\xi \sin(\tau - \tau_0).$$

$$(53)$$

According to (50), it follows that $\cos(k\tau + \varphi_k) = \cos(k(\tau - \tau_0) + \xi)$; that is,

$$\cos(k\tau + \varphi_k) = \cos\xi \cos k(\tau - \tau_0) - \sin\xi \sin k(\tau - \tau_0).$$

$$(54)$$

Furthermore, substitution of (54) and (53) into (47) leads to

$$T_k(\tau) = \left[1 - \cos(\tau - \tau_0)\right]\left[1 + \lambda_k \cos\xi\right]$$

$$- \lambda_k \left[1 - \cos k(\tau - \tau_0)\right]\cos\xi$$

$$+ \lambda_k \left[k\sin(\tau - \tau_0) - \sin k(\tau - \tau_0)\right]\sin\xi.$$

$$(55)$$

According to (A.2) and (A.4) (see Appendices), there is common factor$[1-\cos(\tau-\tau_0)]$for all terms in (55). Consequently, (55) can be written in the form (38), where

$$r_k(\tau) = -\cos\xi + \left[\frac{1 - \cos k(\tau - \tau_0)}{1 - \cos(\tau - \tau_0)}\right]\cos\xi$$

$$- \left[\frac{k\sin(\tau - \tau_0) - \sin k(\tau - \tau_0)}{1 - \cos(\tau - \tau_0)}\right]\sin\xi.$$

$$(56)$$

From (56), by using (A.2), (A.4), and $\cos\xi\cos(\tau - \tau_0) - \sin\xi\sin(\tau - \tau_0) = \cos(n(\tau - \tau_0) + \xi)$, we obtain (39).

In what follows we are going to prove that (40) also holds. According to (38), (τ) is nonnegative if and only if

$$\lambda_k \max_\tau r_k(\tau) \le 1.$$

$$(57)$$

Let us first show that position of global maximum of (τ) belongs to the interval $|\tau - \tau_0| \le 2\pi/k$. Relation (56) can be rewritten as

$$r_k(\tau) = r_k\left(\tau_0 - \frac{2\xi}{k}\right) + q_k(\tau),$$

$$(58)$$

Where

$$r_k\left(\tau_0 - \frac{2\xi}{k}\right) = (k - 1)\cos\xi + k\frac{\sin(\xi - \xi/k)}{\sin(\xi/k)},$$

$$(59)$$

$$q_k(\tau) = \frac{1}{1 - \cos(\tau - \tau_0)}$$

$$\cdot \left[\cos\xi - \cos\left(k\left(\tau - \tau_0 + \frac{\xi}{k}\right)\right)\right.$$

$$\left. + \frac{k\sin\xi}{\sin(\xi/k)}\left(\cos\left(\tau - \tau_0 + \frac{\xi}{k}\right) - \cos\left(\frac{\xi}{k}\right)\right)\right].$$

$$(60)$$

For$|\xi| < \pi$, relation $\sin \xi / \sin(\xi/k) > 0$ obviously holds. From $\cos t > \cos t'$ for $|t| \leq \pi/k < |t'| \leq \pi$ it follows that position of global maximum of the function of type $[c \cos t - \cos(kt)]$ for $c>0$ belongs to interval $|t| \leq \pi/k$. Therefore position of global maximum of the expression in the square brackets in (60) for $|\xi| < \pi$ belongs to interval $|\tau - \tau_0 + \xi/k| \leq \pi/k$. This inequality together with $|\xi| < \pi$ leads to $|\tau - \tau 0| < 2\pi/k$. Since $[1 - \cos(\tau - \tau_0)]^{-1}$ decreases with increasing $|\tau - \tau_0| \leq \pi$, it follows that $qk(\tau)$ for $|\xi| < \pi$ has global maximum on interval $|\tau - \tau_0| < 2\pi/k$. For$|\xi| = \pi$, it is easy to show that $\max_\tau (\tau) = qk(\tau_0 \pm 2\pi/k) = 0$. Since $(\tau) - qk(\tau)$ is constant (see (58)), it follows from previous considerations that $r_k(\tau)$ has global maximum on interval $|\tau - \tau_0| \leq 2\pi/k$.

To find $\max_\tau (\tau)$, let us consider first derivative of $r_k(\tau)$ with respect to τ. Starting from (56), first derivative of (τ) can be expressed in the following form:

$$\frac{dr_k(\tau)}{d\tau} = -s(\tau) \cdot \sin\left(\frac{k(\tau - \tau_0)}{2} + \xi\right),$$

(61)

where

$$s(\tau) = \left[\sin\left(\frac{k(\tau - \tau_0)}{2}\right) \cos\left(\frac{\tau - \tau_0}{2}\right)\right.$$
$$\left. - k \cos\left(\frac{k(\tau - \tau_0)}{2}\right) \sin\left(\frac{\tau - \tau_0}{2}\right)\right]$$
$$\cdot \sin^{-3}\left(\frac{\tau - \tau_0}{2}\right).$$

(62)

Using (A.6) (see Appendices), (62) can be rewritten as

$$s(\tau) = 2 \sum_{n=1}^{k-1} n(k - n) \cos\left(\frac{(k - 2n)(\tau - \tau_0)}{2}\right).$$

(63)

From $(k - n) > 0$ and $|k - 2n| < k$, $n = 1, \ldots, (k - 1)$, it follows that all summands in (63) decrease with increasing $|\tau - \tau_0|$ providing that $|\tau - \tau_0| \leq 2\pi/k$. Therefore $(\tau) \geq (\tau_0 \pm 2\pi/k) = k/\sin^2(\pi/k) > 0$ for $|\tau - \tau_0| \leq 2\pi/k$. Consequently, $d(\tau)/d\tau = 0$ and $|\tau - \tau 0| \leq 2\pi/k$ imply that $\sin(k(\tau - \tau_0)/2 + \xi) = 0$.

From $|\xi| \leq \pi$, $|\tau - \tau_0| \leq 2\pi/k$, and $\sin(k(\tau - \tau_0)/2 + \xi) = 0$ it follows that $\tau - \tau 0 + \xi/k = -\xi/k$ or$|\tau - \tau_0 + \xi/k| = (2\pi - |\xi|)/k$, and therefore $\cos(k(\tau - \tau_0 + \xi/k)) = \cos \xi$. Since $\cos(\xi/k) \geq \cos(2\pi - |\xi|)/k$, it follows that $\max_\tau q_k(\tau)$ is attained for $\tau = \tau_0 - 2\xi/k$. Furthermore, from (60) it follows that $\max_\tau(\tau) = q_k(\tau_0 - 2\xi/k) = 0$, which together with (58)-(59) leads to

$$\max_{\tau} r_k(\tau) = r_k\left(\tau_0 - \frac{2\xi}{k}\right)$$

$$= (k-1)\cos\xi + k\frac{\sin(\xi - \xi/k)}{\sin(\xi/k)}. \tag{64}$$

Both terms on the right hand side of (64) are even functions of ξ and decrease with increase of $|\xi|$, $|\xi| \leq \pi$. Therefore, $\max_{\tau}(\tau)$ attains its lowest value for $|\xi| = \pi$. It is easy to show that right hand side of (64) for $|\xi| = \pi$ is equal to 1, which further implies that

$$\max_{\tau} r_k(\tau) \geq 1. \tag{65}$$

From (65), it follows that (57) can be rewritten as $\lambda_k \leq [\max_{\tau}(\tau)]^{-1}$. Finally, substitution of (64) into $\lambda_k \leq [\max_{\tau}(\tau)]^{-1}$ leads to (40), which completes the proof.

Nonnegative Waveforms with Two Zeros

Nonnegative waveforms of type (35) with two zeros always possess two global minima. Such nonnegative waveforms are therefore related to the conflict set.

In this subsection we provide general description of nonnegative waveforms of type (35) for $k \geq 2$ and exactly two zeros. According to Remark 7, $\lambda_k = 1$ implies $|\xi| = \pi$ and $(\tau) = 1 - \cos k(\tau - \tau_0)$. Number of zeros of $(\tau) = 1 - \cos k(\tau - \tau_0)$ on fundamental period equals k, which is greater than two for $k > 2$ and equal to two for $k = 2$. In the following proposition we exclude all waveforms with $\lambda_k = 1$ (the case when $k = 2$ and $\lambda_2 = 1$ is going to be discussed in Remark 10).

Proposition 9: Every nonnegative waveform of type (35) with exactly two zeros can be expressed in the following form:

$$T_k(\tau) = \lambda_k \left[1 - \cos(\tau - \tau_0)\right]\left[1 - \cos\left(\tau - \tau_0 + \frac{2\xi}{k}\right)\right]$$

$$\cdot \left[c_0 + 2\sum_{n=1}^{k-2} c_n \cos n\left(\tau - \tau_0 + \frac{\xi}{k}\right)\right], \tag{66}$$

where

$$c_n = \left[\sin\left(\xi - \frac{n\xi}{k}\right) \cos\left(\frac{\xi}{k}\right) \right.$$

$$\left. - (k-n)\cos\left(\xi - \frac{n\xi}{k}\right) \sin\left(\frac{\xi}{k}\right) \right]$$

$$\cdot \sin^{-3}\left(\frac{\xi}{k}\right),$$

(67)

$$\lambda_k = \left[(k-1)\cos\xi + k\frac{\sin(\xi - \xi/k)}{\sin(\xi/k)} \right]^{-1},$$

(68)

$$0 < |\xi| < \pi.$$

(69)

Remark 10: For $k=2$ waveforms with $\lambda_2 = 1$ also have exactly two zeros. These waveforms can be included in above proposition by substituting (69) with $0 < |\xi| \le \pi$.

Remark 11: Apart from nonnegative waveforms of type (35) with two zeros, there are another two types of nonnegative waveforms which can be obtained from (66)–(68). These are

1.   nonnegative waveforms with k zeros (corresponding to $|\xi| = \pi$) and

2.   maximally flat nonnegative waveforms (corresponding to $\xi=0$).

Notice that nonnegative waveforms of type (35) with $\lambda_k = 1$ can be obtained from (66)–(68) by setting $|\xi| = \pi$. Substitution of $\lambda_k = 1$ and $|\xi| = \pi$ into (66), along with execution of all multiplications and usage of (A.2) (see Appendices), leads to $(\tau) = 1 - \cos k(\tau - \tau_0)$.

Also, maximally flat nonnegative waveforms (they have only one zero [21]) can be obtained from (66)–(68) by setting $\xi=0$. Thus, substitution of $\xi=0$ into (66)–(68) leads to the following form of maximally flat nonnegative waveform of type (35):

$$T_k(\tau) = \frac{[1 - \cos(\tau - \tau_0)]^2}{3(k^2 - 1)}$$

$$\cdot \left[k(k^2 - 1) \right.$$

$$\left. + 2\sum_{n=1}^{k-2} (k-n)\left((k-n)^2 - 1\right)\cos n(\tau - \tau_0) \right].$$

(70)

Maximally flat nonnegative waveforms of type (35) for $k \leq 4$ can be expressed as

$$T_2(\tau) = \frac{2}{3} \left[1 - \cos(\tau - \tau_0) \right]^2 ,$$

$$T_3(\tau) = \frac{1}{2} \left[1 - \cos(\tau - \tau_0) \right]^2 \left[2 + \cos(\tau - \tau_0) \right] ,$$

$$T_4(\tau) = \frac{4}{15} \left[1 - \cos(\tau - \tau_0) \right]^2$$

$$\cdot \left[5 + 4\cos(\tau - \tau_0) + \cos 2(\tau - \tau_0) \right]. \tag{71}$$

Remark 12: Every nonnegative waveform of type (35) with exactly one zero at nondegenerate critical point can be described as in Proposition 6 providing that symbol "$\leq$" in relation (40) is replaced with "$<$".This is an immediate consequence of Propositions 6 and 9 and Remark 11.

Remark 13: Identity $[1 - \cos(\tau - \tau_0)][1 - \cos(\tau - \tau_0 + 2\xi/k)] = [\cos \xi/k - \cos(\tau - \tau_0 + \xi/k)]2$ implies that (66) can be also rewritten as

$$T_k(\tau) = \lambda_k \left[\frac{\cos \xi}{k} - \cos\left(\tau - \tau_0 + \frac{\xi}{k} \right) \right]^2$$

$$\cdot \left[c_0 + 2 \sum_{n=1}^{k-2} c_n \cos n \left(\tau - \tau_0 + \frac{\xi}{k} \right) \right]. \tag{72}$$

Furthermore, substitution of (67) into (72) leads to

$$T_k(\tau)$$

$$= \lambda_k \left[\frac{\cos \xi}{k} - \cos\left(\tau - \tau_0 + \frac{\xi}{k} \right) \right]$$

$$\cdot \left[\frac{(k-1)\sin \xi}{\sin(\xi/k)} - 2 \sum_{n=1}^{k-1} \frac{\sin(\xi - n\xi/k)}{\sin(\xi/k)} \cos n \left(\tau - \tau_0 + \frac{\xi}{k} \right) \right]. \tag{73}$$

Remark 14: According to (A.6) (see Appendices), it follows that coefficients (67) can be expressed as

$$c_n = 2 \sum_{m=1}^{k-n-1} m(k-n-m) \cos\left(\frac{(k-n-2m)\xi}{k} \right). \tag{74}$$

Furthermore, from (74) it follows that coefficients c_{k-2}, c_{k-3}, c_{k-4} and c_{k-5} are equal to

$$c_{k-2} = 2, \tag{75}$$

$$c_{k-3} = 8 \cos\left(\frac{\xi}{k}\right), \tag{76}$$

$$c_{k-4} = 8 + 12 \cos\left(\frac{2\xi}{k}\right), \tag{77}$$

$$c_{k-5} = 24 \cos\left(\frac{\xi}{k}\right) + 16 \cos\left(\frac{3\xi}{k}\right). \tag{78}$$

For example, for $k=2$, (75) and (68) lead to $c_0 = 2$ and $\lambda_2 = 1/(2 + \cos \xi)$, respectively, which from (72) further imply that

$$T_2(\tau) = \frac{2\left[\cos(\xi/2) - \cos(\tau - \tau_0 + \xi/2)\right]^2}{[2 + \cos \xi]}. \tag{79}$$

Also for $k=3$, (75), (76), and (68) lead to $c_1 = 2$, $c_0 = 8 \cos(\xi/3)$, and $\lambda_3 = [2(3 \cos(\xi/3) + \cos \xi)]^{-1}$, respectively, which from (72) further imply that

$$T_3(\tau) = \frac{2\left[\cos(\xi/3) - \cos(\tau - \tau_0 + \xi/3)\right]^2}{[3 \cos(\xi/3) + \cos \xi]}$$

$$\cdot \left[2 \cos\left(\frac{\xi}{3}\right) + \cos\left(\tau - \tau_0 + \frac{\xi}{3}\right)\right]. \tag{80}$$

Remark 15: According to (A.5) (see Appendices), relation (68) can be rewritten as

$$\lambda_k = \left[(k-1) \cos \xi + k \sum_{n=1}^{k-1} \cos\left(\frac{(k - 2n)\xi}{k}\right)\right]^{-1}. \tag{81}$$

Clearly, amplitude λ_k of kth harmonic of nonnegative waveform of type (35) with exactly two zeros is even function of ξ. Since $\cos((k - 2n)\xi/k)$, $n = 0, \ldots, (k - 1)$, decreases with increase of $|\xi|$ on interval $0 \leq |\xi| \leq \pi$, it follows that λk monotonically increases with increase of $|\xi|$. Right hand side of (68) is equal to $1/(k^2 - 1)$ for $\xi = 0$ and to one for $|\xi| = \pi$. Therefore, for nonnegative waveforms of type (35) with exactly two zeros, the following relation holds:

$$\frac{1}{k^2 - 1} < \lambda_k < 1.$$

(82)

The left boundary in (82) corresponds to maximally flat nonnegative waveforms (see Remark 11). The right boundary in (82) corresponds to nonnegative waveforms with k zeros (also see Remark 11).

Amplitude of kth harmonic of nonnegative waveform of type (35) with two zeros, as a function of parameter ξ for $k \leq 5$, is presented in Figure 5.

Remark 16: Nonnegative waveform of type (35) with two zeros can be also expressed in the following form:

$$T_k(\tau) = 1 - \lambda_k \frac{k \sin \xi}{\sin(\xi/k)} \cos\left(\tau - \tau_0 + \frac{\xi}{k}\right)$$
$$+ \lambda_k \cos\left(k(\tau - \tau_0) + \xi\right),$$

(83)

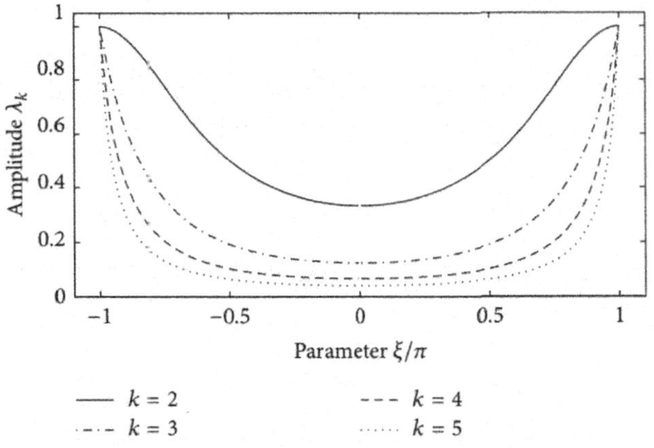

Figure 5: Amplitude of kth harmonic of nonnegative waveform with two zeros as a function of parameter ξ.

where λ_k is given by (68) and $0 < |\xi| < \pi$. From (83) it follows that coefficients of fundamental harmonic of nonnegative waveform of type (35) with two zeros are

$$a_1 = -\lambda_1 \cos\left(\tau_0 - \frac{\xi}{k}\right), \qquad b_1 = -\lambda_1 \sin\left(\tau_0 - \frac{\xi}{k}\right),$$

(84)

where λ_1 is amplitude of fundamental harmonic:

$$\lambda_1 = \frac{k \sin \xi}{\sin(\xi/k)} \lambda_k.$$

(85)

Coefficients of kth harmonic are given by (45)-(46). Notice that (68) can be rewritten as

$$\lambda_k = \left[\cos\left(\frac{\xi}{k}\right) \frac{k \sin\xi}{\sin(\xi/k)} - \cos\xi \right]^{-1}.$$

(86)

By introducing new variable,

$$x = \cos\left(\frac{\xi}{k}\right),$$

(87)

and using the Chebyshev polynomials (e.g., see Appendices), relations (85) and (86) can be rewritten as

$$\lambda_1 = k\lambda_k U_{k-1}(x),$$

(88)

$$\lambda_k = \frac{1}{kxU_{k-1}(x) - V_k(x)},$$

(89)

where $V_k(x)$ and $U_k(x)$ denote the Chebyshev polynomials of the first and second kind, respectively. From (89) it follows that

$$\lambda_k \left[kxU_{k-1}(x) - V_k(x) \right] - 1 = 0,$$

(90)

which is polynomial equation of kth degree in terms of variable x. From $0 < |\xi| < \pi$ and (87) it follows that

$$\cos\left(\frac{\pi}{k}\right) < x < 1.$$

(91)

Since λ_k is monotonically increasing function of $|\xi|$, $0 < |\xi| < \pi$, it follows that λ_k is monotonically decreasing function of x. This further implies that (90) has only one solution that satisfies (91). (For $k=2$ expression (91) reads $\cos(\pi/2) \le x < 1$) This solution for x (which can be obtained at least numerically), according to (88), leads to amplitude $\lambda 1$ of fundamental harmonic.

For $k \le 4$, solutions of (90) and (91) are

$$x = \sqrt{\frac{1-\lambda_2}{2\lambda_2}}, \quad \frac{1}{3} < \lambda_2 \le 1,$$

$$x = \frac{1}{2\sqrt[3]{\lambda_3}}, \quad \frac{1}{8} < \lambda_3 < 1,$$

$$x = \sqrt{\frac{1}{6}\left(1 + \sqrt{\frac{5\lambda_4 + 3}{2\lambda_4}}\right)}, \quad \frac{1}{15} < \lambda_4 < 1.$$

(92)

Insertion of (92) into (88) leads to the following relations between amplitude λ_1 of fundamental and amplitude λ_k of kth harmonic, $k \leq 4$:

$$\lambda_1 = \sqrt{8\lambda_2(1 - \lambda_2)}, \quad \frac{1}{3} < \lambda_2 \leq 1, \tag{93}$$

$$\lambda_1 = 3\left(\sqrt[3]{\lambda_3} - \lambda_3\right), \quad \frac{1}{8} < \lambda_3 < 1, \tag{94}$$

$$\lambda_1 = \sqrt{\frac{32}{27}\left(\sqrt{2\lambda_4(3 + 5\lambda_4)^3} - 2\lambda_4(9 + 7\lambda_4)\right)},$$

$$\frac{1}{15} < \lambda_4 < 1. \tag{95}$$

Proof of Proposition 9: As it has been shown earlier (see Proposition 6), nonnegative waveform of type (35) with at least one zero can be represented in form (38). Since we exclude nonnegative waveforms with $\lambda_k = 1$, according to Remark 7, it follows that we exclude case $|\xi| = \pi$. Therefore in the quest for nonnegative waveforms of type (35) having two zeros we will start with waveforms of type (38) for $|\xi| < \pi$. It is clear that nonnegative waveforms of type (38) have two zeros if and only if

$$\lambda_k = \left[\max_\tau r_k(\tau)\right]^{-1}, \tag{96}$$

and $\max_\tau r_k(\tau) \neq r_k(\tau_0)$. According to (64), $\max_\tau(\tau) \neq r_k(\tau_0)$ implies $|\xi| \neq 0$. Therefore, it is sufficient to consider only the interval (69).

Substituting (96) into (38) we obtain

$$T_k(\tau) = \frac{[1 - \cos(\tau - \tau_0)]\left[\max_\tau r_k(\tau) - r_k(\tau)\right]}{\max_\tau r_k(\tau)}. \tag{97}$$

Expression $\max_\tau(\tau) - r_k(\tau)$, according to (64) and (39), equals

$$\max_\tau r_k(\tau) - r_k(\tau) = k\frac{\sin((k - 1)\xi/k)}{\sin(\xi/k)}$$

$$- 2\sum_{n=1}^{k-1}(k - n)\cos(n(\tau - \tau_0) + \xi). \tag{98}$$

Comparison of (97) with (66) yields

$$\max_{\tau} r_k(\tau) - r_k(\tau) = \left[1 - \cos\left(\tau - \tau_0 + \frac{2\xi}{k} \right) \right]$$

$$\cdot \left[c_0 + 2 \sum_{n=1}^{k-2} c_n \cos n \left(\tau - \tau_0 + \frac{\xi}{k} \right) \right], \tag{99}$$

where coefficients c_n, $n = 0, \ldots, k - 2$, are given by (67). In what follows we are going to show that right hand sides of (98) and (99) are equal.

From (67) it follows that

$$c_0 - c_1 \cos\left(\frac{\xi}{k} \right) = k \frac{\sin(\xi - \xi/k)}{\sin(\xi/k)}. \tag{100}$$

Also, from (67) for $n = 1, \ldots, k-3$ it follows that the following relations hold:

$$(c_{n-1} + c_{n+1}) \cos\left(\frac{\xi}{k} \right) - 2c_n = 2(k - n) \cos\left(\xi - \frac{n\xi}{k} \right),$$

$$(c_{n-1} - c_{n+1}) \sin\left(\frac{\xi}{k} \right) = 2(k - n) \sin\left(\xi - \frac{n\xi}{k} \right). \tag{101}$$

From (99), by using (75), (76), (100)-(101), and trigonometric identities

$$\cos\left(\tau - \tau_0 + \frac{2\xi}{k} \right) = \cos\left(\frac{\xi}{k} \right) \cos\left(\tau - \tau_0 + \frac{\xi}{k} \right)$$

$$- \sin\left(\frac{\xi}{k} \right) \sin\left(\tau - \tau_0 + \frac{\xi}{k} \right),$$

$$\cos\left(\xi - \frac{n\xi}{k} \right) \cos\left(n\left(\tau - \tau_0 + \frac{\xi}{k} \right) \right)$$

$$- \sin\left(\xi - \frac{n\xi}{k} \right) \sin\left(n\left(\tau - \tau_0 + \frac{\xi}{k} \right) \right)$$

$$= \cos\left(n(\tau - \tau_0) + \xi \right), \tag{102}$$

we obtain (98). Consequently (98) and (99) are equal, which completes the proof.

Nonnegative Waveforms with Two Zeros and Prescribed Coefficients of kth Harmonic

In this subsection we show that, for prescribed coefficients a_k and b_k, there are k nonnegative waveforms of type (35) with exactly two zeros. According to(37) and (82), coefficients a_k and b_k of nonnegative waveforms of type (35) with exactly two zeros satisfy the following relation:

$$\frac{1}{k^2 - 1} < \sqrt{a_k^2 + b_k^2} < 1.$$

(103)

According to Remark 16, the value of x (see (87)) that corresponds to $\lambda_k = \sqrt{a_k^2 + b_k^2}$ can be determined from (90)- (91). As we mentioned earlier, (90) has only one solution that satisfies (91). This value of x, according to (88), leads to the amplitude λ_1 of fundamental harmonic (closed form expressions for λ_1 in terms of λ_k and $k \le 4$ are given by (93)- (95)).

On the other hand, from (45)-(46) it follows that

$$k\tau_0 - \xi = \text{atan} \, 2 \left(b_k, a_k\right) + 2q\pi, \quad q = 1, \ldots, (k-1),$$

(104)

where function atan $2(y, x)$ is defined as

$$\text{atan} \, 2 \left(y, x\right) = \begin{cases} \arctan\left(\dfrac{y}{x}\right) & \text{if } x \ge 0, \\[2mm] \arctan\left(\dfrac{y}{x}\right) + \pi & \text{if } x < 0, \, y \ge 0, \\[2mm] \arctan\left(\dfrac{y}{x}\right) - \pi & \text{if } x < 0, \, y < 0, \end{cases}$$

(105)

with the codomain $(-\pi, \pi]$. Furthermore, according to (84) and (104), the coefficients of fundamental harmonic of nonnegative waveforms with two zeros and prescribed coefficients of kth harmonic are equal to

$$a_1 = -\lambda_1 \cos \left[\frac{\text{atan} \, 2 \left(b_k, a_k\right) + 2q\pi}{k}\right],$$

$$b_1 = -\lambda_1 \sin \left[\frac{\text{atan} \, 2 \left(b_k, a_k\right) + 2q\pi}{k}\right],$$

(106)

where $q = 0, \ldots, (k-1)$. For chosen q, according to (104) and (66), positions of zeros are

$$\tau_0 = \frac{1}{k} \left[\xi + \text{atan} \, 2 \left(b_k, a_k\right) + 2q\pi\right],$$

$$\tau_0 - \frac{2\xi}{k} = \frac{1}{k} \left[-\xi + \text{atan} \, 2 \left(b_k, a_k\right) + 2q\pi\right].$$

(107)

From (106) and $q = 0, \ldots, (k-1)$it follows that, for prescribed coefficients a_k and b_k, there are k nonnegative waveforms of type (35) with exactly two zeros.

We provide here an algorithm to facilitate calculation of coefficients a_1 and b_1 of nonnegative waveforms of type (35) with two zeros and prescribed coefficients a_k and b_k, providing that a_k and b_k satisfy (103).

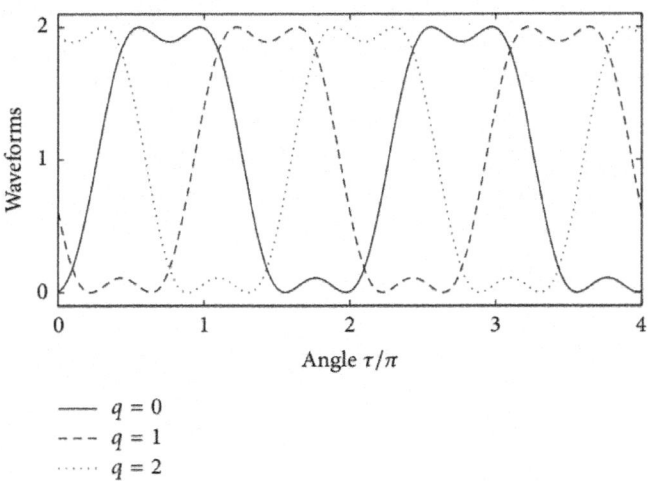

── $q = 0$
--- $q = 1$
······ $q = 2$

Figure 6: Nonnegative waveforms with two zeros for k=3, $a_3 = -0.15$, and $b_3 = -0.2$.

Algorithm 17.

(i)   Calculate $\lambda_k = \sqrt{a_k^2 + b_k^2}$,

(ii)   identify x that satisfies both relations (90) and (91),

(iii)   calculate λ_1 according to (88),

(iv)   choose integer q, such that $0 \leq q \leq k-1$,

(v)   calculate a_1 and b_1 according to (106).

For $k \leq 4$, by using (93) for k=2, (94) for k=3, and (95) for k=4 it is possible to calculate directly λ_1 from λ_k and proceed to step (iv).

For k=2 and prescribed coefficients a_2 and b_2, there are two waveforms with two zeros, one corresponding to $a_1 < 0$ and the other corresponding to $a_1 > 0$ (see also [12]).

Let us take as an input k=3, $a_3 = -0.15$, and $b_3 = -0.2$. Execution of Algorithm 17 on this input yields $\lambda_3 = 0.25$ and $\lambda_1 = 1.1399$ (according to (94)). For q=0 we calculate $a_1 = -0.8432$ and $b_1 = 0.7670$ (corresponding waveform is presented by solid line in Figure 6); for q=1 we calculate $a_1 = -0.2426$ and b_1

$= -1.1138$ (corresponding waveform is presented by dashed line); for $q=2$ we calculate $a_1 = 1.0859$ and $b_1 = 0.3468$ (corresponding waveform is presented by dotted line).

As another example of the usage of Algorithm 17, let us consider case $k=4$ and assume that $a_4 = -0.15$ and $b_4 = -0.2$. Consequently $\lambda_4 = 0.25$ and $\lambda_1 = 0.9861$ (according to (95)). For $q = 0, \ldots, 3$ we calculate the following four pairs (a_1, b_1) of coefficients of fundamental harmonic: $(-0.8388, 0.5184)$ for $q=0$, $(-0.5184, -0.8388)$ for $q=1$, $(0.8388, -0.5184)$ for $q = 2$, and $(0.5184, 0.8388)$ for $q=3$. Corresponding waveforms are presented in Figure 7.

NONNEGATIVE WAVEFORMS WITH MAXIMAL AMPLITUDE OF FUNDAMENTAL HARMONIC

In this section we provide general description of nonnegative waveforms containing fundamental and kth harmonic with maximal amplitude of fundamental harmonic for prescribed amplitude of kth harmonic.

The main result of this section is presented in the following proposition.

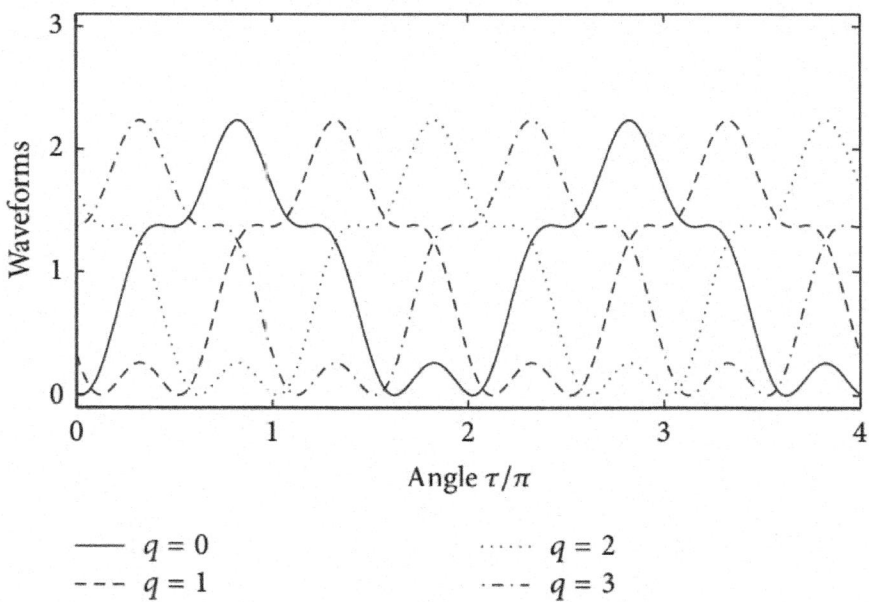

Figure 7: Nonnegative waveforms with two zeros for $k=4$, $a_4 = -0.15$, and $b_4 = -0.2$.

Proposition 18: Every nonnegative waveform of type (35) with maximal amplitude λ_1 of fundamental harmonic and prescribed amplitude λ_k of kth harmonic can be expressed in the following form:

$$T_k(\tau) = \left[1 - \cos(\tau - \tau_0)\right]$$

$$\cdot \left[1 - (k-1)\lambda_k - 2\lambda_k \sum_{n=1}^{k-1}(k-n)\cos(n(\tau - \tau_0))\right], \quad (108)$$

if $0 \le \lambda_k \le 1/(k^2 - 1)$ *or*

$$T_k(\tau) = \lambda_k \left[1 - \cos(\tau - \tau_0)\right]\left[1 - \cos\left(\tau - \tau_0 + \frac{2\xi}{k}\right)\right]$$

$$\cdot \left[c_0 + 2\sum_{n=1}^{k-2} c_n \cos n\left(\tau - \tau_0 + \frac{\xi}{k}\right)\right], \quad (109)$$

if $1/(k^2 - 1) \le \lambda_k \le 1$, providing that c_n, $n = 0, \ldots, k - 2$, and λ_k are related to ξ via relations (67) and (68), respectively, and $|\xi| \le \pi$.

Remark 19: Expression (108) can be obtained from (38) by setting $\xi{=}0$. Furthermore, insertion of $\xi{=}0$ into (43)–(46) leads to the following expressions for coefficients of waveform of type (108):

$$a_1 = -(1 + \lambda_k)\cos\tau_0, \qquad b_1 = -(1 + \lambda_k)\sin\tau_0,$$

$$a_k = \lambda_k \cos(k\tau_0), \qquad b_k = \lambda_k \sin(k\tau_0). \quad (110)$$

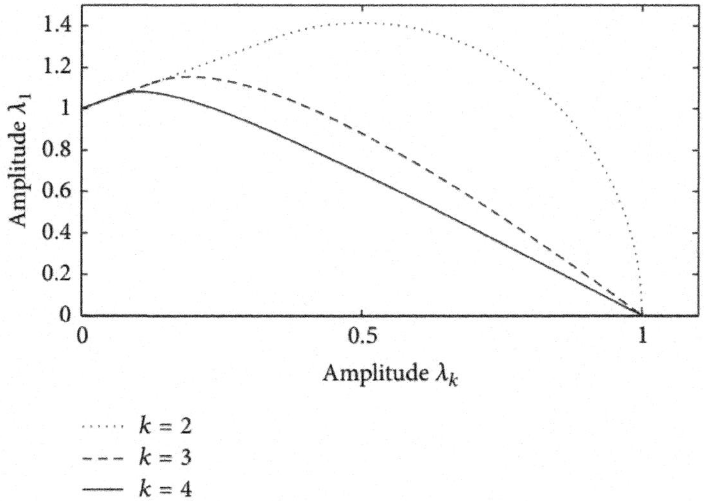

...... $k = 2$
- - - $k = 3$
—— $k = 4$

Figure 8: Maximal amplitude of fundamental harmonic as a function of amplitude of kth harmonic.

On the other hand, (109) coincides with (66). Therefore, the expressions for coefficients of (109) and (66) also coincide. Thus, expressions for coefficients of fundamental harmonic of waveform (109) are given by (84), where λ_1 is given by (85), while expressions for coefficients of kth harmonic are given by (45)-(46).

Waveforms described by (108) have exactly one zero, while waveforms described by (109) for $1/(k^2 - 1) < \lambda_k < 1$ have exactly two zeros. As we mentioned earlier, waveforms (109) for $\lambda_k = 1$ have k zeros.

Remark 20: Maximal amplitude of fundamental harmonic of nonnegative waveforms of type (35) for prescribed amplitude of kth harmonic can be expressed as

$$\lambda_1 = 1 + \lambda_k, \tag{111}$$

if $0 \leq \lambda k \leq 1/(k^2 - 1)$, or

$$\lambda_1 = \frac{k \sin \xi}{k \sin \xi \cos (\xi/k) - \cos \xi \sin (\xi/k)}, \tag{112}$$

if $1/(k^2 - 1) \leq \lambda_k \leq 1$, where ξ is related to λ_k via (68) (or (86)) and $|\xi| \leq \pi$.

From (110) it follows that (111) holds. Substitution of (86) into (85) leads to (112).

Notice that $\lambda_k = 1/(k^2 - 1)$ is the only common point of the intervals $0 \leq \lambda_k \leq 1/(k^2 - 1)$ and $1/(k^2 - 1) \leq \lambda_k \leq 1$. According to (111), $\lambda_k = 1/(k^2 - 1)$ corresponds to $\lambda_1 = k^2/(k^2 - 1)$. It can be also obtained from (112) by setting $\xi = 0$. The waveforms corresponding to this pair of amplitudes are maximally flat nonnegative waveforms.

Maximal amplitude of fundamental harmonic of nonnegative waveform of type (35) for $k \leq 4$, as a function of amplitude of kth harmonic, is presented in Figure 8.

Remark 21: Maximum value of amplitude of fundamental harmonic of nonnegative waveform of type (35) is

$$\lambda_{1,\max} = \frac{1}{\cos (\pi/(2k))}. \tag{113}$$

This maximum value is attained for $|\xi| = \pi/2$ (see (112)). The corresponding value of amplitude of kth harmonic is $\lambda_k = (1/k)\tan(\pi/(2k))$. Nonnegative waveforms of type (35) with $\lambda_1 = \lambda_{1,\max}$ have two zeros at $\tau 0$ and $\tau_0 - \pi/k$ for $\xi = \pi/2$, or at $\tau 0$ and $\tau_0 + \pi/k$ for $\xi = -\pi/2$.

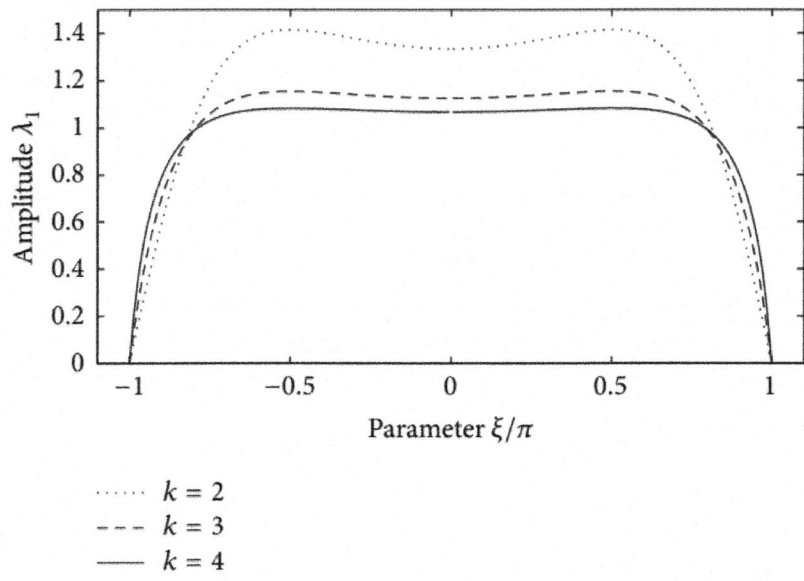

Figure 9: Maximal amplitude of fundamental harmonic as a function of parameter ξ.

To prove that (113) holds, let us first show that the following relation holds for $k \geq 2$:

$$\cos\left(\frac{\pi}{2k}\right) < 1 - \frac{1}{k^2}.$$

(114)

From $k \geq 2$, it follows that $\mathrm{sinc}(\pi/(4k)) > \mathrm{sinc}(\pi/4)$, where $\mathrm{sinc}\ x = (\sin x)/x$, and therefore $\sin(\pi/(4k)) > 1/(\sqrt{2}k)$. By using trigonometric identity $\cos 2x = 1 - 2\sin^2 x$, we immediately obtain (114).

According to (111) and (112), it is clear that λ_1 attains its maximum value on the interval $1/(k^2 - 1) \leq \lambda_k \leq 1$. Since λ_k is monotonic function of $|\xi|$ on interval $|\xi| \leq \pi$ (see Remark 15), it follows that $d\lambda_k/d\xi = 0$ for $0 < |\xi| < \pi$. Therefore, to find critical points of λ_1 as a function of λ_k it is sufficient to find critical points of λ_1 as a function of $|\xi|$, $0 < |\xi| < \pi$, and consider its values at the end points $\xi = 0$ and $|\xi| = \pi$. Plot of $\lambda 1$ as a function of parameter ξ for $k \leq 4$ is presented in Figure 9. According to (112), first derivative of λ_1 with respect to ξ is equal to zero if and only if $(k \cos \xi \sin(\xi/k) - \sin \xi \cos(\xi/k)) \cos \xi = 0$. On interval $0 < |\xi| < \pi$, this is true if and only if $|\xi| = \pi/2$. According to (112), $\lambda 1$ is equal to $k^2/(k^2 - 1)$ for $\xi = 0$, equal to zero for $|\xi| = \pi$, and equal to $1/\cos(\pi/(2k))$ for $|\xi| = \pi/2$. From (114) it follows that $k^2/(k^2 - 1) < 1/\cos(\pi/(2k))$ and therefore maximum value of λ_1 is given by (113). Moreover, maximum value of λ_1 is attained for $|\xi| = \pi/2$.

According to above consideration, all nonnegative waveforms of type (35) having maximum value of amplitude of fundamental harmonic can be obtained from (109) by setting $|\xi| = \pi/2$. Three of them corresponding to $k=3$, $\xi = \pi/2$, and three different values of τ_0 (0, $\pi/6$, and $\pi/3$) are presented in Figure 10. Dotted line corresponds to $\tau_0 = 0$ (coefficients of corresponding waveform are $a_1 = -1$, $b_1 = 1/\sqrt{3}$, $a_3 = 0$, and $b_3 = -\sqrt{3}/9$), solid line to $\tau_0 = \pi/6$ ($a_1 = -2/\sqrt{3}$, $b_1 = 0$, $a_3 = \sqrt{3}/9$, and $b_3 = 0$), and dashed line to $\tau_0 = \pi/3$ ($a_1 = -1$, $b_1 = -1/\sqrt{3}$, $a_3 = 0$, and $b_3 = \sqrt{3}/9$).

Proof of Proposition 18: As it has been shown earlier (Proposition 6), nonnegative waveform of type (35) with at least

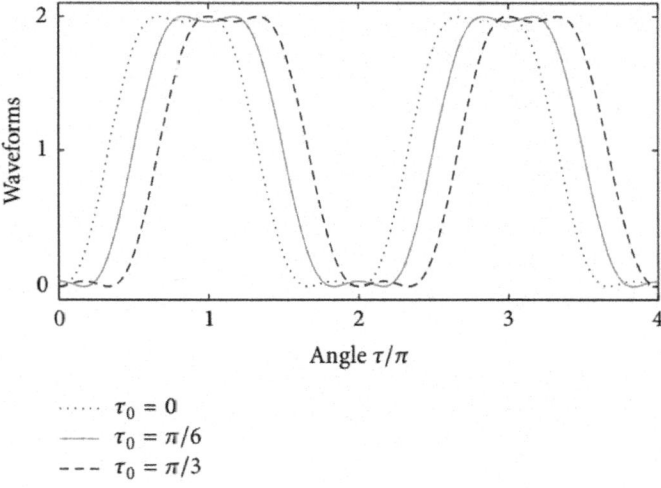

$$\cdots\cdots \quad \tau_0 = 0$$
$$\text{------} \quad \tau_0 = \pi/6$$
$$\text{- - -} \quad \tau_0 = \pi/3$$

Figure 10: Nonnegative waveforms with maximum amplitude of fundamental harmonic for $k=3$ and $\xi = \pi/2$.

one zero can be represented in form (38). According to (43), (44), and (36), for amplitude $\lambda 1$ of fundamental harmonic of waveforms of type (38) the following relation holds:

$$\lambda_1 = \sqrt{\left(1 + \lambda_k \cos \xi\right)^2 + k^2 \lambda_k^2 \sin^2 \xi}, \tag{115}$$

where λ_k satisfy (40) and $|\xi| \leq \pi$.

Because of (40), in the quest of finding maximal λ_1 for prescribed λ_k, we have to consider the following two cases:

(Case i) $\lambda_k < [(k-1)\cos \xi + k \sin(\xi - \xi/k)/\sin(\xi/k)]^{-1}$.

(Case ii) $\lambda_k = [(k-1)\cos \xi + k \sin(\xi - \xi/k)/\sin(\xi/k)]^{-1}$.

Case i: Since $\lambda_k < [(k-1)\cos\xi + k\sin(\xi-\xi/k)/\sin(\xi/k)]-1$ implies $\lambda_k \neq 1$, according to (115), it follows that $\lambda_1 \neq 0$. Hence $d\lambda_1/d\xi = 0$ implies

$$2\lambda_k \sin\xi\left[1-\left(k^2-1\right)\lambda_k\cos\xi\right]=0.$$

(116)

Therefore $d\lambda_1/d\xi = 0$ if $\lambda_k = 0$ (Option 1) or $\sin\xi = 0$ (Option 2) or $(k^2-1)\cos\xi = 1$ (Option 3).

Option 1: According to (115), $\lambda_k = 0$ implies $\lambda_1 = 1$ (notice that this implication shows that $\lambda 1$ does not depend on ξ and therefore we can set ξ to zero value).

Option 2: According to (115), $\sin\xi = 0$ implies $\lambda_1 = 1 + \lambda_k\cos\xi$, which further leads to the conclusion that λ_1 is maximal for $\xi = 0$. For $\xi = 0$, $\lambda_k < [(k-1)\cos\xi + k\sin(\xi-\xi/k)/\sin(\xi/k)]-1$ becomes $\lambda_k < 1/(k^2-1)$.

Option 3: This option leads to contradiction. To show that, notice that $(k^2-1)\cos\xi = 1$ and $\lambda_k < [(k-1)\cos\xi + k\sin(\xi-\xi/k)/\sin(\xi/k)]-1$ imply that $(k-1)\cos\xi > \sin(\xi-\xi/k)/\sin(\xi/k)$. Using (A.5) (see Appendices), the latest inequality can be rewritten as $\sum_{n=1}^{k-1}[\cos\xi - \cos((k-2n)\xi/k)] > 0$. But, from $|k-2n| < k, n = 1, \ldots, (k-1)$, and $|\xi| \leq \pi$ it follows that all summands are not positive and therefore $(k-1)\cos\xi > \sin(\xi-\xi/k)/\sin(\xi/k)$ does not hold for $|\xi| \leq \pi$.

Consequently, Case i implies $\xi = 0$ and $\lambda_k < 1/(k^2-1)$. Finally, substitution of $\xi = 0$ into (38) leads to (108), which proves that (108) holds for $\lambda_k < 1/(k^2-1)$.

Case ii: Relation $\lambda_k = [(k-1)\cos\xi + k\sin(\xi-\xi/k)/\sin(\xi/k)]-1$, according to Proposition 9 and Remark 11, implies that corresponding waveforms can be expressed via (66)–(68) for $|\xi| \leq \pi$. Furthermore, $\lambda_k = [(k-1)\cos\xi + k\sin(\xi-\xi/k)/\sin(\xi/k)]-1$ and $|\xi| \leq \pi$ imply $1/(k^2-1) \leq \lambda_k \leq 1$. This proves that (109) holds for $1/(k^2-1) \leq \lambda_k \leq 1$.

Finally, let us prove that (108) holds for $\lambda_k = 1/(k^2-1)$. According to (68) (see also Remark 11), this value of λ_k corresponds to $\xi = 0$. Furthermore, substitution of $\lambda_k = 1/(k^2-1)$ and $\xi = 0$ into (109) leads to (70), which can be rewritten as

$$T_k(\tau) = \frac{[1-\cos(\tau-\tau_0)]}{(1-k^2)}$$

$$\cdot\left[k(k-1)-2\sum_{n=1}^{k-1}(k-n)\cos(n(\tau-\tau_0))\right].$$

(117)

Waveform (117) coincides with waveform (108) for $\lambda_k = 1/(1 - k^2)$. Consequently, (108) holds for $\lambda_k = 1/(1 - k^2)$, which completes the proof.

NONNEGATIVE WAVEFORMS WITH MAXIMAL ABSO-LUTE VALUE OF THE COEFFICIENT OF COSINE TERM OF FUNDAMENTAL HARMONIC

In this section we consider general description of nonnegative waveforms of type (35) with maximal absolute value of coefficient a_1 for prescribed coefficients of kth harmonic. This type of waveform is of particular interest in PA efficiency analysis. In a number of cases of practical interest either current or voltage waveform is prescribed. In such cases, the problem of finding maximal efficiency of PA can be reduced to the problem of finding nonnegative waveform with maximal coefficient a_1 for prescribed coefficients of kth harmonic (see also Section 7).

In Section 5.1 we provide general description of nonnegative waveforms of type (35) with maximal absolute value of coefficient $a1$ for prescribed coefficients of kth harmonic. In Section 5.2 we illustrate results of Section 5.1 for particular case $k=3$.

Nonnegative Waveforms with Maximal Absolute Value of Coefficient a_1 for $k \geq 2$

Waveforms (τ) of type (35) with $a_1 \geq 0$ can be derived from those with $a_1 \leq 0$ by shifting by π, and therefore we can assume without loss of generality that $a_1 \leq 0$. Notice that if k is even, then shifting (τ) by π produces the same result as replacement of a_1 with $-a_1$ (a_k remains the same). On the other hand, if k is odd, then shifting (τ) by π produces the same result as replacement of a_1 with $-a_1$ and a_k with $-a_k$.

According to (37), coefficients of kth harmonic can be expressed as

$$a_k = \lambda_k \cos\delta, \qquad b_k = \lambda_k \sin\delta, \tag{118}$$

Where

$$|\delta| \leq \pi. \tag{119}$$

Conversely, for prescribed coefficients a_k and b_k, δ can be determined as

$$\delta = \operatorname{atan} 2\left(b_k, a_k\right), \tag{120}$$

where definition of function atan $2(y, x)$ is given by (105).

The main result of this section is stated in the following proposition.

Proposition 22: Every nonnegative waveform of type (35) with maximal absolute value of coefficient $a_1 \leq 0$ for prescribed coefficients a_k and b_k of kth harmonic can be represented as

$$T_k(\tau)$$

$$= [1 - \cos \tau]$$

$$\cdot \left[1 - (k-1)\, a_k - 2 \sum_{n=1}^{k-1} (k-n)\, (a_k \cos n\tau + b_k \sin n\tau) \right], \qquad (121)$$

if $k\lambda_k[\sin \delta / \sin(\delta/k)] \cos(\delta/k) \leq 1 + a_k$, where $\delta = \mathrm{atan}\, 2(b_k, a_k)$, or

$$T_k(\tau) = \lambda_k \left[1 - \cos\left(\tau - \frac{(\delta + \xi)}{k} \right) \right]$$

$$\cdot \left[1 - \cos\left(\tau - \frac{(\delta - \xi)}{k} \right) \right]$$

$$\cdot \left[c_0 + 2 \sum_{n=1}^{k-2} c_n \cos n\left(\tau - \frac{\delta}{k} \right) \right], \qquad (122)$$

if $k\lambda_k[\sin \delta / \sin(\delta/k)] \cos(\delta/k) \geq 1 + a_k$, where c_n, $n = 0, \ldots, k-2$, and $\lambda_k = \sqrt{a_k^2 + b_k^2}$ are related to ξ via relations (67) and (68), respectively, and $|\xi| \leq \pi$.

Remark 23: Expression (121) can be obtained from (38) by setting $\tau_0 = 0$ and $\xi = -\delta$ and then replacing $\lambda_k \cos \delta$ with a_k (see (118)) and $\lambda_k \cos(n\tau - \delta)$ with $a_k \cos n\tau + b_k \sin n\tau$ (see also (118)). Furthermore, insertion of $\tau_0 = 0$ and $\xi = -\delta$ into (43)–(46) leads to the following relations between fundamental and kth harmonic coefficients of waveform (121):

$$a_1 = -(1 + a_k), \qquad b_1 = -kb_k. \qquad (123)$$

On the other hand, expression (122) can be obtained from (66) by replacing $\tau_0 - \xi/k$ with δ/k. Therefore, substitution of $\tau_0 - \xi/k = \delta/k$ in (84) leads to

$$a_1 = -\lambda_1 \cos\left(\frac{\delta}{k} \right), \qquad b_1 = -\lambda_1 \sin\left(\frac{\delta}{k} \right), \qquad (124)$$

where λ_1 is given by (85).

The fundamental harmonic coefficients a_1 and b_1 of waveform of type (35) with maximal absolute value of coefficient $a_1 \leq 0$ satisfy both relations (123) and (124) if a_k and b_k satisfy $1+a_k = k\lambda_k[\sin \delta/ \sin(\delta/k)] \cos(\delta/k)$. For such waveforms, relations $\tau_0 = 0$ and $\xi = -\delta$ also hold.

Remark 24: Amplitude of kth harmonic of nonnegative waveform of type (35) with maximal absolute value of coefficient $a_1 \leq 0$ and coefficients a_k, b_k satisfying $1+a_k = k\lambda_k[\sin \delta/ \sin(\delta/k)] \cos(\delta/k)$, is

$$\lambda_k = \frac{\sin (\delta/k)}{k \sin \delta \cos (\delta/k) - \cos \delta \sin (\delta/k)}.$$

(125)

To show that, it is sufficient to substitute $a_k = \lambda_k \cos \delta$ (see (118)) into $1+a_k = k[\sin \delta/ \sin(\delta/k)] \cos(\delta/k)$.

Introducing new variable,

$$y = \cos\left(\frac{\delta}{k}\right),$$

(126)

and using the Chebyshev polynomials (e.g., see Appendices), relations $a_k = \lambda_k \cos \delta$ and (125) can be rewritten as

$$a_k = \lambda_k V_k (y),$$

(127)

$$\lambda_k = \frac{1}{kyU_{k-1} (y) - V_k (y)},$$

(128)

where $V_k(y)$ and $U_k(y)$ denote the Chebyshev polynomials of the first and second kind, respectively. Substitution of (128) into (127) leads to

$$a_k kyU_{k-1} (y) - (1 + a_k) V_k (y) = 0,$$

(129)

which is polynomial equation of kth degree in terms of variable y. From $|\delta| \leq \pi$ and (126) it follows that

$$\cos\left(\frac{\pi}{k}\right) \leq y \leq 1.$$

(130)

In what follows we show that a_k is monotonically increasing function of y on the interval (130). From $\xi = -\delta$ (see Remark 23) and (81) it follows that $\lambda_k^{-1} = (k - 1) \cos \delta + k \sum_{n=1}^{k-1} \cos((k - 2n)\delta/k) \geq 1$ and therefore $a_k = \lambda_k \cos \delta$ can be rewritten as

$$a_k = \frac{\cos \delta}{(k - 1) \cos \delta + k \sum_{n=1}^{k-1} \cos ((k - 2n) \delta/k)}.$$

(131)

Obviously a_k is even function of δ and all cosines in (131) are monotonically decreasing functions of $|\delta|$ on the interval $|\delta| \le \pi$. It is easy to show that $\cos((k - 2n)\delta/k)$, $n = 1, \ldots, (k - 1)$, decreases slower than $\cos \delta$ when $|\delta|$ increases. This implies that denominator of the right hand side of (131) decreases slower than numerator. Since denominator is positive for $|\delta| \le \pi$ it further implies that a_k is decreasing function of $|\delta|$ on interval $|\delta| \le \pi$. Consequently, a_k is monotonically increasing function of y on the interval (130). Thus we have shown that ak is monotonically increasing function of y on the interval (130) and therefore (129) has only one solution that satisfies (130).

According to (128), the value of y obtained from (129) and (130), either analytically or numerically, leads to amplitude λ_k of kth harmonic.

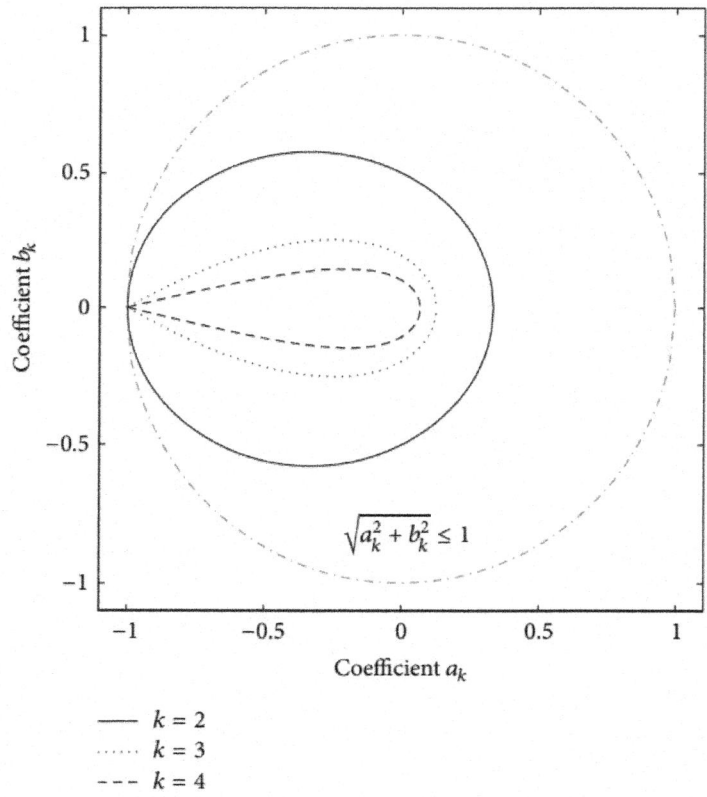

Figure 11: Plot of (a_k, b_k) satisfying $1+a_k = k[\sin \delta/ \sin(\delta/ k)] \cos(\delta/k)$ for $k{\le}4$.

By solving (129) and (130) for $k{\le}4$ we obtain

$$y = \sqrt{\frac{1 + a_2}{2(1 - a_2)}}, \quad -1 \le a_2 \le \frac{1}{3},$$

$$y = \sqrt{\frac{3}{4(1 - 2a_3)}}, \quad -1 \le a_3 \le \frac{1}{8},$$

$$y = \sqrt{\frac{\sqrt{2 - 4a_4 + 10a_4^2} - 2(1 - a_4)}{4(1 - 3a_4)}}, \quad -1 \le a_4 \le \frac{1}{15}. \tag{132}$$

Insertion of (132) into (128) leads to the following explicit expressions for the amplitude λ_k, $k \le 4$:

$$\lambda_2 = \frac{1}{2}(1 - a_2), \quad -1 \le a_2 \le \frac{1}{3}, \tag{133}$$

$$\lambda_3^2 = \left[\frac{1}{3}(1 - 2a_3)\right]^3, \quad -1 \le a_3 \le \frac{1}{8}, \tag{134}$$

$$\lambda_4 = \frac{1}{4}\left(-1 - a_4 + \sqrt{2 - 4a_4 + 10a_4^2}\right), \quad -1 \le a_4 \le \frac{1}{15}. \tag{135}$$

Relations (133)–(135) define closed lines (see Figure 11) which separate points representing waveforms of type (121) from points representing waveforms of type (122). For given k, points inside the corresponding curve refer to nonnegative waveforms of type (121), whereas points outside curve (and $\sqrt{a_k^2 + b_k^2} \le 1$) correspond to nonnegative waveforms of type(122). Points on the respective curve correspond to the waveforms which can be expressed in both forms (121) and (122).

Remark 25: The maximum absolute value of coefficient $a1$ of nonnegative waveform of type (35) is

$$|a_1|_{max} = \frac{1}{\cos(\pi/(2k))}. \tag{136}$$

This maximum value is attained for $|\xi| = \pi/2$ and $\delta = 0$ (see (124)). Notice that $|a_1|_{max}$ is equal to the maximum value $\lambda_{1,max}$ of amplitude of fundamental harmonic (see (113)). Coefficients of waveform with maximum absolute value of coefficient a_1, $a_1 < 0$, are

$$a_1 = -\frac{1}{\cos\left(\pi/(2k)\right)}, \qquad a_k = \frac{1}{k}\tan\left(\frac{\pi}{(2k)}\right),$$

$$b_1 = b_k = 0. \tag{137}$$

Waveform described by (137)is cosine waveform having zeros at $\pi/(2k)$ and $-\pi/(2k)$.

In the course of proving (136), notice first that $|a_1|$max $\le \lambda_1$,max holds. According to (123) and (124), maximum of $|a_1|$ occurs for $k[\sin \delta/\sin(\delta/k)]$ $\cos(\delta/k) \ge 1 + a_k$. From (124) it immediately follows that maximum value of $|a_1|$ is attained if and only if $\lambda_1 = \lambda_{1,\text{max}}$ and $\delta=0$, which because of $\delta/k = \tau_0 - \xi/k$ further implies $\tau_0 = \xi/k$. Since maximum value of λ_1 is attained for $|\xi| = \pi/2$, it follows that corresponding waveform has zeros at $\pi/(2k)$ and $-\pi/(2k)$.

Proof of Proposition 22: As it was mentioned earlier in this section, we can assume without loss of generality that $a_1 \le 0$. We consider waveforms (τ) of type (35)such that $T_k(\tau) \ge 0$ and$T_k(\tau) = 0$ for some τ_0. From assumption that nonnegative waveform (τ) of type (35) has at least one zero, it follows that it can be expressed in form (38).

Let us also assume that τ_0 is position of nondegenerate critical point. Therefore $(\tau_0)=0$ implies $T_k'(\tau_0)=0$ and $T_k''(\tau_0)>0$. According to (55), second derivative of (τ) at τ_0 can be expressed as $T_k''(\tau_0) = 1-\lambda_k(k^2 - 1)\cos\xi$. Since $T_k''(\tau_0)>0$ it follows immediately that

$$1 - \lambda_k\left(k^2 - 1\right)\cos\xi > 0. \tag{138}$$

Let us further assume that (τ) has exactly one zero. The problem of finding maximum absolute value of a_1 is connected to the problem of finding maximum of the minimum function (see Section 2.1). If waveforms possess unique global minimum at nondegenerate critical point then corresponding minimum function is a smooth function of parameters [13]. Consequently, assumption that (τ) has exactly one zero at nondegenerate critical point leads to the conclusion that coefficient a_1 is differentiable function of τ_0. First derivative of a_1 (see (43)) with respect to τ_0, taking into account that $\partial\xi/\partial\tau_0 = k$ (see (50)), can be expressed in the following factorized form:

$$\frac{\partial a_1}{\partial \tau_0} = \sin\tau_0\left[1 - \lambda_k\left(k^2 - 1\right)\cos\xi\right]. \tag{139}$$

From (138) and (139), it is clear that $\partial a_1/\partial\tau_0 = 0$ if and only if $\sin\tau_0 = 0$. According to Remark 12, assumption that(τ) has exactly one zero implies λ_k

< 1. From (51), (48), and $\lambda_k < 1$ it follows that $a_1 \cos \tau_0 + b_1 \sin \tau_0 < 0$, which, together with $\sin \tau_0 = 0$, implies that $a_1 \cos \tau_0 < 0$. Assumption $a_1 \leq 0$, together with relations $a_1 \cos \tau_0 < 0$ and $\sin \tau_0 = 0$, further implies $a_1 = 0$ and

$$\tau_0 = 0. \tag{140}$$

Insertion of $\tau_0 = 0$ into (38) leads to

$$T_k(\tau)$$

$$= [1 - \cos \tau]$$

$$\cdot \left[1 - (k-1) \lambda_k \cos \xi - 2\lambda_k \sum_{n=1}^{k-1} (k-n) \cos(n\tau + \xi) \right]. \tag{141}$$

Substitution of $\tau_0 = 0$ into (45) and (46) yields $a_k = \lambda_k \cos \xi$ and $b_k = -\lambda_k \sin \xi$, respectively. Replacing $\lambda_k \cos \xi$ with a_k and $\lambda_k \cos(n\tau + \xi)$ with $(a_k \cos n\tau + b_k \sin n\tau)$ in (141) immediately leads to (121).

Furthermore, $a_k = \lambda_k \cos \xi$, $b_k = -\lambda_k \sin \xi$, and (118) imply that

$$\delta = -\xi. \tag{142}$$

According to (38)–(40) and (142), it follows that (141) is nonnegative if and only if

$$\lambda_k \left[(k-1) \cos \delta + \frac{k \sin(\delta - \delta/k)}{\sin(\delta/k)} \right] < 1. \tag{143}$$

Notice that $a_k = \lambda_k \cos \delta$ implies that the following relation holds:

$$\lambda_k \left[(k-1) \cos \delta + \frac{k \sin(\delta - \delta/k)}{\sin(\delta/k)} \right]$$

$$= -a_k + k\lambda_k \frac{\sin \delta}{\sin(\delta/k)} \cos\left(\frac{\delta}{k}\right). \tag{144}$$

Finally, substitution of (144) into (143) leads to $k[\sin \delta / \sin(\delta/k)] \cos(\delta/k) < 1 + a_k$, which proves that (121) holds when $k\lambda_k[\sin \delta / \sin(\delta/k)] \cos(\delta/k) < 1 + a_k$.

Apart from nonnegative waveforms with exactly one zero at nondegenerate critical point, in what follows we will also consider other types of nonnegative waveforms with at least one zero. According to Proposition 9 and Remark 11, these waveforms can be described by (66)–(68) providing that $0 \leq |\xi| \leq \pi$.

According to (35), (0) $\geq$ 0 implies $1 + a_1 + a_k \geq 0$. Consequently, $a_1 \leq 0$ implies that $|a_1| \leq 1 + a_k$. On the other hand, according to (123), $|a_1| = 1 + a_k$ holds for waveforms of type (121). The converse is also true; $a_1 \leq 0$ and $|a_1| = 1 + a_k$ imply $a1 = -1 - a_k$, which further from (35) implies (0) = 0. Therefore, in what follows it is enough to consider only nonnegative waveforms which can be described by (66)–(68) and $0 \leq |\xi| \leq \pi$, with coefficients ak and bk satisfying $k\lambda_k[\sin \delta / \sin(\delta/k)] \cos(\delta/k) \geq 1 + a_k$.

For prescribed coefficients a_k and b_k, the amplitude $\lambda_k = \sqrt{a_k^2 + b_k^2}$ of kth harmonic is also prescribed. According to Remark 15 (see also Remark 16), λ_k is monotonically decreasing function of $x = \cos(\xi/k)$. The value of x can be obtained by solving (90) subject to the constraint $\cos(\pi/k) \leq x \leq 1$. Then λ_1 can be determined from (88). From (106) it immediately follows that maximal absolute value of $a_1 \leq 0$ corresponds to $q=0$, which from (104) and (120) further implies that

$$\delta = k\tau_0 - \xi.$$
(145)

Furthermore $q=0$, according to (107), implies that waveform zeros are

$$\tau_0 = \frac{(\delta + \xi)}{k}, \qquad \tau_0' = \tau_0 - \frac{2\xi}{k} = \frac{(\delta - \xi)}{k}.$$
(146)

Substitution of $\tau_0 = (\delta + \xi)/k$ into (66) yields (122), which proves that (122) holds when $k[\sin \delta / \sin(\delta/k)] \cos(\delta/k) \geq 1 + a_k$.

In what follows we prove that (121) also holds when $k[\sin \delta / \sin(\delta/k)] \cos(\delta/k) = 1 + a_k$. Substitution of $a_k = \lambda_k \cos \delta$ into $k[\sin \delta / \sin(\delta/k)] \cos(\delta/k) = 1 + a_k$ leads to

$$\lambda_k \left[(k-1) \cos \delta + \frac{k \sin (\delta - \delta/k)}{\sin (\delta/k)} \right] = 1.$$
(147)

As we mentioned earlier, relation (142) holds for all waveforms of type (121). Substituting (142) into (147) we obtain

$$\lambda_k \left[(k-1) \cos \xi + k \frac{\sin (\xi - \xi/k)}{\sin (\xi/k)} \right] = 1.$$
(148)

This expression can be rearranged as

$$\lambda_k \frac{k \sin ((k-1) \xi/k)}{\sin \xi/k} = 1 - (k-1) \lambda_k \cos \xi.$$
(149)

On the other hand, for waveforms of type (122), according to (68), relations (148) and (149) also hold. Substitution of $\tau_0 = (\delta + \xi)/k$ (see (145)) and (67) into (122) leads to

$$T_k(\tau)$$

$$= \lambda_k \left[1 - \cos(\tau - \tau_0)\right]$$

$$\cdot \left[\frac{k\sin((k-1)\xi/k)}{\sin \xi/k} - 2\sum_{n=1}^{k-1}(k-n)\cos(n(\tau-\tau_0)+\xi)\right].$$

(150)

Furthermore, substitution of (142) into (145) implies that $\tau_0 = 0$. Finally, substitution of $\tau_0 = 0$ and (149) into (150) leads to (141). Therefore (141) holds when $k[\sin \delta/\sin(\delta/k)]\cos(\delta/k) = 1 + a_k$, which in turn shows that (121) holds when $k\lambda_k[\sin \delta/\sin(\delta/k)]\cos(\delta/k) = 1 + a_k$. This completes the proof.

Nonnegative Waveforms with Maximal Absolute Value of Coefficient a_1 for $k=3$.

Nonnegative waveform of type (35) for $k=3$ is widely used in PA design (e.g., see [10]). In this subsection we illustrate results of Section 5.1 for this particular case. The case $k=2$ is presented in detail in [12].

Coefficients of fundamental harmonic of nonnegative waveform of type (35) with $k=3$ and maximal absolute value of coefficient $a_1 \leq 0$ for prescribed coefficients a_3 and b_3 ($\lambda_3 = \sqrt{a_3^2 + b_3^2}$), according to (123), (124), (134), (94), and (120), are equal to

$$a_1 = -1 - a_3, \qquad b_1 = -3b_3,$$

(151)

if $\lambda_3^2 \leq [(1 - 2a_3)/3]^3$,

$$a_1 = -\lambda_1 \cos\left(\frac{\delta}{3}\right), \qquad b_1 = -\lambda_1 \sin\left(\frac{\delta}{3}\right),$$

(152)

where $\lambda_1 = 3(\sqrt[3]{\lambda_3} - \lambda_3)$ and $\delta = \text{atan } 2(b_3, a_3)$, if $[(1 - 2a_3)/3]^3 \leq \lambda_3^2 \leq 1$. The line $\lambda_3^2 = [(1-2a_3)/3]^3$ (see case $k=3$ in Figure 11) separates points representing waveforms with coefficients satisfying (151) from points representing waveforms with coefficients satisfying (152). Waveforms described by (151) for $\lambda_3^2{}_3 < [(1 - 2a_3)/3]^3$ have exactly one zero at $\tau_0 = 0$. Waveforms described by (151) and (152) for $\lambda_3^2 = [(1 - 2a3)/3]^3$ also have zero at $\tau_0 = 0$. These waveforms as a rule have exactly two zeros. However there are two exceptions: one related to the maximally flat nonnegative waveform with coefficients $a_1 = -9/8$, $a_3 = 1/8$, and $b_1 = b_3 = 0$, which has only one zero, and the other related to the waveform with coefficients $a_1 = 0$, $a_3 = -1$, and $b_1 = b_2 = 0$, which has three zeros. Waveforms described by (152) for $[(1-2a_3)/3]^3 < \lambda_3^2 < 1$ have two zeros. Waveforms with $\lambda_3 = 1$ have only third harmonic (fundamental harmonic is zero).

Plot of contours of maximal absolute value of coefficient a_1, $a_1 \leq 0$, for prescribed coefficients a_3 and b_3 is presented in Figure 12. According to Remark 25, the waveform with maximum absolute value of $a_1 \leq 0$ is fully described with the following coefficients: $a_1 = -2/\sqrt{3}$, $a_3 = \sqrt{3}/9$ and $b_1 = b_3 = 0$. This waveform has two zeros at $\pm\pi/6$.

Two examples of nonnegative waveforms for $k=3$ and maximal absolute value of coefficient a_1, $a_1 \leq 0$, with prescribed coefficients a_3 and b_3 are presented in Figure 13. One waveform corresponds to the case $\lambda_3^2 < [(1 - 2a_3)/3]^3$ (solid line) and the other to the case $\lambda_3^2 > [(1 - 2a_3)/3]^3$ (dashed line). The waveform represented by solid line has one zero and its coefficients are $a_3 = -0.1$, $b_3 = 0.1$, $a_1 = -0.9$, and $b_1 = -0.3$. Dashed line corresponds to the waveform having two zeros with coefficients $a_3 = -0.1$, $b_3 = 0.3$, $a_1 = -0.8844$, and $b_1 = -0.6460$ (case $\lambda_3^2 > [(1 - 2a_3)/3]^3$).

NONNEGATIVE COSINE WAVEFORMS WITH AT LEAST ONE ZERO

Nonnegative cosine waveforms have proved to be of importance for waveform modelling in PA design (e.g., see [10]). In this section we consider nonnegative cosine waveforms containing fundamental and kth harmonic with at least one zero.

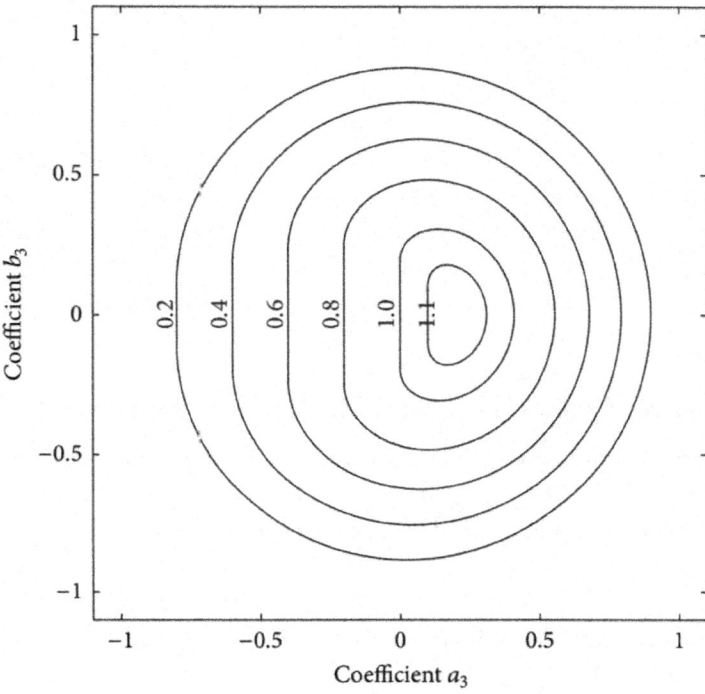

Figure 12: Contours of maximal absolute value of coefficient a_1, $a_1 \leq 0$, as a function of a_3 and b_3.

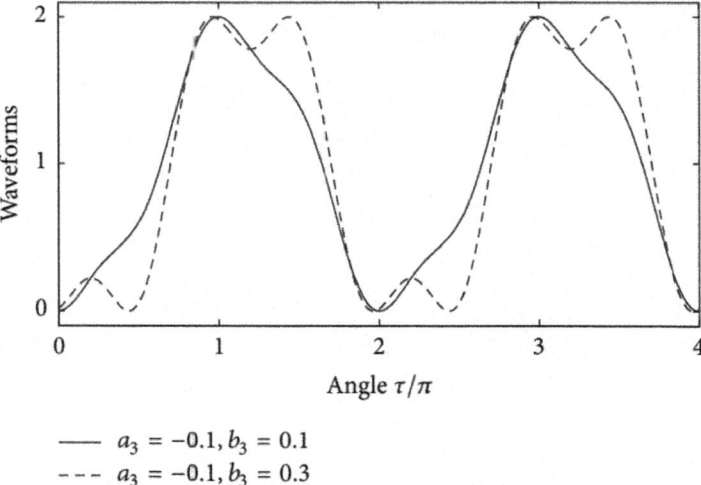

$$\text{---} \quad a_3 = -0.1, b_3 = 0.1$$
$$\text{- - -} \quad a_3 = -0.1, b_3 = 0.3$$

Figure 13: Nonnegative waveforms for $k=3$ and maximal absolute value of a_1, $a_1 \leq 0$, with prescribed coefficients a_3 and b_3.

Cosine waveform with dc component, fundamental and kth harmonic, can be obtained from (35) by setting $b_1 = b_k = 0$; that is,

$$T_k(\tau) = 1 + a_1 \cos \tau + a_k \cos k\tau. \tag{153}$$

In Section 6.1 we provide general description of nonnegative cosine waveforms of type (153) with at least one zero. We show that nonnegative cosine waveforms with at least one zero coincide with nonnegative cosine waveforms with maximal absolute value of coefficient a_1 for prescribed coefficient a_k. In Section 6.2 we illustrate results of Section 6.1 for particular case $k=3$.

Nonnegative Cosine Waveforms with at Least One Zero for $k \geq 2$

Amplitudes of fundamental and kth harmonic of cosine waveform of type (153) are $\lambda_1 = |a_1|$ and $\lambda_k = |a_k|$, respectively. According to (42), for nonnegative cosine waveforms of type (153) the following relation holds:

$$-1 \leq a_k \leq 1. \tag{154}$$

This explains why kth harmonic coefficient a_k in Proposition 26 goes through interval $[-1, 1]$.

Waveforms (153) with $a_1 \geq 0$ can be obtained from waveforms with $a_1 \leq 0$ by shifting by π, and therefore, without loss of generality, we can assume that $a_1 \leq 0$.

Proposition 26: Each nonnegative cosine waveform of type (153) with $a_1 \leq 0$ and at least one zero can be represented as

$$T_k(\tau) = [1 - \cos \tau] \left[1 - (k-1) a_k - 2a_k \sum_{n=1}^{k-1} (k-n) \cos n\tau \right], \tag{155}$$

$$if -1 \leq a_k \leq 1/(k^2 - 1), \text{ or}$$

$$T_k(\tau) = a_k [1 - \cos(\tau - \tau_0)][1 - \cos(\tau + \tau_0)]$$

$$\cdot \left[c_0 + 2 \sum_{n=1}^{k-2} c_n \cos n\tau \right], \tag{156}$$

where

$$c_n = \frac{\sin((k-n)\tau_0) \cos \tau_0 - (k-n) \cos((k-n)\tau_0) \sin \tau_0}{\sin^3 \tau_0}, \tag{157}$$

$$a_k = \frac{\sin \tau_0}{k \sin (k\tau_0) \cos \tau_0 - \cos (k\tau_0) \sin \tau_0},$$ (158)

$$|\tau_0| \le \frac{\pi}{k},$$ (159)

if $1/(k^2 - 1) \le a_k \le 1$.

Remark 27: Identity $[1-\cos(\tau-\tau_0)][1-\cos(\tau+\tau_0)] = [\cos \tau_0 - \cos \tau]^2$ implies that (156) can be rewritten as

$$T_k (\tau) = a_k \left[\cos \tau_0 - \cos \tau\right]^2 \left[c_0 + 2\sum_{n=1}^{k-2} c_n \cos n\tau \right].$$ (160)

Furthermore, substitution of (157) into (160) leads to

$$T_k (\tau) = a_k \left[\cos \tau_0 - \cos \tau\right]$$

$$\cdot \left[\frac{(k - 1) \sin k\tau_0}{\sin \tau_0} - 2\sum_{n=1}^{k-1} \frac{\sin ((k - n) \tau_0)}{\sin \tau_0} \cos n\tau \right].$$ (161)

Remark 28: All nonnegative cosine waveforms of type (153) with at least one zero and $a_1 \le 0$, except one of them, can be represented either in form (155) or form (156). This exception is maximally flat cosine waveform with $a_1 < 0$ which can be obtained from (155) for $a_k = 1/(k^2 - 1)$ or from (156) for $\tau_0 = 0$. Maximally flat cosine waveform with $a_1 < 0$ can also be obtained from (70) by setting $\tau_0 = 0$. Furthermore, setting $\tau_0 = 0$ in (71) leads to maximally flat cosine waveforms for $k \le 4$ and $a_1 < 0$.

Remark 29: Nonnegative cosine waveform of type (155) with $a_1 < 0$ and $-1 < a_k \le 1/(k^2 - 1)$ has exactly one zero at $\tau=0$. Nonnegative cosine waveform described by (156) with $a_1 < 0$ and $1/(k^2 - 1) < a_k < 1$ has two zeros at $\pm\tau_0$, where $0 < |\tau_0| < \pi/k$. For $a_k = \mp 1$, nonnegative cosine waveform of type (153) reduces to $(\tau) = 1 \mp \cos k\tau$ (clearly, these two waveforms both have k zeros).

Remark 30: Transformation of (155) into an additive form leads to the following relation:

$$a_1 = -1 - a_k,$$ (162)

where $-1 \le a_k \le 1/(k^2 - 1)$. Similarly, transformation of (156) leads to the following relation:

$$a_1 = -a_k \frac{k \sin k\tau_0}{\sin \tau_0},$$

(163)

where a_k is given by (158), $1/(k^2 -1) \leq a_k \leq 1$, and $|\tau_0| \leq \pi/k$. Notice that coefficients of maximally flat cosine waveform, namely, $a_k = 1/(k^2 - 1)$ and $a_1 = -k^2/(k^2 - 1)$, satisfy relation (162). They also satisfy relation (163) for $\tau_0 = 0$.

Remark 31: Nonnegative cosine waveforms of type (153) with at least one zero coincide with nonnegative cosine waveforms with maximal absolute value of coefficient a_1 for prescribed coefficient a_k.

In proving that Remark 31 holds, notice that expression (155) can be obtained from (121) by setting $b_k = 0$. Furthermore, if $a_k \geq 0$, then $\lambda_k = a_k$, which together with $b_k = 0$ and (118) implies $\delta=0$. In this case $k[\sin \delta/ \sin(\delta/k)]$ $\cos(\delta/k) \leq 1 + a_k$ becomes $k_2 \, a_k \leq 1 + a_k$. On the other hand, if $a_k < 0$, then $\lambda_k = -a_k$, which together with $bk = 0$ and (118) implies $|\delta| = \pi$. In this case $k[\sin \delta/ \sin(\delta/k)] \cos(\delta/k) \leq 1 + ak$ becomes $0 \leq 1 + a_k$. Therefore, every nonnegative cosine waveform of type (155) has maximal absolute value of coefficient a_1 for prescribed coefficient a_k, when $-1 \leq a_k \leq 1/(k^2 - 1)$.

Let us now show that expression (156) can be obtained from (122) by setting $b_k = 0$ and $a_k > 0$. For waveforms of type (122), according to (118), $b_k = 0$ and $a_k > 0$ imply $\delta=0$ and $\lambda_k = a_k$. Substitution of $\lambda_k = ak$ and $\delta=0$ into $k[\sin \delta/ \sin(\delta/k)] \cos(\delta/k) \geq 1 + a_k$ leads to $a_k \geq 1/(k^2 - 1)$. Furthermore, substitution of $\delta=0$ into (145) yields $\tau_0 = \xi/k$. Insertion of $\lambda_k = a_k$, $\delta=0$, and $\tau 0 = \xi/k$ into (122) leads to (156). Therefore, every nonnegative cosine waveform of type (156) has maximal absolute value of coefficient a_1 for prescribed coefficient a_k, when $1/(k^2 - 1) \leq a_k \leq 1$.

Proof of Proposition 26: Let us start with nonnegative cosine waveform of type (153) with $\lambda_k = |a_k| = 1$. According to Remark 7, $\lambda_k = |a_k| = 1$ implies that $\lambda_1 = |a_1| = 0$. Substitution of $a_k = -1$ into (155) and using (A.2) (see Appendices) lead to $(\tau) = 1 - \cos k\tau$. Consequently, (155) holds for $a_k = -1$. On the other hand, substitution of $a_k = 1$ into (158) yields $|\tau_0| = \pi/k$. Furthermore, substitution of $a_k = 1$ and $\tau_0 = \pi/k$ (or $\tau_0 = -\pi/k$) into (156), along with performing all multiplications and using (A.2), leads to $(\tau) = 1 + \cos k\tau$. Consequently, (156)–(158) hold for $a_k = 1$ and $|\tau_0| = \pi/k$.

It is easy to see that $\lambda_k = |a_k| < 1$ and $(\tau_0)=0$ for some τ_0 imply $\lambda_1 = |a_1| \neq 0$. Therefore in what follows we assume that $|a_k| \neq 1$ and $a_1 < 0$.

Cosine waveforms are even functions of τ. Therefore, if nonnegative cosine waveform has exactly one zero it has to be either at 0 or at π. On the other hand, if nonnegative cosine waveform with $a_1 \neq 0$ has exactly two zeros then these zeros are placed at $\pm\tau_0$, such that τ_0 is neither 0 nor π.

In order to prove that (155) holds for $-1 < a_k \leq 1/(1-k^2)$, let us start by referring to the description (38) of nonnegative waveforms with at least one zero. As we mentioned earlier, for nonnegative cosine waveform with exactly one zero (denoted by τ_0) it is either $\tau_0 = 0$ or $\tau_0 = \pi$. Therefore in both cases $\sin \tau_0 = 0$. Substitution of $\sin \tau_0 = 0$ into (43), together with $a_1 \neq 0$ and $\lambda_k = |a_k| < 1$, leads to

$$\tau_0 = 0. \tag{164}$$

Clearly $\tau_0 = 0$, $b_1 = 0$, and $b_k = 0$, according to (44) and (46), imply $\lambda_k \sin \xi = 0$. Since $\lambda_k = |a_k|$ it follows that $|a_k|\sin \xi = 0$ also holds, which further implies $\lambda_k = a_k = 0$ or $\sin \xi = 0$. In the case when $\lambda_k = a_k = 0$, from (164) and (43) we obtain $a_1 = -1$, which further implies that $(\tau) = 1 - \cos \tau$. Consequently (155) holds for $a_k = 0$. In the case when $\sin \xi = 0$, from (164) and (45) we obtain $a_k = \lambda_k$ if $\xi = 0$, or $a_k = -\lambda_k$ if $\xi = \pi$. Relations $a_k = \lambda_k$ and $\xi = 0$, according to (40), imply that $0 \leq a_k \leq 1/(1 - k^2)$. Substitution of $\xi = 0$, $\lambda_k = a_k$, and (164) into (38) leads to (155), which proves that (155) holds for $0 \leq a_k \leq 1/(1 - k2)$. On the other hand, relations $a_k = -\lambda_k$ and $\xi = \pi$, according to (40), imply that $-1 < a_k \leq 0$. Substitution of $\xi = \pi$, $\lambda_k = -a_k$, and (164) into (38) also leads to (155), which proves that (155) also holds for $-1 < a_k \leq 0$. Consequently (155) holds for $-1 < a_k \leq 1/(1 - k^2)$.

In what follows we first prove that (156)-(157) hold for $1/(1 - k^2) < 1$. For this purpose let us start with nonnegative waveforms with two zeros described by (66). As we mentioned before, nonnegative cosine waveforms with two zeros have zeros at τ_0 and $-\tau_0$, such that $\tau_0 \neq 0$ and $\tau0 \neq$. Relations $a_1 < 0$ and $b_1 = 0$, according to (84), imply $\cos(\tau_0 - \xi/k) = 1$ and therefore

$$\frac{\xi}{k} = \tau_0. \tag{165}$$

From $\xi/k = \tau_0$ and $0 < |\xi| < \pi$ it follows that $0 < |\tau_0| < \pi/k$. Insertion of $\xi/k = \tau_0$ into (45) yields $ak = \lambda_k$. Relations $a_k = \lambda_k$ and (82) imply that $1/(1 - k^2) < 1$. Substitution of $\lambda_k = a_k$ and $\xi/k = \tau_0$ into (66)-(68) leads to (156)-(158), which proves that (156)-(158) hold for $1/(1 - k^2) < 1$ and $0 < |\tau_0| < \pi/k$.

Finally, substitution of $a_k = 1/(1 - k^2)$ and $\tau_0 = 0$ into (161) leads to

$$T_k(\tau) = \frac{[1 - \cos \tau]}{(1 - k^2)} \left[k(k-1) - 2\sum_{n=1}^{k-1}(k-n)\cos n\tau \right]. \tag{166}$$

Waveform (166) coincides with waveform (155) for $a_k = 1/(1 - k^2)$, which in turn proves that (156) holds for $a_k = 1/(1 - k^2)$ and $\tau_0 = 0$. This completes the proof.

Nonnegative Cosine Waveforms with at Least One Zero for $k=3$

In this subsection we consider nonnegative cosine waveforms with at least one zero for $k=3$ (for case $k=2$ see [12]).

Cosine waveform with fundamental and third harmonic reads

$$T_3(\tau) = 1 + a_1 \cos \tau + a_3 \cos 3\tau. \tag{167}$$

For $a_1 \leq 0$ and $-1 \leq a_3 \leq 1/8$, according to (155), nonnegative cosine waveform of type (167) with at least one zero can be expressed as

$$T_3(\tau) = (1 - \cos \tau)\left[1 - 2a_3(1 + 2\cos \tau + \cos 2\tau)\right]. \tag{168}$$

From $T_3(\tau + \pi) = 2 - T_3(\tau)$, it immediately follows that, for $a_1 \geq 0$ and $-1/8 \leq a_3 \leq 1$, $T_3(\tau)$ can be expressed as

$$T_3(\tau) = (1 + \cos \tau)\left[1 + 2a_3(1 - 2\cos \tau + \cos 2\tau)\right]. \tag{169}$$

For $a_1 \leq 0$ and $1/8 \leq a_3 \leq 1$, from (158) it follows that $a_3 = [8\cos^3 \tau_0]^{-1}$. This relation, along with (160) and (157), further implies that $T_3(\tau)$ can be expressed as

$$T_3(\tau) = \frac{\left[\cos \tau_0 - \cos \tau\right]^2 \left[2\cos \tau_0 + \cos \tau\right]}{2\cos^3 \tau_0}, \tag{170}$$

providing that $|\tau_0| \leq \pi/3$. From $T_3(\tau + \pi) = 2 - T_3(\tau)$, it follows that (170) also holds for $a_1 \geq 0$ and $-1 \leq a_3 \leq -1/8$, providing that $\tau_0 \in [2\pi/3, 4\pi/3]$.

Maximally flat nonnegative cosine waveform of type (167) with $a_1 < 0$ (minimum at $\tau_0 = 0$) reads $T_3(\tau) = [1 - \cos \tau]^2 [1 + (1/2) \cos \tau]$. Dually, maximally flat nonnegative cosine waveform with $a_1 > 0$ (minimum at $\tau_0 = \pi$) reads $T_3(\tau) = [1 + \cos \tau]^2 [1 - (1/2) \cos \tau]$.

In what follows we provide relations between coefficients $a1$ and $a3$ of nonnegative cosine waveforms of type (167) with at least one zero.

For $a_1 \leq 0$, conversion of (168) into an additive form immediately leads to the following relation:

$$a_1 = -1 - a_3 \quad \text{for} \quad -1 \leq a_3 \leq \frac{1}{8}. \tag{171}$$

Conversion of (170) into an additive form leads to $a_1 = -3a_3(2 \cos 2\tau_0 + 1)$, which can be also expressed as $a_1 = -3a_3(4\cos^2 \tau 0 - 1)$. For $a_1 \leq 0$, relations $|\tau_0| \leq \pi/3$, $a_1 = -3a_3(4\cos^2 \tau_0 - 1)$, and $a_3 = [8\cos^3 \tau_0]^{-1}$ lead to

$$a_1 = -3 \left[\sqrt[3]{a_3} - a_3 \right] \quad \text{for } \frac{1}{8} \leq a_3 \leq 1.$$

(172)

Similarly for $a_1 \geq 0$, conversion of (169) into an additive form leads to the following relation:

$$a_1 = 1 - a_3 \quad \text{for } -\frac{1}{8} \leq a_3 \leq 1.$$

(173)

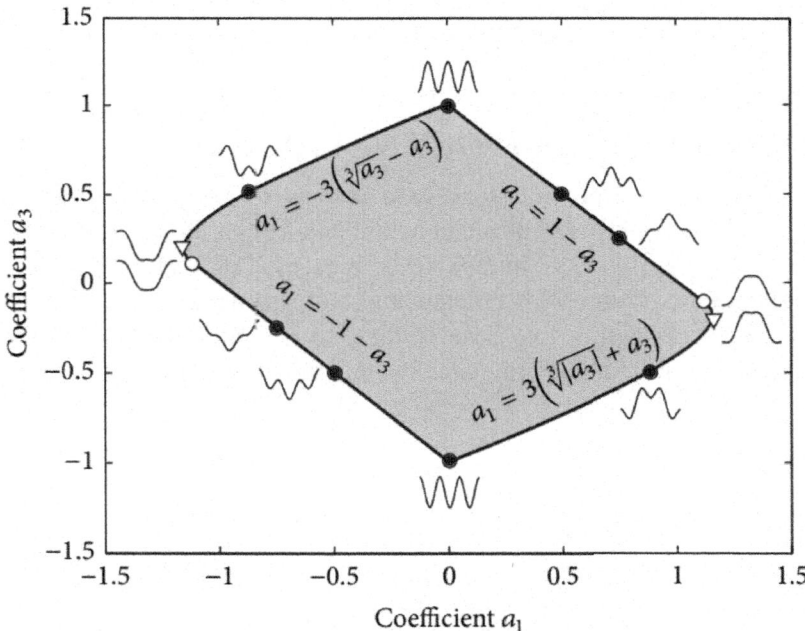

Figure 14: Parameter space of cosine waveforms for $k=3$.

For waveform of type (170) with $a_1 \geq 0$, relations $\tau_0 \in [2\pi/3, 4\pi/3]$, $a_1 = -3a3(4\cos^2 \tau_0 - 1)$, and $a_3 = [8\cos^3 \tau_0]^{-1}$ lead to

$$a_1 = 3 \left[\sqrt[3]{|a_3|} + a_3 \right] \quad \text{for} -1 \leq a_3 \leq -\frac{1}{8}.$$

(174)

Every cosine waveform of type (167) corresponds to a pair of real numbers $(a1, a3)$ and vice versa. Points (a_1, a_3) in grey area in Figure 14 correspond to nonnegative cosine waveforms for $k=3$. The points at the boundary of grey area correspond to nonnegative cosine waveforms with at least one zero. A number of shapes of nonnegative cosine waveforms with $k=3$ and at least one zero, plotted on interval $[-\pi, \pi]$, are also presented in Figure 14. The boundary of grey area in Figure 14 consists of four line segments described by relations (171)–(174). The common point of line segments (172) and (173) is cusp point with coordinates $a_1 = 0$ and $a_3 = 1$. Another cusp point, with coordinates $a_1 = 0$ and $a_3 = -1$, is the common point of line segments (171) and (174). The common point of line segments (171)-(172) has coordinates $(-9/8, 1/8)$ and common point of line segments (173)-(174) has coordinates $(9/8, -1/8)$. These points are represented by white circle dots and they correspond to maximally flat cosine waveforms (e.g., see [21]). White triangle dots with coordinates $(2/\sqrt{3}, -\sqrt{3}/9)$ and $(-2/\sqrt{3}, \sqrt{3}/9)$ refer to the nonnegative cosine waveforms with maximum value of amplitude of fundamental harmonic.

FOUR CASE STUDIES OF USAGE OF NONNEGATIVE WAVEFORMS IN PA EFFICIENCY ANALYSIS

In this section we provide four case studies of usage of description of nonnegative waveforms with fundamental and kth harmonic in PA efficiency analysis. In first two case studies, to be presented in Section 7.1, voltage is nonnegative waveform with fundamental and second harmonic with at least one zero. In remaining two case studies, to be considered in Section 7.2, voltage waveform contains fundamental and third harmonic.

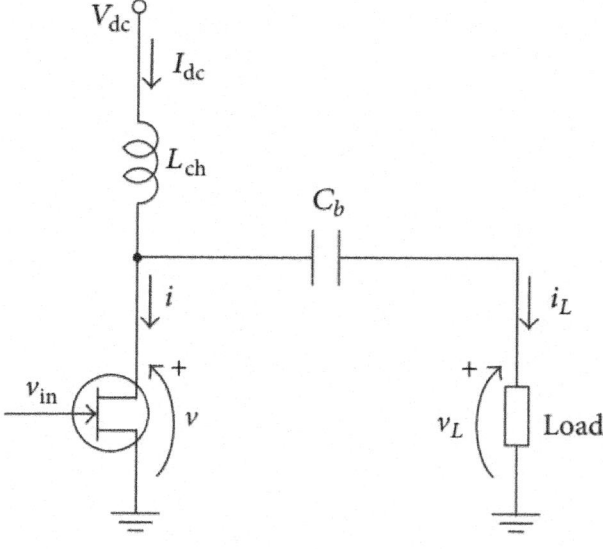

Figure 15: Generic PA circuit diagram.

Let us consider generic PA circuit diagram, as shown in Figure 15. We assume here that voltage and current waveforms at the transistor output are

$$v(\theta) = 1 + a_{1v} \cos\theta + b_{1v} \sin\theta + a_{kv} \cos k\theta + b_{kv} \sin k\theta,$$

$$i(\theta) = 1 + a_{1i} \cos\theta + \sum_{n=2}^{\infty} a_{ni} \cos n\theta, \tag{175}$$

where θ stands for ωt. Both waveforms are normalized in the sense that dc components of voltage and current are $V_{dc} = 1$ and $I_{dc} = 1$, respectively. Under assumption that blocking capacitor C_b behaves as short-circuit at the fundamental and higher harmonics, current and voltage waveforms at the load are

$$v_L(\theta) = a_{1v} \cos\theta + b_{1v} \sin\theta + a_{kv} \cos k\theta + b_{kv} \sin k\theta,$$

$$i_L(\theta) = -a_{1i} \cos\theta - \sum_{n=2}^{\infty} a_{ni} \cos n\theta. \tag{176}$$

In terms of coefficients of voltage and current waveforms, the load impedance at fundamental harmonic is $z_1 = -(a_{1v} - jb_{1v})/a_{1i}$, whereas load impedance at kth harmonic is $z_k = -(a_{kv} - jb_{kv})/a_{ki}$. All other harmonics are short-circuited ($z_n = 0$ for $n \neq 1$ and $n \neq k$). Time average output power of PA (e.g., see [10]) with waveform pair (175) at fundamental frequency can be expressed as

$$P_1 = -\frac{a_{1i}a_{1v}}{2}. \tag{177}$$

For normalized waveforms (175) with $V_{dc} = 1$ and $I_{dc} = 1$, dc power is $P_{dc} = 1$. Consequently, PA efficiency $\eta = P_1/P_{dc}$ (e.g., see, [10, 26]) is equal to

$$\eta = -\frac{a_{1i}a_{1v}}{2}. \tag{178}$$

Thus, time average output power $P1$ of PA with pair of normalized waveform (175) is equal to efficiency (178).

Power utilization factor (PUF) is defined [26] as "the ratio of power delivered in a given situation to the power delivered by the same device with the same supply voltage in Class A mode." Since the output power in class-A mode is $P_{1,\text{class-A}} = \max[V(\theta)] \cdot \max[i(\theta)]/8$ (e.g., see [9]), it follows that power utilization factor PUF $= P_1/P_{1,\text{class-A}}$ for PA with pair of normalized waveforms (175) can be expressed as

$$PUF = \frac{8\eta}{\max\left[v\left(\theta\right)\right] \cdot \max\left[i\left(\theta\right)\right]}. \tag{179}$$

Nonnegative Waveforms for $k=2$ in PA Efficiency Analysis

In this subsection we provide two case studies of usage of description of nonnegative waveforms with fundamental and second harmonic ($k=2$) in PA efficiency analysis. For more examples of usage of descriptions of nonnegative waveforms with fundamental and second harmonic in PA efficiency analysis see [12].

Case Study

In this case study we consider efficiency of PA for given second harmonic impedance, providing that voltage is nonnegative waveform with fundamental and second harmonic and current is "half-sine" waveform frequently used in efficiency analysis of classical PA operation (e.g., see [10]).

Standard model of current waveform for classical PA operation has the form (e.g., see [10, 26])

$$i_D\left(\theta\right) = \begin{cases} I_D\left[\cos\theta - \cos\left(\dfrac{\alpha}{2}\right)\right], & |\theta| \le \dfrac{\alpha}{2} \\ 0, & \dfrac{\alpha}{2} \le |\theta| \le \pi, \end{cases} \tag{180}$$

where α is conduction angle and $I_D > 0$. Since (θ) is even function, it immediately follows that its Fourier series contains only dc component and cosine terms:

$$i_D\left(\theta\right) = I_{dc} + \sum_{n=1}^{\infty} I_n \cos n\theta. \tag{181}$$

The dc component of the waveform (180) is

$$I_{dc} = \frac{I_D \alpha}{2\pi}\left[\operatorname{sinc}\left(\frac{\alpha}{2}\right) - \cos\left(\frac{\alpha}{2}\right)\right], \tag{182}$$

where $\operatorname{sinc} x = (\sin x)/x$. The coefficient of the fundamental harmonic component reads

$$I_1 = \frac{I_D \alpha}{2\pi}\left(1 - \operatorname{sinc}\alpha\right), \tag{183}$$

and the coefficient of nth harmonic component can be written in the form

$$I_n = \frac{I_D}{n\pi} \left[\frac{\sin\left((n-1)\,\alpha/2\right)}{(n-1)} - \frac{\sin\left((n+1)\,\alpha/2\right)}{(n+1)} \right], \quad n \geq 2. \tag{184}$$

For "half-sine" current waveform, conduction angle is equal to π (class-B conduction angle). According to (182), this further implies that $I_{dc} = I_D/\pi$. To obtain normalized form of waveform (180), we set $I_{dc} = 1$ which implies that $I_D = \pi$. Furthermore, substitution of $\alpha=\pi$ and $I_D = \pi$ in (180) leads to

$$i\,(\theta) = \begin{cases} \pi\cos\theta, & |\theta| < \dfrac{\pi}{2}, \\[2mm] 0, & \dfrac{\pi}{2} < |\theta| \leq \pi. \end{cases} \tag{185}$$

Similarly, substitution of $I_D = \pi$ and $\alpha=\pi$ into (183) and (184) leads to the coefficients of waveform (185). Coefficients of fundamental and second harmonic, respectively, are

$$a_{1i} = \frac{\pi}{2}, \qquad a_{2i} = \frac{2}{3}. \tag{186}$$

On the other hand, voltage waveform of type (35) for $k = 2$ reads

$$v\,(\theta) = 1 + a_{1v}\cos\theta + b_{1v}\sin\theta + a_{2v}\cos 2\theta + b_{2v}\sin 2\theta. \tag{187}$$

This waveform contains only fundamental and second harmonic, and therefore all harmonics of order higher than two are short-circuited ($\underline{Z}_n = 0$ for $n>2$). For current voltage pair (185) and (187), load impedance at fundamental harmonic is $\underline{Z}_1 = -(a_{1v} - jb_{1v})/a_{1i}$, whereas load impedance at second harmonic is $\underline{Z}_2 = -(a_{2v} - jb_{2v})/a_{2i}$. According to our assumption, the load is passive and therefore $\mathrm{Re}\{\underline{Z}_1\}>0$ and $\mathrm{Re}\{\underline{Z}_2\}\geq 0$, which further imply $a_{1i}a_{1v} < 0$ and $a_{2i}a_{2v} \leq 0$, respectively.

It is easy to see that problem of finding maximal efficiency of PA with current-voltage pair (185) and (187) for prescribed second harmonic impedance can be reduced to the problem of finding voltage waveform of type (187) with maximal coefficient $|a_{1v}|$, for prescribed coefficients of second harmonic (see Section 5).

The following algorithm (analogous to Algorithm 22 presented in [12]) provides the procedure for calculation of maximal efficiency with current-voltage pair (185) and (187) for prescribed second harmonic impedance. The definition of function $\mathrm{atan}\,2(y, x)$, which appears in the step (iii) of the following algorithm, is given by (105).

Algorithm 32

1. Choose $\underline{z}_2 = r_2 + jx_2$ such that $|\underline{z}_2| \leq 1/|a_{2i}|$,

2. calculate $a_{2v} - jb_{2v} = -\underline{z}_2 a_{2i}$ and $\lambda_{2v} = \sqrt{a_{2v}^2 + b_{2v}^2}$,

3. if $\quad 2\lambda_{2v} \leq 1 - a_{2v} \quad$ then $\quad$ calculate $\quad a_{1v} = -1 - a_{2v} \quad$ and $b_{1v} = -2b_{2v}$; else, calculate $\lambda_{1v} = \sqrt{8\lambda_{2v}(1 - \lambda_{2v})}$, $\theta_{0v} - \xi_v/2 = (1/2)\mathrm{atan2}(b_{2v}, a_{2v})$, $a_{1v} = -\lambda_{1v}\cos(\theta_{0v} - \xi_v/2)$, and $b_{1v} = -\lambda_{1v}\sin(\theta_{0v} - \xi_v/2)$,

4. calculate efficiency $\eta = -a_{1i}a_{1V}/2$,

5. calculate $\underline{z}_1 = -(a_{1v} - jb_{1v})/a_{1i}$ and $\underline{z}_{2n} = \underline{z}_2/\mathrm{Re}\{\underline{z}_1\}$.

In this case study, coefficients of fundamental and second harmonic of current waveform are given by (186). Maximal efficiency of PA associated with the waveform pair (185) and (187), as a function of normalized second harmonic impedance $\underline{z}_{2n} = \underline{z}_2/\mathrm{Re}\{\underline{z}_1\}$, is presented in Figure 16(a). As can be seen from Figure 16(a), efficiency of 0.78 is achieved at the edge of Smith chart, where second harmonic impedance has small resistive part. Corresponding PUF calculated according to (179) is presented in Figure 16(b). Peak efficiency $\eta = \pi/4 = 0.7854$ and peak value of PUF = 1 are attained when second harmonic is short-circuited (which corresponds to ideal class-B operation [10, 26]).

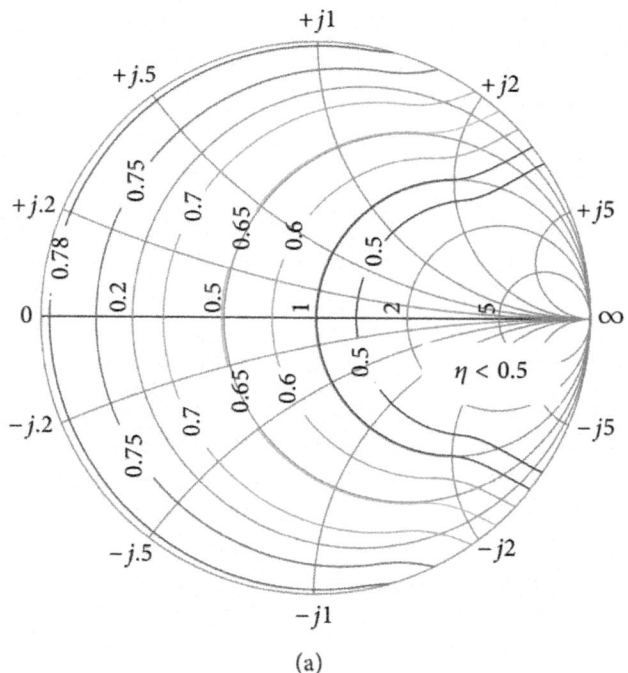

(a)

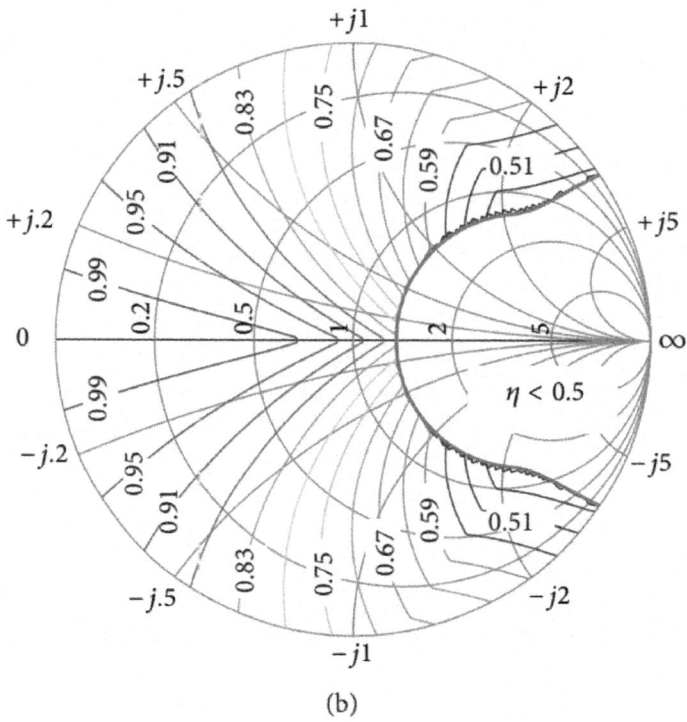

(b)

Figure 16: (a) Contours of maximal efficiency of PA and (b) contours of corresponding PUF, associated with the waveform pair (185) and (187), as functions of normalized second harmonic impedance $z_{2n} = z_2/\operatorname{Re}\{z_1\}$.

For example, for second harmonic impedance $z_2 = 0.1 - j0.5$ and current waveform (185), from Algorithm 32 it follows that $2\lambda_{2V} \leq 1 - a_{2V}$. Furthermore, according to step (iii) of above algorithm, maximal efficiency of PA is attained with voltage waveform of type (187) with coefficients $a_{2V} = -0.0667$, $b_{2V} = -0.3333$, $a_{1V} = -0.9333$, and $b_{1V} = 0.6667$ (see Figure 17). Corresponding efficiency, PUF, and normalized second harmonic impedance are $\eta = 0.7330$, PUF = 0.7572, and $z_{2n} = 0.1683 - j0.8415$, respectively.

On the other hand, for second harmonic impedance $z_2 = 0.1 - j0.8$ and current waveform (185), from Algorithm 32 it follows that $2\lambda_{2V} > 1 - a_{2V}$. Then, according to step (iii) of above algorithm, maximal efficiency is attained with voltage waveform of type (187) with coefficients $a_{2V} = -0.0667$, $b_{2V} = -0.5333$, $a_{1V} = -0.9333$, and $b_{1V} = 1.0572$ (see Figure 18). Efficiency, PUF, and normalized second harmonic impedance are $\eta = 0.7330$, PUF = 0.6332, and $z_{2n} = 0.1683 - j1.3465$, respectively.

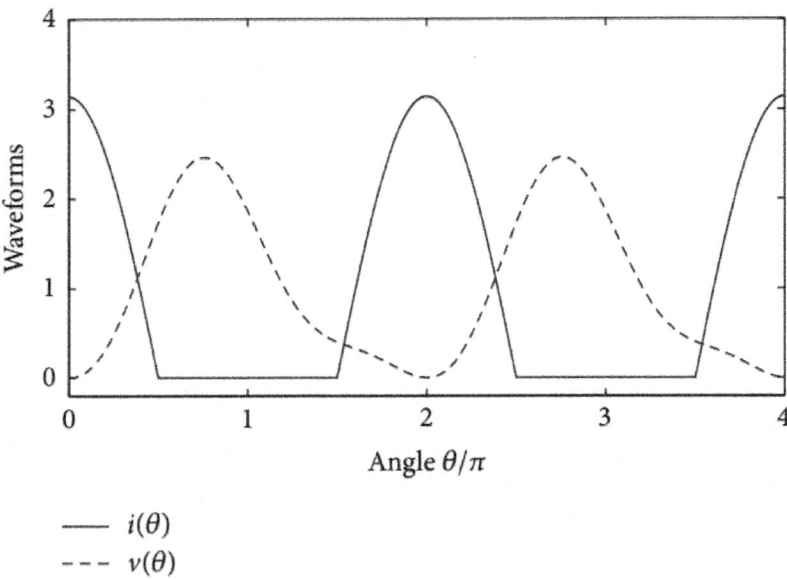

Figure 17: Waveform pair (185) and (187) that provides maximal efficiency for $z_2 = 0.1 - j0.5$.

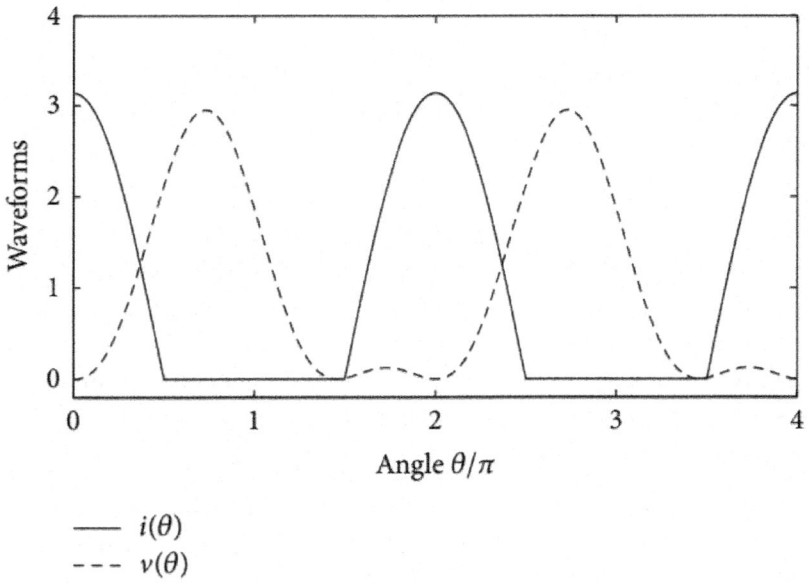

Figure 18: Waveform pair (185) and (187) that provides maximal efficiency for $z_2 = 0.1 - j0.8$.

Case Study

As another case study, let us consider the efficiency of PA, providing that current waveform is nonnegative cosine waveform up to third harmonic with maximum value of amplitude of fundamental harmonic [22] (see also [8]):

$$i(\theta) = 1 + \frac{1 + \sqrt{5}}{2} \cos\theta + \frac{2\sqrt{5}}{5} \cos 2\theta + \frac{5 - \sqrt{5}}{10} \cos 3\theta,$$

(188)

and voltage waveform is nonnegative waveform of type (187). Load impedances at fundamental, second, and third harmonic are $z_1 = -(a_{1v} - jb_{1v})/a_{1i}$, $z_2 = -(a_{2v} - jb_{2v})/a_{2i}$, and $z_3 = 0$, respectively. According to our assumption, the load is passive and therefore $\mathrm{Re}\{z_1\} > 0$ and $\mathrm{Re}\{z_2\} \geq 0$, which further imply $a_{1i}a_{1v} < 0$ and $a_{2i}a_{2v} \leq 0$, respectively.

Because current waveform (188) contains only cosine terms and voltage waveform is the same as in previous case study, the procedure for calculation of maximal efficiency of PA with waveform pair (187)-(188) is the same as presented in Algorithm 32.

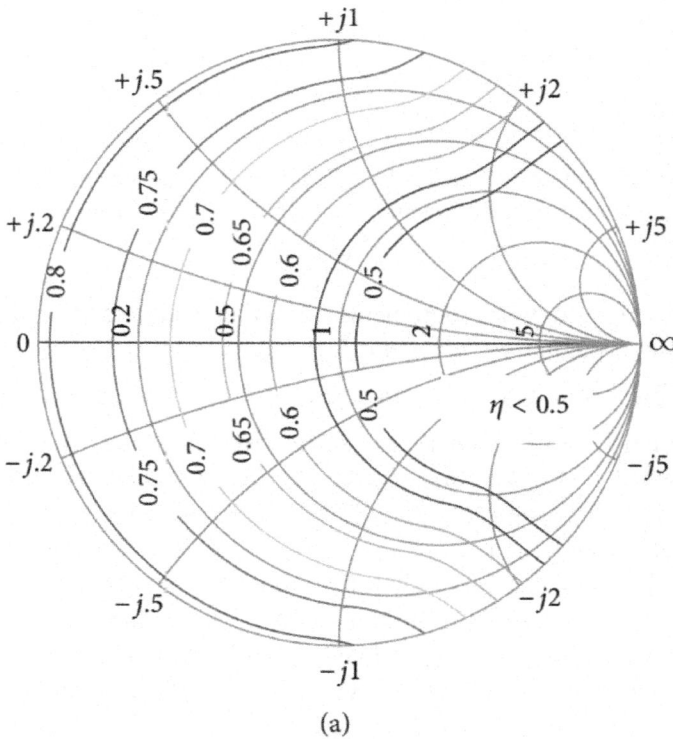

(a)

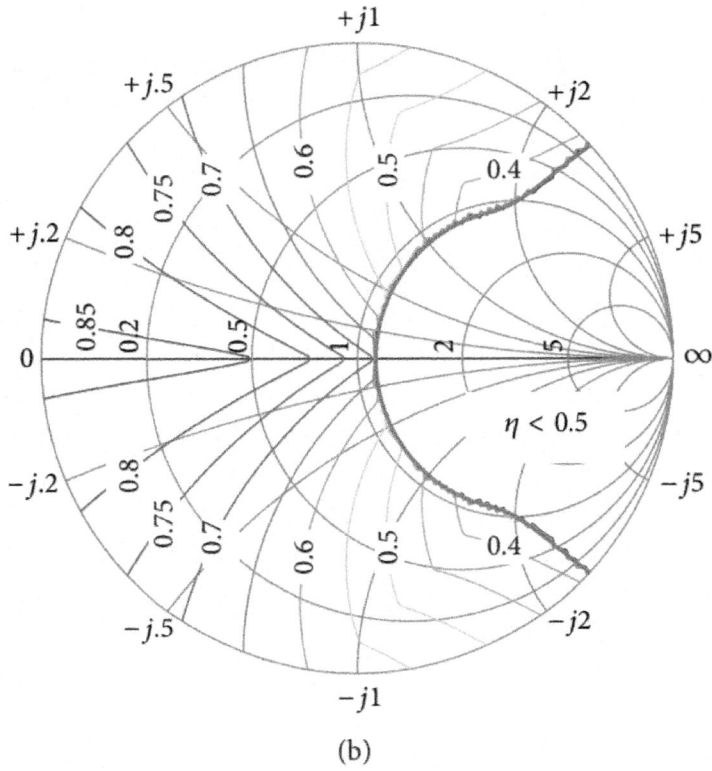

(b)

Figure 19: (a) Contours of maximal efficiency of PA and (b) contours of corresponding PUF, associated with the waveform pair (187)-(188) as functions of normalized second harmonic impedance $\underline{z}_{2n} = \underline{z}_2 / \operatorname{Re}\{\underline{z}_1\}$.

In this case study the coefficients of fundamental and second harmonic of current waveform are $a_{1i} = (1 + \sqrt{5})/2$ and $a_{2i} = 2\sqrt{5}/5$, respectively.

Maximal efficiency of PA associated with the waveform pair(187)-(188), as a function of normalized second harmonic impedance $\underline{z}_{2n} = \underline{z}_2 / \operatorname{Re}\{\underline{z}_1\}$, is presented in Figure 19(a). Efficiency of 0.8 is achieved at the edge of Smith chart, where second harmonic impedance has small resistive part. The theoretical upper bound $\eta = (1 + \sqrt{5})/4 \approx 0.8090$ is attained when second harmonic is short-circuited. When this upper bound is reached, both second and third harmonic are short-circuited which implies that we are dealing with finite harmonic class-C [6, 8], or dually, when current and voltage interchange their roles, with finite harmonic inverse class-C [6, 9]. Corresponding PUF, calculated according to (179), is presented in Figure 19(b). Peak value of PUF ≈ 0.8541 is attained when second harmonic is short-circuited.

For example, for second harmonic impedance $z2 = 0.07 - j0.4$ and current waveform (188), from Algorithm 32 it follows that $2\lambda_{2v} \leq 1 - a_{2v}$. Furthermore, according to step (iii) of Algorithm 32, maximal efficiency of PA is attained with voltage waveform of type (187) with coefficients $a_{2v} = -0.0626$, $b_{2v} = -0.3578$, $a_{1V} = -0.9374$, and $b_{1V} = 0.7155$ (see Figure 20). Corresponding efficiency, PUF, and normalized second harmonic impedance are $\eta = 0.7584$, PUF = 0.6337, and $z_{2n} = 0.1208 - j0.6904$, respectively.

On the other hand, for $z_2 = 0.05 - j0.7$ and current waveform (187) it follows that $2\lambda_{2v} > 1 - a_{2v}$. Then, according to step (iii) of Algorithm 32, the maximal efficiency is attained with voltage waveform of type (187) with coefficients $a_{2V} = -0.0447$, $b_{2v} = -0.6261$, $a_{1V} = -0.9318$, and $b_{1V} = 1.0007$ (see Figure 21). Efficiency, PUF, and normalized second harmonic impedance are $\eta = 0.7538$, PUF = 0.5314, and $z_{2n} = 0.0868 - j1.2156$, respectively.

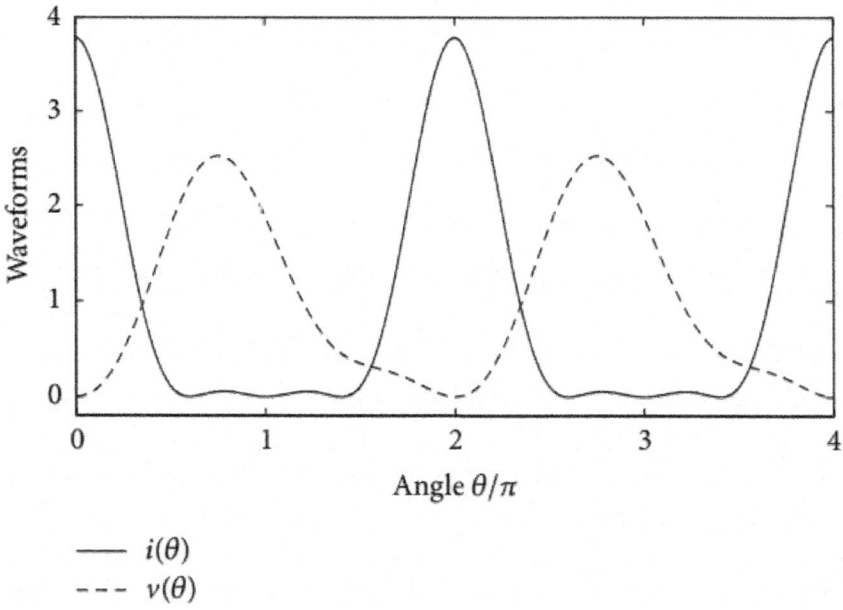

Figure 20: Waveform pair (187)-(188) that provides maximal efficiency for $z_2 = 0.07 - j0.4$.

Nonnegative Waveforms for $k=3$ in PA Efficiency Analysis

In this subsection we provide another two case studies of usage of description of nonnegative waveforms in PA efficiency analysis, this time with fundamental and third harmonic ($k=3$).

Case Study

Let us consider current-voltage pair such that voltage is nonnegative waveform with fundamental and third harmonic:

$$v(\theta) = 1 + a_{1v}\cos\theta + b_{1v}\sin\theta + a_{3v}\cos 3\theta + b_{3v}\sin 3\theta, \tag{189}$$

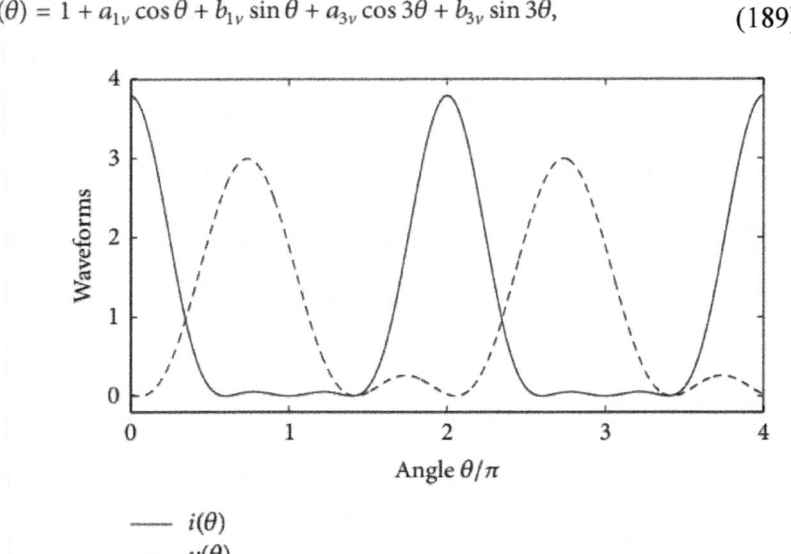

—— $i(\theta)$

--- $v(\theta)$

Figure 21: Waveform pair (187)-(188) that provides maximal efficiency for $\underline{Z}_2 = 0.05 - j0.7$.

and current is nonnegative cosine waveform given by (188). Load impedances at fundamental, second, and third harmonic are $\underline{Z}_1 = -(a_{1v} - jb_{1v})/a_{1i}$, $\underline{Z}_2 = 0$, and $\underline{Z}_3 = -(a_{3v} - jb_{3v})/a_{3i}$, respectively. According to our assumption, the load is passive and therefore Re$\{\underline{Z}_1\}$>0 and Re$\{\underline{Z}_3\}$≥0, which further imply $a_{1i}a_{1v} < 0$ and $a_{3i}a_{3v} \leq 0$.

In this subsection we consider the problem of finding maximal efficiency of PA with waveform pair (188)-(189) for given third harmonic impedance. As we mentioned earlier, problem of finding maximal efficiency of PA with current-voltage pair (188)-(189) for prescribed third harmonic impedance can be reduced to the problem of finding voltage waveform of type (189) with maximal coefficient $|a_{1v}|$, for prescribed coefficients of third harmonic (see Section 5.2).

The following algorithm provides the procedure for calculation of maximal efficiency with current-voltage pair (188)-(189). The definition of function atan $2(y, x)$, which appears in step (iii) of the following algorithm, is given by (105).

Algorithm 33

(i) Choose $\underline{z}_3 = r_3 + jx_3$ such that $|\underline{z}_3| \leq 1/|a_{3i}|$,

(ii) calculate $a_{3v} - jb_{3v} = -\underline{z}_3 a_{3i}$ and $\lambda_{3v} = \sqrt{a_{3v}^2 + b_{3v}^2}$,

(iii) if $27\lambda_{3v}^2 \leq (1 - 2a_{3v})^3$ then calculate $a_{1v} = -1 - a_{3v}$
and $b_{1v} = -3b_{3v}$; else, calculate $\lambda_{1v} = 3(\sqrt[3]{\lambda_{3v}} - \lambda_{3v})$, $\theta_{0v} - \xi_v/3 = (1/3)\mathrm{atan}\,2(b_{3v}, a_{3v})$, $a_{1v} = -\lambda_{1v}\cos(\theta_{0v} - \xi_v/3)$, and $b_{1v} = -\lambda_{1v}\sin(\theta_{0v} - \xi_v/3)$,

(iv) calculate efficiency $\eta = -a_{1i}a_{1v}/2$,

(v) calculate $\underline{z}_1 = -(a_{1v} - jb_{1v})/a_{1i}$ and $\underline{z}_{3n} = \underline{z}_3/\,\mathrm{Re}\{\underline{z}_1\}$.

In this case study coefficients of fundamental and third harmonic of current waveform are $a_{1i} = (1 + \sqrt{5})/2$ and $a_{3i} = (5 - \sqrt{5})/10$, respectively. For the waveform pair (188)-(189), maximal efficiency of PA as a function of normalized third harmonic impedance $\underline{z}_{3n} = \underline{z}_3 / \mathrm{Re}\{\underline{z}_1\}$ is presented in Figure 22. Efficiency of 0.8 is reached when third harmonic impedance has small resistive part. Peak efficiency $\eta = (1 + \sqrt{5})/4 \approx 0.8090$ is achieved when third harmonic is short circuited.

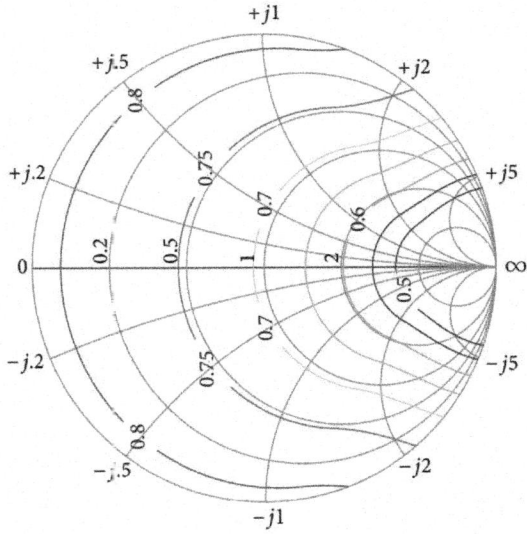

Figure 22: Contours of maximal efficiency of PA, associated with the waveform pair (188)-(189) as a function of normalized third harmonic impedance $\underline{z}_{3n} = \underline{z}_3/\mathrm{Re}\{\underline{z}_1\}$.

For the present case study, in what follows we show that power utilization factor is proportional to efficiency. For voltage waveform of type (189) it is easy to see that $V(\theta + \pi) = 2 - V(\theta)$ holds. This relation along with the fact that waveform $V(\theta)$ that provides maximal efficiency has at least one zero implies that $\max[V(\theta)] = 2$. On the other hand, current waveform (188) is cosine waveform with positive coefficients and therefore $\max[i(\theta)] = i(0) = 2 + 4/\sqrt{5}$. Consequently, according to (179), the following relation holds:

$$PUF_{\text{Case study }7.3} = 2\left(5 - 2\sqrt{5}\right)\eta = 1.0557\eta.$$

(190)

Clearly, the ratio PUF/η is constant and therefore in this case study PUF can be easily calculated from the corresponding efficiency. Accordingly, peak efficiency and peak value of PUF Case study $= 3\sqrt{5}/2 - 5/2 = 0.8541$ are attained for the same voltage waveform (when third harmonic is shortcircuited).

In the first example, current waveform (188) and $\underline{z}_3 = 0.2 - j0.5$ imply that $27\lambda_{3v}^2 \le (1 - 2a_{3v})^3$. Then, according to Algorithm 33, the voltage waveform of type (189) that provides maximal efficiency has the following coefficients: $a_{3v} = -0.0553$, $b_{3v} = -0.1382$, $a_{1v} = -0.9447$, and $b_{1v} = 0.4146$ (see Figure 23). Efficiency, PUF, and normalized third harmonic impedance are $\eta = 0.7643$, PUF $= 0.8069$, and $\underline{z}_{3n} = 0.3425 - j0.8564$, respectively.

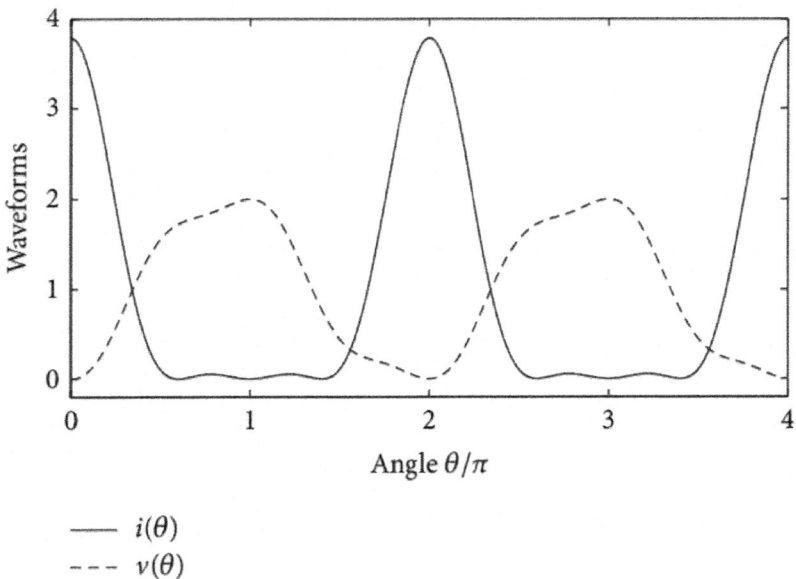

Figure 23: Waveform pair (188)-(189) that provides maximal efficiency for $\underline{z}_3 = 0.2 - j0.5$.

In the second example, current waveform (188) and $\underline{Z}_3 = 0.1 - j1.1$ imply that $27\lambda_{3v}^2 > (1 - 2a_{3v})^3$. Then, according to Algorithm 33, the voltage waveform of type (189) that provides maximal efficiency has the following coefficients: $a_{3v} = -0.0276$, $b_{3v} = -0.3040$, $a_{1v} = -0.9391$, and $b_{1v} = 0.5807$ (see Figure 24). Efficiency, PUF, and normalized third harmonic impedance are $\eta = 0.7598$, PUF $= 0.8021$, and $\underline{Z}_{3n} = 0.1723 - j1.8952$, respectively.

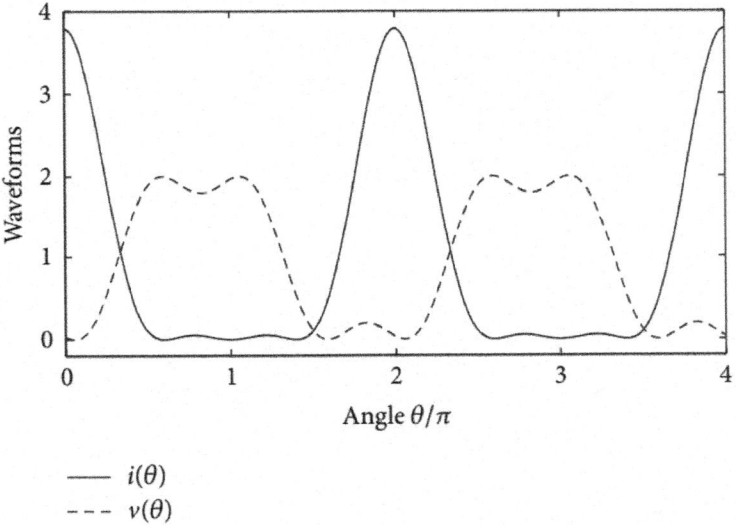

Angle θ/π

—— $i(\theta)$
--- $v(\theta)$

Figure 24: Waveform pair (188)-(189) that provides maximal efficiency for $\underline{Z}_3 = 0.1 - j1.1$.

Case Study 7.4: In this case study let us consider currentvoltage pair, where current is normalized waveform of type (180) with conduction angle $\alpha = 1.15\pi$ (207°) and voltage is nonnegative waveform of type (189). Substitution of $\alpha = 1.15\pi$ and $I_{dc} = 1$ into (182) leads to $I_D = 2.2535$. Furthermore, substitution of $\alpha = 1.15\pi$ and $I_D = 2.2535$ into (180) leads to

$$i(\theta)$$

$$= \begin{cases} 2.2535 \left[\cos\theta - \cos\left(\dfrac{1.15\pi}{2}\right) \right], & |\theta| \leq \dfrac{1.15\pi}{2}, \\ 0, & \dfrac{1.15\pi}{2} \leq |\theta| \leq \pi. \end{cases} \quad (191)$$

Similarly, substitution of $\alpha = 1.15\pi$ and $I_D = 2.2535$ into (183) and (184) for $n=3$ yields coefficients of fundamental and third harmonic of waveform (191):

$$a_{1i} = 1.4586, \qquad a_{3i} = -0.1026.$$

$$(192)$$

Because current waveform (191) contains only cosine terms and voltage waveform is the same as in previous case study, the procedure for calculation of maximal efficiency of PA with waveform pair (189)–(191) is the same as that presented in Algorithm 33. In this case study the coefficients of fundamental and third harmonic of current waveform are given by (192).

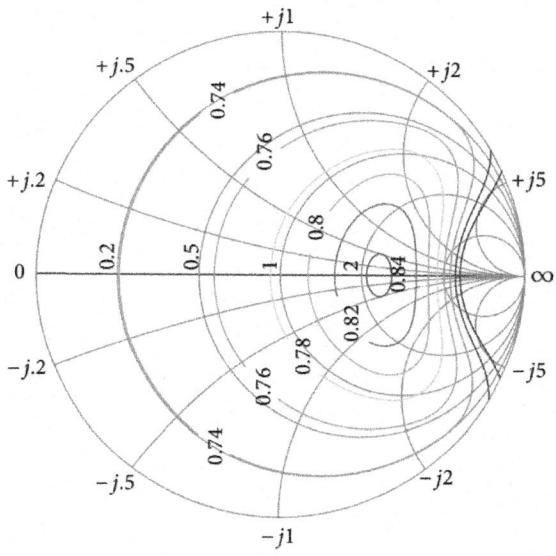

Figure 25: Contours of maximal efficiency of PA, associated with the waveform pair (189) and (191) as a function of normalized third harmonic impedance $\underline{z}_{3n} = \underline{z}_3 / \text{Re}\{\underline{z}_1\}$.

For the waveform pair (189) and (191), maximal efficiency of PA as a function of normalized third harmonic impedance $\underline{z}_{3n} = \underline{z}_3 / \text{Re}\{\underline{z}_1\}$ is presented in Figure 25. Efficiency of 0.84 is obtained in vicinity of $\underline{z}_{3n} = 2.3685$ (corresponding to $\underline{z}_3 = 1.8750$). Peak efficiency $\eta \approx 0.8421$ is achieved for voltage waveform of type (189) with coefficients $a_{1V} = -2/\sqrt{3}$, $a_{3V} = \sqrt{3}/9$, and $b_{1V} = b_{3V} = 0$.

In the course of finding power utilization factor, notice that current waveform of type (191) attains its maximum value for $\theta=0$. Insertion of $\max[i(\theta)] = i(0) = 2.78$ and $\max[V(\theta)] = 2$ for voltage waveform of type (189) into (179) leads to

$$PUF_{\text{Case study 7.4}} = 1.439\eta.$$

$$(193)$$

Again, the ratio PUF/η is constant and PUF can be easily calculated from the corresponding efficiency. Accordingly, peak value of $\text{PUF}_{\text{Case study 7.4}} \approx 1.2118$ and peak efficiency are attained for the same voltage waveform.

In the first example, current waveform (191) and $\underline{z}_3 = 1 - j0.2$ imply that $27\lambda_{3v}^2 \leq (1 - 2a_{3v})^3$. Then, according to Algorithm 33, voltage waveform of type (189) which provides maximal efficiency has coefficients: $a_{3V} = 0.1026$, $b_{3V} = 0.0205$, $a_{1V} = -1.1026$, and $b_{1V} = -0.0616$ (see Figure 26). Efficiency, PUF, and normalized third harmonic impedance are $\eta = 0.8042$, PUF $= 1.1572$, and $\underline{z}_{3n} = 1.3228 - j0.2646$, respectively.

In second example, current waveform (191) and $\underline{z}_3 = 1.5 - j1.2$ imply that $27\lambda_{3v}^2 > (1 - 2a_{3v})^3$. Then, according to Algorithm 33, voltage waveform of type (189) which provides maximal efficiency has coefficients: $a_{3V} = 0.1540$, $b_{3V} = 0.1232$, $a_{1V} = -1.1255$, and $b_{1V} = -0.2575$ (see Figure 27). Efficiency, PUF, and normalized third harmonic impedance are $\eta = 0.8208$, PUF $= 1.1812$, and $\underline{z}_{3n} = 1.9439 - j1.5552$, respectively.

CONCLUSION

In this paper we consider a problem of finding general descriptions of various classes of nonnegative waveforms with fundamental and kth harmonic. These classes include nonnegative waveforms with at least one zero, nonnegative waveforms with maximal amplitude of fundamental harmonic for prescribed amplitude of kth harmonic, nonnegative waveforms with maximal coefficient of cosine part of fundamental harmonic for prescribed coefficients of kth harmonic, and nonnegative cosine waveforms with at least one zero. Main results are stated in six propositions (Propositions 1, 6, 9, 18, 22, and 26), four corollaries (Corollaries 2–5), twenty remarks, and three algorithms. Four case studies of usage of closed form descriptions of nonnegative waveforms in PA efficiency analysis are considered in detail in Section 7.

APPENDICES

Here we provide a list of finite sums of trigonometric functions used in this paper (Appendix A) and brief account of the Chebyshev polynomials (Appendix B).

List of Some Finite Sums of Trigonometric Functions

Dirichlet kernel (e.g., see [27]) is as follows:

$$D_{k-1}(\tau) = 1 + 2\sum_{n=1}^{k-1} \cos n\tau = \frac{\sin((2k-1)\tau/2)}{\sin(\tau/2)}.$$

(A.1)

Fejer kernel (e.g., see [´27]) can be expressed in the following equivalent forms:

$$F_{k-1}(\tau) = \frac{1}{k}\sum_{n=0}^{k-1} D_n(\tau) = 1 + \frac{2}{k}\sum_{n=1}^{k-1}(k-n)\cos n\tau$$

$$= \frac{(1 - \cos k\tau)}{k(1 - \cos \tau)}.$$

(A.2)

Lagrange's trigonometric identity (e.g., see [28]) is as follows:

$$S_1(\tau) = \sum_{n=1}^{k-1} \sin n\tau = \frac{\sin(k\tau/2)\sin((k-1)\tau/2)}{\sin(\tau/2)}.$$

(A.3)

In what follows we show that the following three trigonometric identities also hold:

$$2\sum_{n=1}^{k-1}(k-n)\sin n\tau = \frac{k\sin\tau - \sin k\tau}{1 - \cos\tau},$$

(A.4)

$$\sum_{n=1}^{k-1} \cos(k-2n)\tau = \frac{\sin(k-1)\tau}{\sin\tau},$$

(A.5)

$$\sum_{n=1}^{k-1} n(k-n)\cos(k-2n)\tau$$

$$= \frac{\sin(k\tau)\cos\tau - k\cos(k\tau)\sin\tau}{2\sin^3\tau}.$$

(A.6)

Denote

$$S_2(\tau) = 2\sum_{n=1}^{k-1}(k-n)\sin n\tau, \; S_3(\tau) = \sum_{n=1}^{k-1} \cos(k-$$

$2n)\tau$, and $S_4(\tau) = \sum_{n=1}^{k-1} n(k-n)\cos(k-2n)\tau.$

Notice that $S_2(\tau) = 2kS_1(\tau) + dD_{k-1}(\tau)/d\tau,$, which immediately leads to (A.4).

Identity (A.5) can be obtained as follows:

$$\frac{\sin (k-1)\,\tau}{\sin \tau} = \frac{e^{j(k-1)\tau} - e^{-j(k-1)\tau}}{e^{j\tau} - e^{-j\tau}}$$

$$= e^{jk\tau}\frac{e^{-2j\tau} - e^{-2j(k-1)\tau}}{1 - e^{-2j\tau}}$$

$$= e^{jk\tau}\sum_{n=1}^{k-1} e^{-2jn\tau} = \sum_{n=1}^{k-1} e^{j(k-2n)\tau}$$

$$= \sum_{n=1}^{k-1} \cos (k-2n)\,\tau.$$

$$(A.7)$$

From $4(k-n) = k^2 - (k-2n)^2$ it follows that $4S_4(\tau) = k^2 S_3(\tau) + d^2 S_3(\tau)/d\tau^2$, which leads to (A.6).

THE CHEBYSHEV POLYNOMIALS

The Chebyshev polynomials of the first kind (x) can be defined by the following relation (e.g., see [29]):

$$V_n (x) = \cos n\tau, \quad \text{when } x = \cos \tau.$$

$$(B.1)$$

The Chebyshev polynomials of the second kind (x) can be defined by the following relation (e.g., see [29]):

$$U_n (x) = \frac{\sin (n+1)\,\tau}{\sin \tau}, \quad \text{when } x = \cos \tau.$$

$$(B.2)$$

The Chebyshev polynomials satisfy the following recurrence relations (e.g., see [29]):

$$V_0 (x) = 1, \qquad V_1 (x) = x,$$

$$V_{n+1} (x) = 2xV_n (x) - V_{n-1} (x),$$

$$U_0 (x) = 1, \qquad U_1 (x) = 2x,$$

$$U_{n+1} (x) = 2xU_n (x) - U_{n-1} (x).$$

$$(B.3)$$

The first few Chebyshev polynomials of the first and second kind are $V_2(x)$ $= 2x^2 - 1$, $V_3(x) = 4x^3 - 3x$, $V_4(x) = 8x^4 - 8x^2 + 1$, $U_2(x) = 4x^2 - 1$, $U_3(x) = 8x^3 - 4x$, and $U_4(x) = 16x^4 - 12x^2 + 1$.

ACKNOWLEDGMENT

This work is supported by the Serbian Ministry of Education, Science and Technology Development, as a part of Project TP32016.

REFERENCES

1. V. I. Arnol'd, V. S. Afrajmovich, Y. S. Il'yashenko, and L. P. Shil'nikov, Dynamical Systems V, Bifurcation Theory and Catastrophe Theory, Springer, Berlin, Germany, 1994.

2. E. Polak, "On the mathematical foundations of nondifferentiable optimization in engineering design," SIAM Review, vol. 29, no. 1, pp. 21–89, 1987.

3. N. S. Fuzik, "Biharmonic modes of a tuned RF power amplifier," Radiotehnika, vol. 25, no. 7, pp. 62–71, 1970 (Russian).

4. P. Colantonio, F. Giannini, G. Leuzzi, and E. Limiti, "Class G approach for low-voltage, high-efficiency PA design," International Journal of RF and Microwave Computer-Aided Engineering, vol. 10, no. 6, pp. 366–378, 2000.

5. F. H. Raab, "Maximum efficiency and output of class-F power amplifiers," IEEE Transactions on Microwave Theory and Techniques, vol. 49, no. 6, pp. 1162–1166, 2001.

6. F. H. Raab, "Class-E, class-C, and class-F power amplifiers based upon a finite number of harmonics," IEEE Transactions on Microwave Theory and Techniques, vol. 49, no. 8, pp. 1462–1468, 2001.

7. J. D. Rhodes, "Output universality in maximum efficiency linear power amplifiers," International Journal of Circuit Theory and Applications, vol. 31, no. 4, pp. 385–405, 2003.

8. A. Juhas and L. A. Novak, "Comments on 'Class-E, class-C, and classF power amplifier based upon a finite number of harmonics'," IEEE Transactions on Microwave Theory and Techniques, vol. 57, no. 6, pp. 1623–1625, 2009.

9. M. Roberg and Z. Popovic, "Analysis of high-efficiency power ´ amplifiers with arbitrary output harmonic terminations," IEEE Transactions on Microwave Theory and Techniques, vol. 59, no. 8, pp. 2037–2048, 2011.

10. A. Grebennikov, N. O. Sokal, and M. J. Franco, Switchmode RF Power Amplifiers, Elsevier/Academic Press, San Diego, Calif, USA, 2nd edition, 2012.

11. T. Canning, P. J. Tasker, and S. C. Cripps, "Continuous mode power amplifier design using harmonic clipping contours: theory and practice,"

IEEE Transactions on Microwave Theory and Techniques, vol. 62, no. 1, pp. 100–110, 2014.

12. A. Juhas and L. A. Novak, "General description of nonnegative waveforms up to second harmonic for power amplifier modelling," Mathematical Problems in Engineering, vol. 2014, Article ID 709762, 18 pages, 2014.

13. V. I. Arnol'd, V. V. Goryunov, O. V. Lyashko, and V. A. Vasil'ev, Dynamical Systems VIII—Singularity Theory II, Applications, Springer, Berlin, Germany, 1993.

14. D. Siersma, "Properties of conflict sets in the plane," Banach Center Publications, Polish Academy of Sciences, vol. 50, no. 1, pp. 267–276, 1999, Proceedings of the Banach Center Symposium on Geometry and Topology of Caustics (Caustics '98), Warsaw, Poland.

15. M. van Manen, The geometry of conflict sets

16. Dissertation., Universiteit Utrecht, Utrecht, The Netherlands, 2003, http:// igitur-archive.library.uu.nl/dissertations/2003-0912-123058/c4.pdf.

17. Y. L. Sachkov, "Maxwell strata and symmetries in the problem of optimal rolling of a sphere over a plane," Sbornik: Mathematics, vol. 201, no. 7-8, pp. 1029–1051, 2010.

18. I. A. Bogaevsky, "Perestroikas of shocks and singularities of minimum functions," Physica D: Nonlinear Phenomena, vol. 173, no. 1-2, pp. 1–28, 2002.

19. Y. L. Sachkov, "Maxwell strata in the Euler elastic problem," Journal of Dynamical and Control Systems, vol. 14, no. 2, pp. 169–234, 2008.

20. M. Siino and T. Koike, "Topological classification of black holes: generic Maxwell set and crease set of a horizon," International Journal of Modern Physics, D: Gravitation, Astrophysics, Cosmology, vol. 20, no. 6, pp. 1095–1122, 2011.

21. F. H. Raab, "Class-F power amplifiers with maximally flat waveforms," IEEE Transactions on Microwave Theory and Techniques, vol. 45, no. 11, pp. 2007–2012, 1997.

22. A. Juhas and L. A. Novak, "Maximally flat waveforms with finite number of harmonics in class-F power amplifiers," Mathematical Problems in Engineering, vol. 2013, Article ID 169590, 9 pages, 2013.

23. L. Fejer, " ´ Uber trigonometrische polynome," ¨ Journal fur die ¨ Reine und Angewandte Mathematik, vol. 1916, no. 146, pp. 53–82, 1916 (German).

24. S. C. Cripps, "Bessel Waives

25. microwave bytes.," IEEE Microwave Magazine, vol. 10, no. 7, pp. 30–36, 117, 2009.

26. L. N. Bryzgalova, "Singularities of the maximum of parametrically dependent function," Functional Analysis and Its Applications, vol. 11, no. 1, pp. 49–51, 1977.

27. V. I. Arnold, A. A. Davydov, V. A. Vassiliev, and V. M. Zakalyukin, Mathematical Models of Catastrophes, Control of Catastrophic Process, Encyclopedia of Life Support Systems (EOLSS), EOLSS Publishers, Oxford, UK, 2006.

28. S. C. Cripps, RF Power Amplifiers for Wireless Communications, Artech House, Norwood, Mass, USA, 2nd edition, 2006.

29. A. Zygmund, Trigonometric Series, vol. 1, Cambridge University Press, Cambridge, UK, 2nd edition, 1959.

30. A. Jeffrey and H. Dai, Handbook of Mathematical Formulas and Integrals, Elsevier/Academic Press, San Diego, Calif, USA, 4th edition, 2008.

31. J. C. Mason and D. C. Handscomb, Chebyshev Polynomials, Chapman & Hall, CRC Press, Boca Raton, Fla, USA, 2003.

Chapter 2

POWER AMPLIFIERS FOR ELECTRONIC BIO-IMPLANTS

Anthony N. Laskovski and Mehmet R. Yuce

The University of Newcastle Australia

INTRODUCTION

Healthcare systems face continual challenges in meeting their aims to provide quality care to their citizens within tight budgets. Ageing populations in the developed world are perhaps one of the greatest concerns in providing quality healthcare in the future. Figure 1 shows projections from the United Nations, indicating that the median age of citizens in economically developed regions is set to approach 40 years by the year 2050, and reach as high as 55 years in Japan. This trend is likely to lead to strained economies caused by less revenue raised by smaller workforces. Another effect of ageing populations is the need of further care in order to remain healthy.

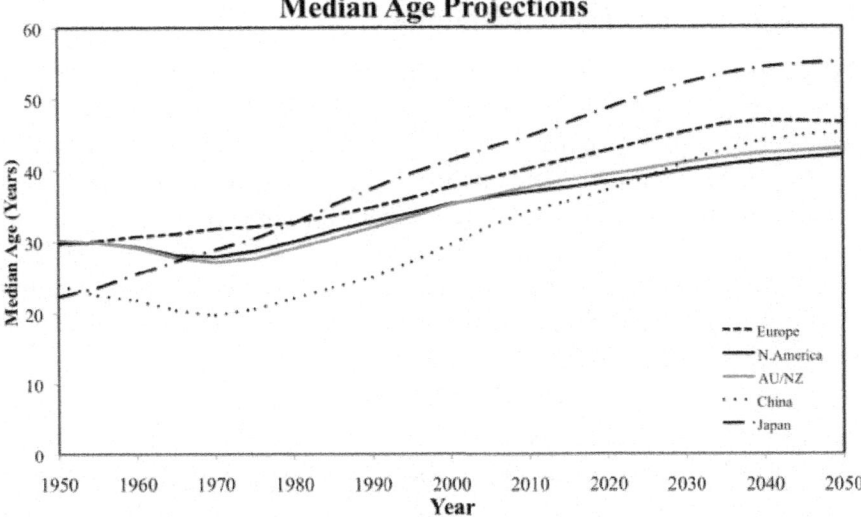

Figure 1: UN Median Age Statistics (UN, 2010).

This care varies from frequent check-ups to condition monitoring, compensation for organ malfunction and serious surgical operations. As a result of these trends, healthcare systems will face the task of servicing more people with more serious and expensive health services, all using less available funds. Effort is being focused on running cheaper and more effective healthcare systems and the development of technology to assist in this process is a natural research priority.

Technology in Medicine

Archaeological evidence shows the application of technology to medicine for rehabilitative, functional and aesthetic purposes as far back as 5000 years ago in ancient Egypt, whereprosthetic devices were designed engineered and constructed with basic materials such as leather and wood. The earliest written evidence exists from ancient India, mentioning a prosthetic iron leg (Thurston, 2007). These materials formed the basics of prosthetics until recently, the only variations being in manufacturing techniques.

The application of titanium alloys to medicine was a significant advance due to their ability to form biological bonds with human tissue (Long and Rack, 1998). Developments in polymer technology led to biocompatible polymers, allowing more precise, detailed and finer implants to be made such as blood vessel reinforcements (Ramakrishna et al., 2001). The latest developments in biomaterial research is in fact designing polymers to allow the body to heal itself (Hench and Polak, 2002).

Electrical and Electronic Technology

The field of electronics has been a relatively recent technological advance in history, and it has seen an escalating rate of sophistication. After the renaissance, serious curiosity in the phenomenon of electrical charge developed, and several fundamental developments were made such as the discovery in 1791 by Galvani, that electricity was the medium through which information was passed to muscles in the body. The voltaic pile was developed in 1800 by Volta, which provided the first reliable source of electrical energy, and other major developments happened such as the recognition of electromagnetism by Orsted and Ampere, and Faraday's electric motor.

Tesla's achievements in the transmission of low frequency wireless power were significant. He proposed to apply the concept of resonance to electrical energy in order to transmit energy wirelessly. Hertz used spark gaps to generate high frequency power and detect it at a receiving end, using parabolic reflectors at the transmitting and receiving ends. These developments were further built upon in the late 1930s with the availability of higher energy microwave power generators. Developments in microwave power transmission escalated during the 20th century due to World War II and the Cold War, resulting in sophisticated satellite communication technologies (Brown, 1984).

The development of quantum theory and semiconductor electronics laid the foundations for rapid technological development in what is now being called 'The Age of Silicon' (Jenkins, 2005). They allowed for the rapid development of integrated circuit technology characterised by Moore's Law, which states that the number of transistors in a given surface area increases exponentially with time (Łukasiak and Jakubowski).

Computer networks developed in the 1970s and led to the eventual creation of internet (Kleinrock, 2008). This has led to a technological and sociological revolution characterising the 21st century as 'The Information Age', with omnipresent networks, small sensors, constant and cheap access to information on increasingly intelligent personal devices that are modestly called 'phones'.

Electronics in Medicine

Galvani's frog experiment showed biology as one of the original phenomena through which human understanding of electricity was developed. Interestingly, knowledge in the field of electronic engineering has since advanced to a stage where it is being used to understand, monitor and even treat biological and medical systems.

Medical imaging is fundamental to the understanding of the human body and diagnosing medical problems. X-Ray technology is widely used to capture two-dimensional details fororthopaedic applications. The rays are created by rapidly decelerating electrons to produce high frequency electromagnetic radiation, which is diffracted and penetrated differently by bones and flesh, allowing the resultant radiation to be recorded on X-ray sensitive polymers to show internal details of the body. Ultrasound is commonly use to provide a real-time image of the body's internal operations, being a popular and safe technology in monitoring various stages of pregnancy. Computed Tomography (CT) scanning and Magnetic Resonance Imaging (MRI) provide three-dimensional images of the body's internal organs, allowing fine differentiation between different types of body tissue. Such types of scans involve powerful computing capability to reconstruct models of internal organs, and have been invaluable to the understanding of the human body in a non-destructive way (Seligman, 1982).

Robotics in medicine has become another exciting field in which the application of intelligent electronics is contributing greatly, to the point where they are used to conduct complex surgery, which is remotely controlled by surgeons. Their ability to move accurately without shaking hands or unstable movements allows minute and delicate operations to take place, while still being controlled by a doctor. The application of robotics to medical prosthesis is another significant advance since the first pneumatically powered hand in 1915 (Childress, 1985). So advanced is this field, that robotic prosthetic arms are being developed and controlled by electrical signals sent by the brain through the body's nervous system.

The cardiac pacemaker is the oldest and perhaps best known implantable prosthetic electronic device. It was first used externally on a patient in 1952 and as the first semiconductor transistors were developed, the possibility to implant led to the first human implant in 1960 (Greatbatch and Holmes, 1991). This was the beginning of several exciting developments in the area of medical prosthetics. Cochlear implants, popularly termed `Bionic Ears' were a major breakthrough in medical prosthesis.

The Cochlea is a part of the ear that converts sound vibrations to electrical signals that are sent via the audio nerve to the brain where they are interpreted. In deaf patients where the Cochlea does not operate properly and the auditory nerve does, cochlear implants are possible. A system was designed and created to replace the Cochlea with an electronic prosthetic device, such that the sound recorded by a microphone is processed by an implanted device and sent to the brain on the audio nerve. Retinal prostheses, popularly termed 'Bionic Eyes' have been the focus of much research.

The concept is similar to Cochlear prosthesis, however this electronic prosthetic device aims to substitute the retina, which is the part of the eye which converts light to electrical signals and sent to the brain via the optic nerve. For patients that have suffered blindness due to macular degeneration, this prosthetic device has the potential to re-introduce sight. Patient monitoring is an important part of medicine in that it assists doctors in understanding the condition of their patients, be it for known issues or as a means of diagnosis. Condition monitoring of patients is also conducted after serious surgical operations, in order to ensure that no complications arise. This is often a major reason for a patient's long stay in hospital after an operation.

Prevention is preferable to treatment, and the ability to monitor vital health indicators such as the electrocardiogram (ECG), body temperature and blood pressure information via medical telemetry may offer adequate tools to view logged or real time data for vulnerable patients, especially the elderly. Growing telecommunications infrastructure with increasing sophistication is opening the possibilities with regards to medical telemetry, making ittheoretically possible for patients to carry out their daily tasks while being monitored remotely by doctors. Implantable medical telemetry is in fact becoming an increasingly important field of research, with the potential to reduce medical risks, lower medical costs and cater for ageing populations.

TELEMETRY

Telemetry is a significant element of health care, involving the measurement and communication of a patient's biological information for interpretation by medical professionals. It is mostly conducted by external medical equipment, however medical telemetry is making its way into the body in the form of implantable monitoring devices, which will potentially be able to measure very detailed body signals. Figure 2 shows a general block diagram of most implantable telemetry systems.

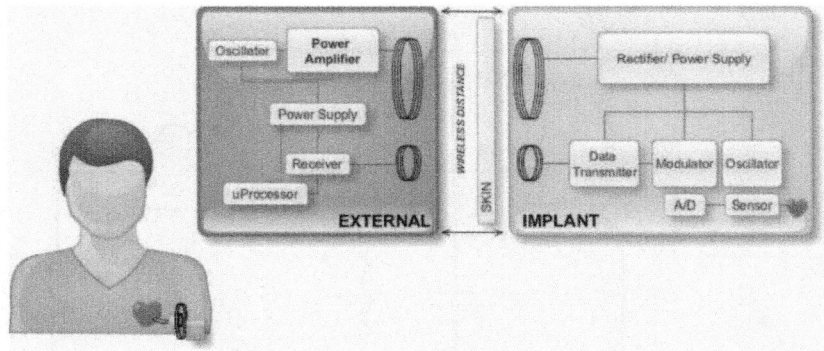

Figure 2: General architecture of telemetry systems.

One important factor to consider when dealing with implantable devices is the supply of power. In order to send power to implantable devices, wireless links are usually employed in the form of inductive links. Inductive power transfer is more efficient at lower frequencies (Vaillancourt et al., 1997). However, lower transmission frequencies use larger circuit components, especially transmission coils. From the perspective of implantable devices, space is important and this has led to a need to design highly efficient transmission circuits at higher frequencies.

Switching power amplifiers have been a popular choice for the transmission of wireless power (Raab et al., 2002). While the most popular choice has been the Class-E amplifier, it is also useful to gain an understanding of other power amplifiers, Class-F and Class-D.

CLASS-F AMPLIFIER

The Class-F power amplifier may be seen as a development from the Class-A and Class-B power amplifier, with a 50% conduction time and the use of harmonic resonators on the load network (Reynaert and Steyaert, 2006). An example of the Class-F amplifier is shown in Figure 3 comprising a transistor, choke inductor and an input source.

The network attached to the output of the transistor is manipulated by harmonics such that the voltage and current are manipulated. The voltage is shaped by odd harmonics of thefundamental frequency such that the voltage appears as a square-wave. The current is 1800 out of phase and shaped to appear as a half sine-wave (Raab, 1997).

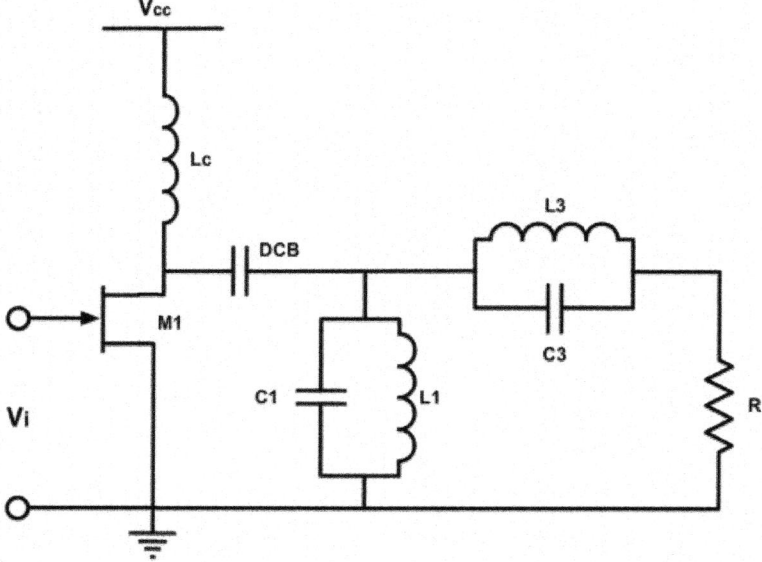

Figure 3: Class-F Amplifier.

The more harmonic frequencies are used to shape the voltage and current curves, the higher the efficiency of the Class-F amplifier. The theoretical efficiency of the amplifier with the use of third harmonics is 88.4%, while the additional use of fifth harmonic resonators produces an efficiency of 92% (Reynaert and Steyaert, 2006).

Inverse Class-F amplifiers also exist where the current curve is shaped to be a square-wave, while the voltage is shaped as a half sine-wave (Young, 2006).

CLASS-D AMPLIFIER

Like the Class-F power amplifier, the Class-D power amplifier is a non-linear amplifier in that the transistors of the amplifier behave as switches such that the output of the transistors is related to the supply or reference voltage, depending on which transistor is turned on at the time. The fact that the output signal of the amplifier is determined by the switching of the amplifier's transistors means that the Class-D power amplifier may be described as a switching power amplifier (Reynaert and Steyaert, 2006).

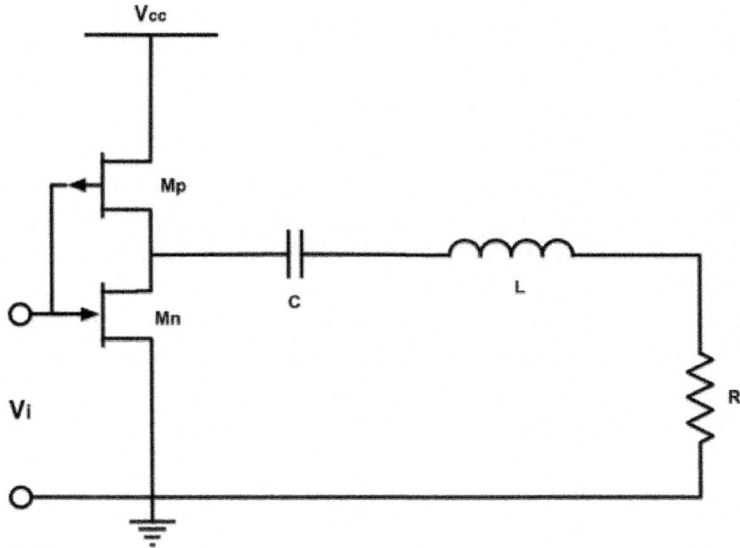

Figure 4: The Class-D amplifier.

Figure 4 shows an example of a Class-D switching power amplifier. It comprises an inverter, which switches two transistors on and off alternatively to generate a square wave. The output of the inverter is connected to a series RLC network as shown in Figure 4, which is resonant at the fundamental frequency of the square-wave, producing a sinusoidal signal at this frequency. Assuming that the series L and C network only allows sinusoidal current to reach the load R, the theoretical efficiency of the Class-D amplifier is 100%.

In reality, circuit elements are not ideal and several losses have been analysed with a focus on parasitic drain-source capacitance in each of the transistors, which becomes significant in higher frequency RF designs. The drain-source capacitance, C_{ds} actually introduces a capacitor where an open circuit should ideally exist. At high frequencies, typical capacitor values are in the order of pico Farads, which means that parasitic capacitance C_{ds} becomes a significant circuit element, which dissipates energy during switching cycles thus decreasing the amplifier's efficiency (El-Hamamsy, 1994, Kiri et al., 2009, Raab et al., 2002).

CLASS-E AMPLIFIER

The Class-E power amplifier was introduced by Sokal et al., shown in Figure 5 (Sokal and Sokal, 1975). Like the Class-D amplifier it is also a switching power amplifier driven by a square-wave input, however rather than two transistors it comprises one transistor and a choke inductor. As a result the signal seen by the load is not hard-switched.

Similar to the Class-D amplifier, the series LC network of the Class-E amplifier only allows a sinusoidal voltage and current to pass to the load. The Class-E amplifier also includes a capacitor C_1 across the transistor terminals and forms a key component of the circuit's high efficiency operation at high frequencies as well as absorbing C_{ds}. The amplifier's high efficiency operation lies in the shape of the voltage across C_1. Circuit elements are chosen such that the voltage at this point is zero when the transistor is switched on such that no stored energy is dissipated from the capacitor. The voltage is shaped such that the rate of change of voltage (dv_{C1}/dt) across this point is also zero. This feature enables robustness to phase or frequency irregularities in practice.

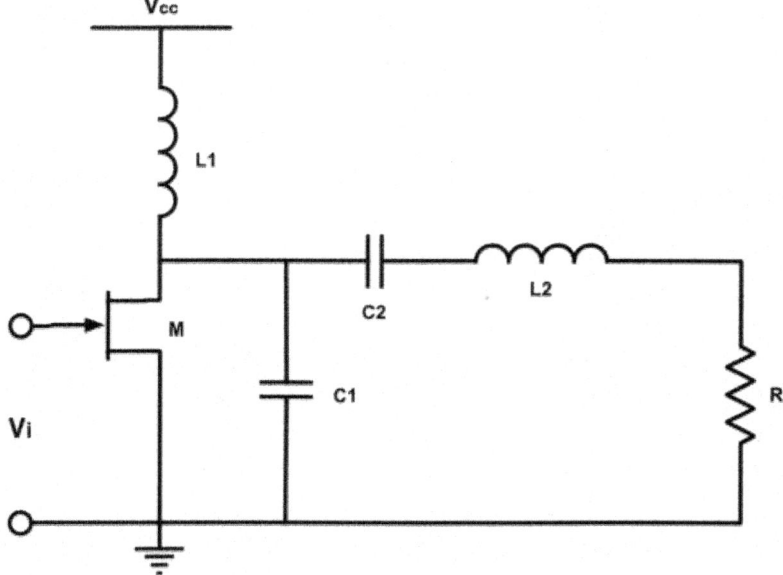

Figure 5: The Class-E amplifier.

Since the amplifier's introduction several analyses have been presented to enhance the design processes of the Class-E amplifier such that it includes more practical considerations. One of the original assumptions of the Class-E amplifier design process was that it has aninfinitely loaded quality factor

(Q). Kazimierczuk et al. presented a design procedure in which the amplifier can be designed at a specific Q and switch duty cycle (Kazimierczuk and Puczko, 1987). Suetsugu et al. presented a design procedure to handle off-nominal operation where the voltage across C_1 is zero but its derivative is not, concluding that a higher C_1 capacitance is required for such conditions (Suetsugu and Kazimierczuk, 2006). The Class-E amplifier has been applied to a number of applications, however its relevance to biomedical engineering came to light with Troyk et al.'s proposal to use the Class-E amplifier as a transmitter to transfer inductive power and data for micro implants, with L2 representing the primary inductive coil (Troyk and Schwan, 1992).

A number of design procedures have been presented in literature, however it is interesting to consider the amplifier in the frequency domain. Given that L1 is considered to be large, the transfer function for the output voltage (across R) is given by (1). This transfer function is a second order system, which implies that it has a resonant frequency ω, damping factor ζ and Q factor, indicated in (2)-(4).

$$\frac{V_{out}(s)}{V_{in}(s)} = \frac{g_m R C_2}{s^2 L_2 C_1 C_2 + sRC_1 C_2 + C_1 + C_2} \tag{1}$$

$$\omega = \frac{1}{\sqrt{L_2 C_1 || C_2}} \tag{2}$$

$$\zeta = \frac{R}{2}\sqrt{\frac{C_1 || C_2}{L_2}} \tag{3}$$

$$Q = \frac{1}{2\zeta} \tag{4}$$

If a Class-E amplifier design was to be conducted for a practical application where the inductive coil's properties (L_2) are known as is the load R and resonant frequency ω, equations (2)-(4) can be re-arranged to select the unknown parameters. Combining (2) and (3) gives (5) and (6).

$$\zeta = \frac{R}{2\omega L_2} \tag{5}$$

$$C_1 || C_2 = \frac{2\zeta}{R\omega} \tag{6}$$

Where:

$$C_1 | | C_2 = \frac{C_1 C_2}{C_1 + C_2}$$

(7)

The damping factor ζ (and therefore Q) is determined in the first step by substituting the known values of R, L_2 and ω into (5). This essentially implies that the quality factor of the amplifier is highly dependant on the coil inductor's quality factor. The ζ value is then used in (6), along with R and ω, which determines the capacitor combination $C_1 || C_2$.

Determining the individual capacitor values C_1 and C_2 is the more complicated step and requires care, given that the voltage between the two capacitors is vital to the circuit's Class E operation. Generally speaking, if C_1 is smaller than C_2, charge across C_1 is dissipated quickly into C2 prior to the transistor's next half-cycle switch. This implies that the voltage and voltage derivative of the capacitor junction is zero during switching.

OSCILLATORS

Power amplifiers require square-wave clock signal inputs, so while they are known to operate efficiently at high frequencies they require a high frequency square-wave input in order to operate effectively, which is often not included in the determination of the efficiency of the amplifiers. These input signals are produced by oscillator circuits.

Oscillators are frequently used to generate high frequency signals, using resonant elements and a form of feedback. The Colpitts oscillator is a popular oscillator topology, which involves an LC network with feedback to a transistor. Other oscillators use crystals as the resonant feedback network rather than inductors and capacitors.

The idea of feeding an oscillated output signal back to the input of the amplifier implies that the circuit becomes self-oscillating- similar to the Colpitts oscillator- while operating with zero switching conditions. This is the concept behind the Class-E Oscillator shown in Figure 6 (Ebert and Kazimierczuk, 1981).

Additional circuit elements are added to form the Class-E oscillator, namely feedback elements C_3, C_4 and L_3. It was designed by Ebert et al. to constructively shift the phase of the feedback point of the oscillator. The diode D_1 is placed at the input of the transistor in order to clip the input signal such that it appears as a square wave, satisfying the requirement of the Class-E circuit to have a square-wave input.

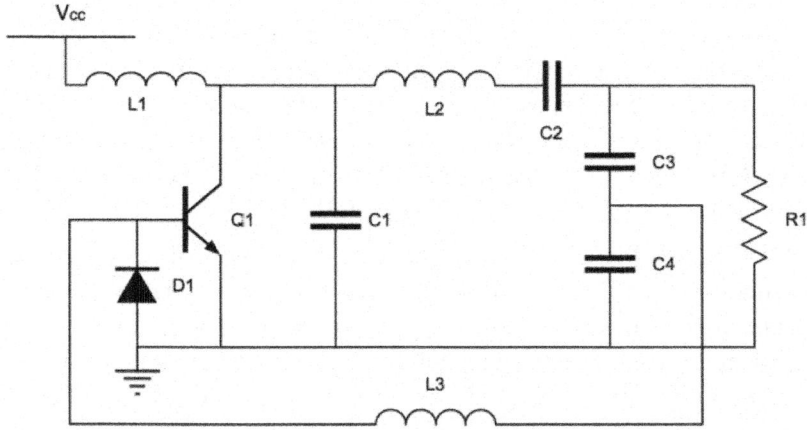

Figure 6: Class-E Oscillator (Ebert and Kazimierczuk, 1981).

Given that low power consumption is advantageous in biomedical systems, it is useful to consider a self-oscillating Class-E oscillator as a wireless power transmitter rather than a Class-E power amplifier. Similar to the power amplifier, the oscillator would transmit energy through L_2. This idea is currently being explored.

Wireless Power Links

The next module of an implanted telemetry system is the wireless power link. As previously mentioned, inductive power transfer has been the most popular means to transfer power wirelessly to implants. Inductive power transfer may be understood by considering two inductive coils L_1 and L_2 shown as the power transmission coils in Figure 2. A time-varying current i1 in L_1 produces a linearly proportional magnetic flux, which passes through L_2 inducing an e.m.f. v_2 in that coil as shown in (8). The symbol M is a combination of the magnetic flux flowing between the two inductors and proportional to the number of turns in L_1 and L_2, and is referred to as the mutual inductance between the coils. A pair of inductors is considered to be strongly coupled if the mutual inductance between them is high in comparison to the respective inductances, as shown in (9), where k is referred to as the coupling coefficient.

$$v_2(t) = M \frac{di_1(t)}{dt}$$

(8)

$$k = \frac{M}{\sqrt{L_1 L_2}}$$

(9)

In the application of inductively powering implantable medical devices, one inductor of the power transmitter circuit forms the primary coil, and a receiving inductor implanted in the body forms the secondary coil. This essentially describes a weakly coupled transformer, the core of which is a combination of air and the layers of human tissue that exist between the two coils (Schuylenbergh and Puers, 2009). Typical coupling coefficients for power transfer in air are 0.17 (Ghovanloo and Atluri, 2007).

It is more efficient to transmit wireless power at lower frequencies (Vaillancourt et al., 1997), and as the complexity of implants increases, data rates are also required to increase. Wang et al. proposed the advantages of biomedical implants operating in dual frequency bands to send power and data, and it has since been the basis of further work in the area (Wang et al., 2006).

Transmission coils are an obvious point of focus, as their design holds the key to how well power is transmitted from the external device and received by the implant. Many biomedical implants employ traditional wire-wound cylindrical inductors for the power transmitting and receiving coils. In some scenarios such as retinal prosthesis wire-wound coils are preferred. Best results are usually obtained with the use of Litz wire, which reduces eddy currents caused by the skin-effect (Yang et al., 2007).

In situations such as pre-clinical monitoring, the issues related to wireless power transfer for implanted devices become more difficult to manage, mainly due to random movement by the subject of the implant. A common pre-clinical scenario involves an enclosure in which the subject is free to move. Zimmerman et al. (Zimmerman et al., 2006) investigated the optimisation of wireless power transfer in such a situation, monitoring the overall transfer efficiency by varying transmission frequency and the number of turns on the secondary coil, which was a distance of 1cm from the primary coil. The system produced 3V at 1.3mA in the implant itself, accounting for a tilting angle of 600. The primary coil was a cylindrical wirewound coil, wrapped around the circumference of the base of the enclosure.

Zeirhofer et al. (Zierhofer and Hochmair, 1996) investigated the enhancement of magnetic coupling between coils using a geometric approach. It was concluded that coupling is enhanced when turns of the coil are distributed across the radii rather than concentrating them at the outer radius of the inductors. A number of subsequent papers have been presented analysing and using planar spiral coils for implantable applications (Harrison, 2007, Jow and Ghovanloo, 2010, Silay et al., 2008, Simons et al., 2004). The theory used to design planar spiral coils is quite involved, with most designers opting for simplified and sometimes empirically derived equations such as (10), where

L is the inductance calculated by the surface area A of a square spiral and the number of turns n within the area (Liao, 1987, Wadell, 1991).

$$L = 8.5\sqrt{A}n^{5/3}$$

(10)

Work is being implemented in the use of stacked spiral coils for use in implantable devices. Stacking spiral coils together allows the advantages of spiral shapes to be combined with space efficiency. An increase in coil capacitance also reduces the self-resonant frequency of these coils making them compact and optimised for lower frequency transmission, which is advantageous for the inductive transfer of power (Laskovski et al., 2009).

DATA CARRIER GENERATION

Implantable biomedical telemetry schemes are moving towards a dual-band approach, meaning that power and data are sent at different frequencies, power at a lower frequency and data at a higher frequency. There are several methods used to generate data carrier frequencies for implantable devices.

Many systems involve the generation of data carrier signals on the external side of the system, leaving only the data recovery, modulation and transmission to the implantable circuitry (Mandal and Sarpeshkar, 2008, Zhou et al., 2006). Ziaie et al. presented a dual band implantable neuromuscular stimulator, the 2MHz data clock of which is recovered from 2MHz power supply (Ziaie et al., 1997), while Wise et al. operated at 4MHz (Wise et al., 2004) as did Sauer et al.'s (Sauer et al., 2005). Generating data carrier signals external to the implanted device allows for the reduction of device complexity and power consumption, however such a system requires a synchronised send/receive protocol as well as an accurate data recovery block.

The other option popularly used is to generate a data carrier frequency from within the implantable device itself. Kocer et al. presented an on-chip LC oscillator for general nonimplantable non-medical telemetry, and other options for implantable devices involve ring oscillators (Ghovanloo and Najafi, 2004).

One idea currently being developed for implantable involves the generation of a data carrier signal within the implant. However, this signal is generated without the use of a dedicated oscillator block. It is generated by using an inverter to turn the incoming power signal into a non-sinusoidal square-wave signal in order to generate harmonics. One of these harmonics is then filtered and used to transmit data (Laskovski and Yuce, 2008).

MODULATION TECHNIQUES

The majority of modulation techniques used to encode biological signals are digital in that the signals are digitised within the implants. Some common forms of modulation include Frequency Shift Keying (FSK), Load Shift Keying (LSK), Amplitude Shift Keying (ASK), Phase Shift Keying (PSK).

FSK involves allocating different frequencies for different bit values. For example, binary FSK translates to bit `0' transmitting at a frequency $f_1$ and bit `1' transmitting at a different frequency f_2. Modulating a signal using FSK involves a switch and the generation of twodifferent carrier frequencies, and bits are usually decoded by bandpass filters (Ghovanloo and Najafi, 2004).

Impedance modulation or LSK involves altering a load in the transmitting circuit according to digital information. Since it usually involves switching one part of the load on and off, the frequency of transmission is varied with each bit, making this scheme very similar to FSK. A number of medical devices make use of this scheme (Chaimanonart and Young, 2006, Mandal and Sarpeshkar, 2008, Wang et al., 2005).

Shifting a carrier frequency's phase occurs to achieve PSK. Depending on the number of symbols in the scheme, the phase shift varies. A popular PSK scheme is Binary PSK (BPSK), where a 1800 phase shift is implemented in order to indicate a particular bit. A number of biomedical and non-biomedical telemetry systems use this scheme (Kocer and Flynn, 2006, Zhou et al., 2006)

ASK modulation is achieved by producing a different amplitude for different bits. A typical and simple example of ASK is called On-Off Keying (OOK), where bits are distinguished by either sending data at a carrier frequency to represent bit `1', or no no signal to represent bit `0'. This type of modulation is very straightforward to implement, being as simple as implementing a data controlled switch in series with an RF transmitter. It can be decoded by rectification and/or a lowpass filter (Ziaie et al., 1997). For low-power implantable circuits, OOK is a simple and space efficient method of modulation.

CONCLUSION

This chapter provided a broad background in the development of biomedical engineering, and the recent contribution of electronics to this field. The role of power amplifiers was explained in the form of three switching power amplifiers, specifically the Class-E amplifier, which included a simple design process. A new idea to use Class-E oscillators was highlighted and is being developed. Basic theory of wireless power transfer was explained and methods in data carrier generation explained. A new method of generating carrier frequencies was briefly explained, which simplifies and reduces the power use of implantable devices. The meaning behind the acronyms of major data modulation schemes were explained, with the features of each described.

REFERENCES

1. BROWN, W. (1984) The History of Power Transmission by Radio Waves. Microwave Theory and Techniques, IEEE Transactions on, 32, 1230 - 1242.

2. CHAIMANONART, N. & YOUNG, D. (2006) Remote RF powering system for wireless MEMS strain sensors. IEEE Sensors Journal,, 6, 484-489.

3. CHILDRESS, D. (1985) Historical aspects of powered limb prostheses. Clinical prosthetics and orthotics.

4. EBERT, J. & KAZIMIERCZUK, M. (1981) Class E high-efficiency tuned power oscillator. IEEE Journal of Solid-State Circuits, , 16, 62 - 66.

5. EL-HAMAMSY, S.-A. (1994) Design of high-efficiency RF Class-D power amplifier. IEEE Transactions on Power Electronics, , 9, 297-308.

6. GHOVANLOO, M. & ATLURI, S. (2007) A Wide-Band Power-Efficient Inductive Wireless Link for Implantable Microelectronic Devices Using Multiple Carriers. IEEE Transactions on Circuits and Systems I: Regular Papers, , 54, 2211 - 2221.

7. GHOVANLOO, M. & NAJAFI, K. (2004) A wideband frequency-shift keying wireless link for inductively powered biomedical implants. IEEE Transactions on Circuits and Systems I: Regular Papers, , 51, 2374 - 2383.

8. GREATBATCH, W. & HOLMES, C. (1991) History of implantable devices. IEEE Engineering in Medicine and Biology Magazine, , 10, 38 - 41.

9. HARRISON, R. (2007) Designing Efficient Inductive Power Links for Implantable Devices. IEEE International Symposium on Circuits and Systems, 2007. ISCAS 2007. , 2080 - 2083.

10. HENCH, L. & POLAK, J. (2002) Third-generation biomedical materials. Science's STKE. JENKINS, T. (2005) A brief history of... semiconductors. Physics education.

11. JOW, U. & GHOVANLOO, M. (2010) Optimization of Data Coils in a Multiband Wireless Link for Neuroprosthetic Implantable Devices. IEEE Transactions on Biomedical Circuits and Systems, , PP, 1 - 1.

12. KAZIMIERCZUK, M. & PUCZKO, K. (1987) Exact analysis of class E tuned power amplifier at any Q and switch duty cycle. IEEE Transactions on Circuits and Systems, , 34, 149 - 159.

13. KIRI, A., OHARA, K., TOMITA, Y., SHUKURI, S., YASUKOUCHI, T. & SUETSUGU, T. (2009) Class D and class E selectable power amplifier. 31st International Telecommunications Energy Conference, 2009. INTELEC 2009. , 1 - 4.

14. KLEINROCK, L. (2008) History of the Internet and its flexible future. IEEE Wireless Communications, , 15, 8 - 18.

15. KOCER, F. & FLYNN, M. (2006) A new transponder architecture with on-chip ADC for long-range telemetry applications. IEEE Journal of Solid-State Circuits,, 41, 1142 - 1148.

16. LASKOVSKI, A. N. & YUCE, M. R. (2008) Harmonics-based bio-implantable telemetry system. 30th Annual International Conference of the IEEE Engineering in Medicine and Biology Society, 2008. EMBS 2008. , 3196 - 3199.

17. LASKOVSKI, A. N., YUCE, M. R. & DISSANAYAKE, T. (2009) Stacked spirals for use in biomedical implants. Asia Pacific Microwave Conference, 2009. APMC 2009., 389 - 392.

18. LIAO, S. Y. (1987) Microwave Circuit Analysis and Amplifier Design, Englewood Cliffs, NJ USA, Prentice-Hall.

19. LONG, M. & RACK, H. (1998) Titanium alloys in total joint replacement-a materials science perspective. Biomaterials.

20. ŁUKASIAK, L. & JAKUBOWSKI, A. History of Semiconductors. itl. waw.pl.

21. MANDAL, S. & SARPESHKAR, R. (2008) Power-Efficient Impedance-Modulation Wireless Data Links for Biomedical Implants. IEEE Transactions on Biomedical Circuits and Systems, , 2, 301 - 315.

22. RAAB (1997) Class-F power amplifiers with maximally flat waveforms. IEEE Transactions on Microwave Theory and Techniques, , 45, 2007 - 2012.

23. RAAB, F., ASBECK, P., CRIPPS, S., KENINGTON, P., POPOVIC, Z., POTHECARY, N., SEVIC, J. & SOKAL, N. (2002) Power amplifiers and transmitters for RF and microwave. IEEE Transactions on Microwave Theory and Techniques, , 50, 814 - 826.

24. RAMAKRISHNA, S., MAYER, J. & WINTERMANTEL, E. (2001) Biomedical applications of polymer-composite materials: a review. Composites Science and ….

25. REYNAERT, P. & STEYAERT, M. (2006) RF Power Amplifiers for Mobile Communications, P.O. Box 17, 3300 AA Dordrecht, The Netherlands, Springer.

26. SAUER, C., STANACEVIC, M., CAUWENBERGHS, G. & THAKOR, N. (2005) Power harvesting and telemetry in CMOS for implanted devices. IEEE Transactions on Circuits and Systems I: Regular Papers,, 52, 2605-2613.

27. SCHUYLENBERGH, K. V. & PUERS, R. (2009) Inductive Powering. Basic Theory and Application to Biomedical Systems, P.O. Box 17, 3300 AA Dordrecht, The Netherlands, Springer.

28. SELIGMAN, L. J. (1982) Physiological Stimulators: From Electric Fish to Programmable Implants. IEEE Transactions on Biomedical Engineering, , BME-29, 270 - 284.

29. SILAY, K., DEHOLLAIN, C. & DECLERCQ, M. (2008) Improvement of power efficiency of inductive links for implantable devices. PRIME 2008. Ph.D. Research in Microelectronics and Electronics, 2008. , 229 - 232.

30. SIMONS, R., HALL, D. & MIRANDA, F. (2004) Spiral chip implantable radiator and printed loop external receptor for RF telemetry in bio-sensor systems. IEEE Radio and Wireless Conference, 2004.

31. SOKAL, N. & SOKAL, A. (1975) Class E-A new class of high-efficiency tuned single-ended switching power amplifiers. IEEE Journal of Solid-State Circuits,, 10, 168-176.

32. SUETSUGU, T. & KAZIMIERCZUK, M. K. (2006) Design procedure of class-E amplifier for off-nominal operation at 50\% duty ratio. IEEE Transactions on Circuits and Systems I: Regular Papers, , 53, 1468 - 1476.

33. THURSTON, A. (2007) Pare and prosthetics: the early history of artificial limbs. ANZ Journal of Surgery.

34. TROYK, P. & SCHWAN, M. (1992) Closed-loop class E transcutaneous power and data link for MicroImplants. IEEE Transactions on Biomedical Engineering, 39, 589-599.

35. UN (2010). United Nations.

36. VAILLANCOURT, P., DJEMOUAI, A., HARVEY, J. & SAWAN, M. (1997) EM radiation behavior upon biological tissues in a radio-frequency power transfer link for a cortical visual implant. Proceedings of the 19th Annual International Conference of the IEEE Engineering in Medicine and Biology Society, 1997. .

37. WADELL, B. C. (1991) Transmission Line Design Handbook, 685 Canton Street Norwood, MA 02062 USA, Artech House.

38. WANG, G., LIU, W., SIVAPRAKASAM, M. & KENDIR, G. (2005) Design and analysis of an adaptive transcutaneous power telemetry for biomedical implants. IEEE Transactions on Circuits and Systems I: Regular Papers, , 52, 2109-2117.

39. WANG, G., LIU, W., SIVAPRAKASAM, M., ZHOU, M., WEILAND, J. & HUMAYUN, M. (2006) A Dual Band Wireless Power and Data Telemetry for Retinal Prosthesis. 28th Annual International Conference of the IEEE Engineering in Medicine and Biology Society, 2006. EMBS ‹06. .

40. WISE, K., ANDERSON, D. J., HETKE, J. F., KIPKE, D. R. & NAJAFI, K. (2004) Wireless implantable microsystems: high-density electronic interfaces to the nervous system. Proceedings of the IEEE, 92, 76 - 97.

41. YANG, Z., LIU, W. & BASHAM, E. (2007) Inductor Modeling in Wireless Links for Implantable Electronics. IEEE Transactions on Magnetics, 43, 3851-3860.

42. YOUNG (2006) Analysis and experiments for high-efficiency class-F and inverse class-F power amplifiers. IEEE Transactions on Microwave Theory and Techniques,, 54, 1969 - 1974.

43. ZHOU, M., LIU, W., WANG, G., SIVAPRAKASAM, M., YUCE, M., WEILAND, J. & HUMAYUN, M. (2006) A Transcutaneous Data Telemetry System Tolerant to

44. Power Telemetry Interference. 28th Annual International Conference of the IEEE Engineering in Medicine and Biology Society, 2006. EMBS ‹06. , 5884 - 5887.

45. ZIAIE, B., NARDIN, M. & COGHLAN (1997) A single-channel implantable microstimulator for functional neuromuscular stimulation. Biomedical Engineering, IEEE Transactions on, 44, 909 - 920.

46. ZIERHOFER, C. & HOCHMAIR, E. (1996) Geometric approach for coupling enhancement of magnetically coupled coils. IEEE Transactions on Biomedical Engineering, , 43, 708 - 714.

47. ZIMMERMAN, M., CHAIMANONART, N. & YOUNG, D. (2006) In Vivo RF Powering for Advanced Biological Research. Engineering in Medicine and Biology Society, 2006. EMBS ‹06. 28th Annual International Conference of the IEEE.

Chapter 3

HYBRID FIBER AMPLIFIER

Inderpreet Kaur[1] and Neena Gupta[2]

[1]Rayat and Bahra Institute of Engineering, Mohali, India

[2]PEC University of Technology (Formally Punjab Engineering College), Chandigarh, India

INTRODUCTION

The advent of telecommunications in 1870s completely revolutionized the world of communications. Metallic cables consisting of twisted wire cables, co-axial cables were the media of choice for many years. These could be used efficiently up to frequencies of 10MHz but the system performance degraded beyond this range. However, with the increasing demand for telephone services, it was necessary to find an alternative medium for telephony to cope up with the high demand. The development of low loss optical fibers gave a solution to this problem and their use revolutionized the speed of telecommunication. Optical fibers have become an unavoidable part of any high speed communication system due to its high information carrying capacity, high bandwidth and extremely low loss. The transmission performance of the optical communication systems is limited by various effects such as attenuation, dispersion, non- linearity, scattering etc, which degrade the level of the signal. To compensate for all these limitations the signals have to be regenerated within the transmission link after some distance. While setting up the transmission link, it is to be ensured that the signal can be retrieved intelligibly at the receiving end. This can be done either by using optoelectronic repeaters or optical amplifiers. In optoelectronic repeaters the optical signal is first converted into an electric signal, then amplified in electric domain and finally converted back into optical signals. Regeneration by making use of repeater is a traditional way to compensate for loss and degradation along the transmission medium. Such regenerators become quite complex and expensive for dense wavelength division multiplexed (DWDM) lightwave systems. This process works well for moderate speed single wavelength operation but it can be fairly complex and expensive for high- speed multi- wavelength systems.

Moreover these so called opto-electronic repeaters once installed into the system can not be upgraded to higher bit rates. Thus a great deal of effort has been spent to develop all optical amplifiers. These devices operate in the optical domain to boost the power level of the signals. In the history of optical fiber communication systems, the advent of optical amplifier was an important milestone. Optical amplifiers can amplify the optical signals directly without requiring its conversion to the electric domain. The development of optical amplifiers started in early eighties and their use for long haul communication systems became widespread during late nineties. Optical amplifiers provided flexibility while upgrading the installed transmission links to higher bit rates. This flexibility of the bit rates allows overcoming the electrical bottleneck of an electric repeater, which was unable to transmit at high bit rates. The opto-electronic repeaters provided with maximum of 40-80 Gbps bit rate.

DWDM Systems

To increase the transmission capacity of a single fiber, DWDM is used. DWDM is a technology, which combines large number of independent information carrying wavelengths onto the same fiber. A characteristic of DWDM is that the discrete wavelengths form an orthogonal set of carriers, which can be separated, routed and switched without interfering with each other. This isolation between channels holds as long as the total optical power intensity is kept sufficiently low to prevent non linear effects e.g. Stimulated Brillouin Scattering (SBS) and Four Wave Mixing processes (FWM) from degrading the link performance. The implementation of DWDM system requires a variety of passive and active devices to combine, distribute, isolate and amplify optical power at different wavelengths. Passive devices require no external control for their operation, so they are less flexible. The wavelength dependent performance of active devices can be controlled electronically, so they provide more flexibility to the network system. Optical amplifiers, tunable filters and tunable sources are integral part of any DWDM system. The key component of DWDM system is optical amplifier. In DWDM system, it is desirable to set a very narrow grid of optical carriers in order to allow more channels in the same optical bandwidth. This not only demands an optical amplifier with high gain but also very broad and flat gain profile to ensure a nearly identical amplification factor in every channel. Figure 1 shows the implementation of active as well as passive components in a typical DWDM system having post amplifier, in-line amplifier and preamplifier [Keiser 2009; Mynbaev 2003].

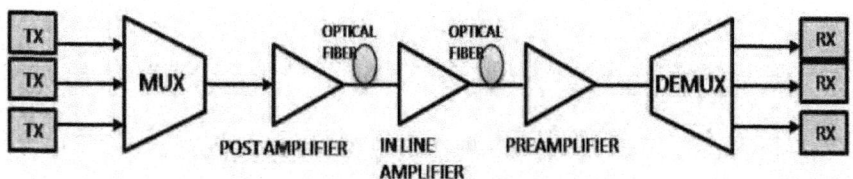

Figure 1: Implementation of A DWDM System Having Various Types of Optical Amplifiers.

REVIEW OF AN OPTICAL AMPLIFIER

An optical amplifier works on the principle of stimulated emission. Optical amplifier increases the level of signal through this process. The mechanism for stimulated emission is same as that for lasers. The operation of laser diodes that are required for the fiber amplifier is similar to the external current injection method (which is used in semiconductor optical amplifiers, SOAs, discussed later). This method is the pumping method used to create population inversion needed for gain mechanism in fiber amplifiers. The sum of injection, stimulated emission and spontaneous recombination rates gives the rate equation that governs the carrier density N (t) in the excited state of both the amplifiers. This carrier density is given by equation (1) [Keiser 2009;Mynbaev 2003].

$$\frac{\partial N(t)}{\partial t} = R_1(t) - R_2(t) - \frac{N(t)}{\tau_r}$$

(1)

where,

$$R_1(t) = \frac{J(t)}{qd}$$

(2)

is the external pumping rate from the injection current density J(t) into an active layer having thickness d, t_r is the combined time constant coming from carrier-recombination mechanisms and spontaneous emission, and

$$R_2(t) \cong g v_g N_p$$

(3)

is the net stimulated emission rate. Here, v_g is the group velocity of incident light, N_p is the photon density and g is the overall gain per unit length. The photon density N_p is dependent on optical signal power, energy of photons, group velocity and dimensions of active area of optical amplifier.

This photon density N_p is given by equation (4),

$$N_p = \frac{P_s}{(hv)(wd)v_g}$$

(4)

In equation (4), P_s is the signal power, v_g is group velocity, w and d are the width and thickness of active area of optical amplifier respectively. The difference between the structure of optical amplifiers and laser diodes is that there is no feedback system in optical amplifiers. So, for boosting an incoming signal optical amplifier requires a pump. The pump supplies energy to the electrons in an active medium, which in turn causes population inversion. An incoming signal photon triggers these excited electrons to drop to lower levels through a stimulated emission process, thereby producing an amplified signal. The amplifier is connected with the optical fiber through a fiber- to- amplifier coupler. The basic components of an optical amplifier are shown in the figure 2) [Keiser 2009;Mynbaev 2003].

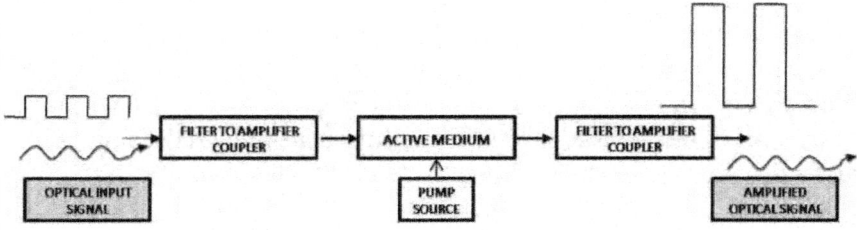

Figure 2: The Basic Structure of an Optical Amplifier.

The optical gain depends on the frequency/ wavelength of the signal. Let us consider a medium of two level systems for demonstrating the dependence of gain on frequency. The gain coefficient of such a medium can be written as below [Agarwal 2003]:

$$g(\omega) = \frac{g_0}{1 + (\omega - \omega_0)^2 T^2 + \dfrac{P}{P_{sat}}}$$

(5)

Where, g_0 is the peak value of the gain, ω is the optical frequency of the incident signal, ω_0 is the atomic transition frequency , P is the optical power of the signal being amplified, P_{sat} is the saturation power and T is the dipole relaxation time. In the unsaturated region, $P_s/P_s \ll 1$. So, the gain coefficient becomes

$$g(\omega) = \frac{g_0}{1+(\omega - \omega_0)^2 T^2}$$

(6)

This equation shows that the gain reaches its maximum when the incident frequency coincides with the atomic transition frequency. Another term associated with optical amplifiers is amplification factor or amplifier gain (G) defined as:-

$$G = \frac{P_{out}}{P_{in}}$$

(7)

Where P_{in} and P_{out} are the input and output powers of the continuous wave signal being amplified.

APPLICATIONS OF OPTICAL AMPLIFIERS

Optical amplifiers have found many applications ranging from ultra long undersea links to short links in access networks[Keiser 2009; Mynbaev 2003; Olsson1989 & Agarwal 2003].

- In-line amplifier
- Pre-amplifier
- Post –amplifier

In-Line Amplifier: This is used as a repeater along the link at intermediate points. It can be used to compensate for transmission loss and increase the distance between regenerative repeaters, as shown in figure 3a.

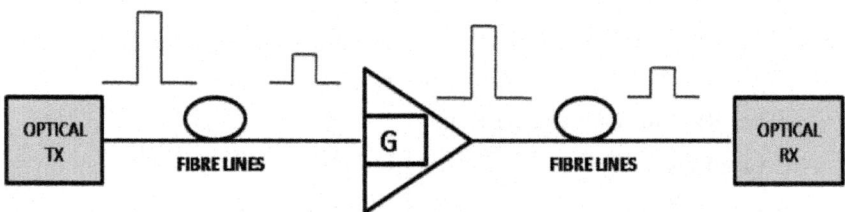

Figure 3a: Optical Amplifier as In-Line Amplifier.

Pre-Amplifier: This is used before the photo detector at the receiver in order to strengthen the weak received signal. This increases the sensitivity of the detector effectively. This configuration is shown in figure 3b.

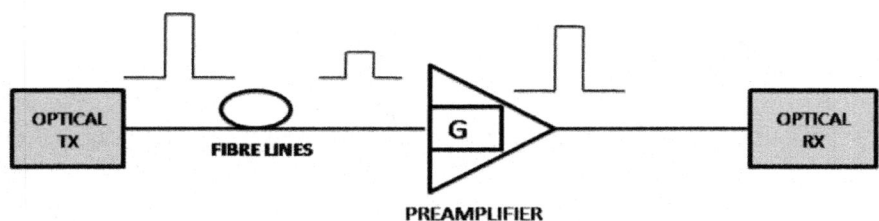

Figure 3b: Optical Amplifier as Preamplifier.

Post -Amplifier: This is used at the transmitting end, after the source and operates near the saturation region. The power launched into the fiber is enhanced and so the repeater span can become large. This serves to increase the transmission distance by 10- 100km depending on the amplifier gain and fiber loss. This configuration is shown in figure 3c.

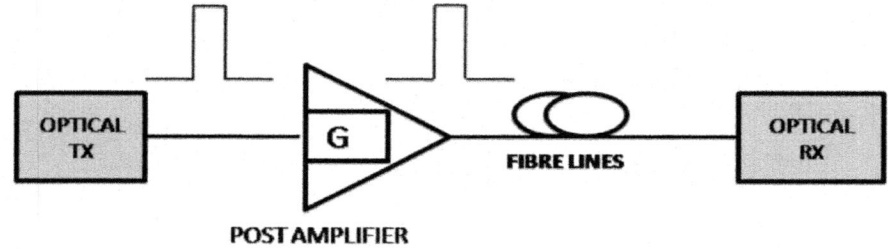

Figure 3c: Optical Amplifier as Post Amplifier.

TYPES OF OPTICAL AMPLIFIERS

The optical amplifiers which find widespread use in communication systems can be classified into three categories:-

1.   Fiber Raman Amplifier (FRA)
2.   Erbium Doped Fiber Amplifier (EDFA)
3.   Semiconductor Optical Amplifier (SOA)

The first two types, Fiber Raman Amplifier (FRA) and Erbium Doped Fiber Amplifier (EDFA) can be efficiently coupled to the transmission fiber by splicing with a minimum coupling loss. Of these two, EDFA requires lesser power for the pump source and the pump power requirements can be easily met by semiconductor laser diodes. Besides, the gain characteristics of EDFA are insensitive to polarization. Semiconductor Optical Amplifier (SOA) has the advantages of smaller size and lower power consumption. Its dimensionalcompatibility with the transmission fiber is obviously not as good as the fiber amplifier. However, SOA is suitable for optoelectronic integrated circuits.Table1 shows the basic difference between the three optical amplifiers and Table 2 shows the comparison of optical amplifiers.

Table 1: Difference of materials and operating bandwidth of three optical amplifiers

Type of optical amplifier	Material required	Operating Working band
Semiconductor Optical Amplifier (SOA)	Semiconductor material from group III and V. *e.g.*phosphorous, gallium, indium and arsenic	O-Band and C-Band
Erbium Doped Fiber Amplifier (EDFA)	Lightly doping silica or tellurite with rare earth element *i.e.* erbium.	O-Band, S-Band, C-Band and L-Band
Fiber Raman Amplifier (FRA)	Raman Lasers	All Operating Bands

Table 2: Comparison of Optical Amplifiers

S. No.	Parameter	Semiconductor Optical Amplifier (SOA)	Erbium Doped Fiber Amplifier (EDFA)	Fiber Raman Amplifier (FRA)
1	Gain(dB)	>30	>40	>25
2	Bandwidth (3dB)	60	30-60	Pump dependent
3	Max. Saturation (dBm)	18	22	0.75 X pump
4	Noise Figure (dB)	8	5	5
5	Pump Power	<400mA	25dBm	>30dBm
6	Wavelength(nm)	1260-1650	1530-1560	1260-1650
7	Time Constant	2×10^{-9} s	10^{-2}s	10^{-15} s
8	Size	Compact	Rack Mounted	Bulk Module
9	Cost factor	Low	Medium	High
10	Polarization Sensitivity	Yes	No	No

Although gain bandwidth of semiconductor laser amplifiers is ideally large, they have several drawbacks like polarization sensitivity, interchannel cross- talk and large coupling losses. Fiber amplifiers are preferable since the coupling loss due to fusion splice is negligible for them. Fiber amplifiers are also insensitive to polarization and have negligible noise for interchannel cross talk, which is one of the main noise sources in multichannel transmission or Dense Wavelength Division Multiplexing (DWDM). These reasons and available gain properties make the fiber amplifiers very suitable for modern optical transmission.

Advantages of EDFA

It is clear that EDFAs are the best choice for optical amplification in present lightwave systems. Erbium (Er: 68) is used as dopant into glass host (fiber) and the 'doped fiber' isused as an amplifying medium. Er-doped fibers give an amplified output around 1550nm [Desurvire 2002; Becker, Olsson & Simpson 1999; Sun et.al.1997]. The EDFA is one of the key devices used for dense wavelength division multiplexed (DWDM) transmission systems. EDFAs are revolutionizing lightwave systems by reducing system costs and enhancing network performance. Some of the advantages offered by EDFAs are:

- High gain (~50dB)
- High output power (>100mW)
- Low noise figure (~4dB)
- Less gain variation
- Wide bandwidth of operating suiting DWDM
- Inherent compatibility to transmission fiber with low insertion loss
- Cross talk immunity in multichannel systems

WORKING PRINCIPLE OF EDFA

The invention of the Erbium Doped Fiber Amplifier (EDFA) in the late eighties was one of the major events in the history of optical communication systems. It provided new life to the research of technologies that allow high bit rate transmission over long distances. EDFA has a narrow high gain peak at 1532nm and a broad peak with a lower centered at 1550nm. The use of an increasing number of channels in the present day DWDM optical networks requires a flat gain spectrum across the whole usable bandwidth. Owing to their versatility, useful gain bandwidth, high pumping efficiency and low intrinsic noise, EDFAs are the amplifier of choice for most of the network applications. They are based on single mode optical fibers with cores that have been doped, typically to a few hundred part per million, with the trivalent erbium ion, Er^{3+}. The gain is provided through stimulated emission, as in laser. The Er^{3+} ion acts mostly as a three level system, in which the main participants are the $^{4}I15/2$ ground state, the $^{4}I13/2$ first excited level and the $^{4}I11/2$second excited level. The energy level diagram of Er^{3+} is shown in figure 4) [Keiser 2009;Mynbaev 2003].

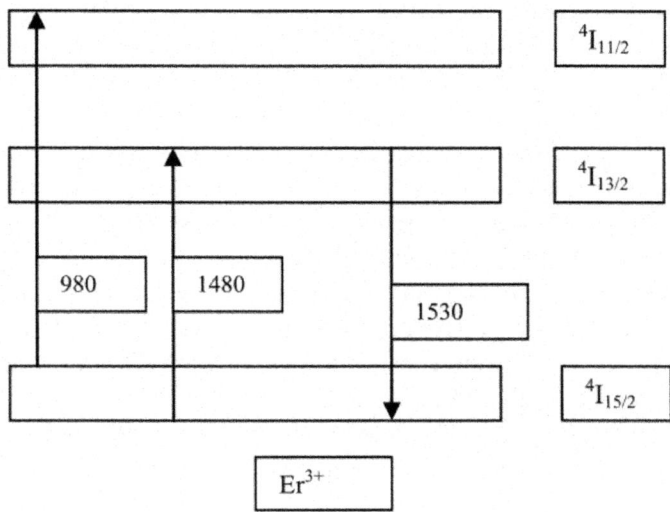

Figure 4: Energy Level Diagram of Er^{3+}.

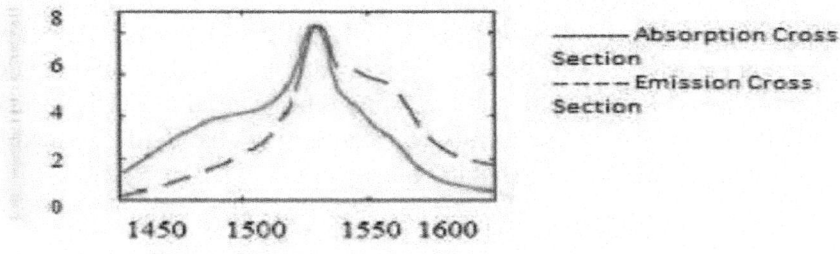

Figure 5: Absorption and Emission Spectra of EDFA.

EDFAs are of particular interest in telecommunications, because their emission spectrum shows a gain of more than 20dB over the range of 1530-1560nm. This is also the third window used in optical communication. The absorption spectrum reveals that good absorption takes place around 380nm, 520nm, 800nm, 980nm, and 1480nm. The absorption bands at shorter wavelengths are not of interest owing to the non- availability of semiconductor laser diodes at these wavelengths. At 980nm and 1480nm, efficient laser diodes are available and therefore used as pump sources.

LIMITATIONS OF EDFA

The main practical limitation of an EDFA stems from the spectral non-uniformity of the amplifier gain. As a result, different channels of a DWDM system are amplified by different amounts. These problems become quite severe in long-haul systems, employing cascaded chain of EDFAs. Secondly, for many EDFA deployments, automatic gain control (AGC) is used to ensure that the output signal power is proportional to the input power. However, there are times when a constant optical signal output, independent of input power, is more desirable, e.g., in an optical preamplifier at an optical receiver [Qiao & Vella 2007]. The figure 6) [Keiser 2009;Mynbaev 2003].Figure 6 shows the gain spectrum of EDFA, from which it is clear that EDFA has peak gain at 1530nm, beyond which the gain reduces slightly and remains flat almost until 1550nm. After that, the gain reduces sharply. Several gain flattening techniques of EDFA are available [Lee et,al 1996; Ono et.al.1997; Kim et.al.1998; Park et.al.1998; Kawai et.al.1999; Yun et al. 1999; Lu & Chu 2000; Pasquale & Federighi 1995; Kemtchou et. al.1996; Hwang et al.2000; .Bakshi et.al.2001; Sohn et.al.2002; Arbore et.al.2002; Kaur & Gupta 2009 and Lobo et.al. 2003]

So, EDFAs are widely used in the C-band (1530-1560nm) for optical communication networks. So, there is a necessity to improve the amplification bandwidth of EDFA (i.e. broadening as well as flattening of gain spectrum). This would help to cater the needs of present day communication systems. In order to overcome this limitation of EDFA, different doping elements are coming into existence. One of such doping material is thulium and the doped fiber amplifier is known as Thulium Doped Fiber Amplifier (TDFA). TDFAs are highly viable alternative to meet out the limitations of EDFAs and have bright future prospects to be used in optical communication systems.

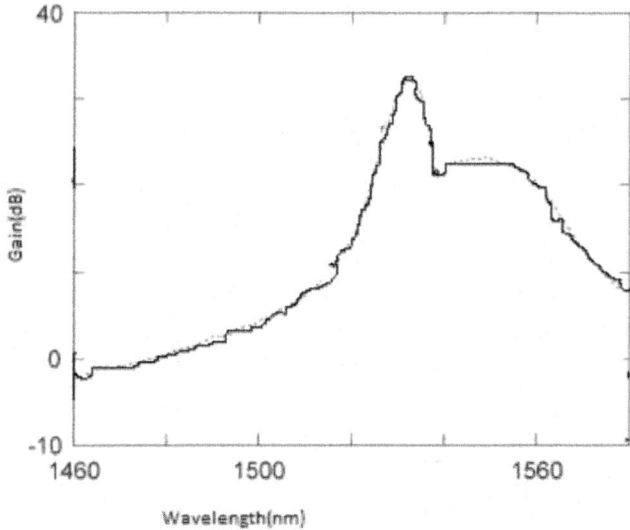

Figure 6: Gain Spectrum of EDFA.

ROLE OF TDFA IN COMMUNICATION SYSTEMS

The optical fiber can be doped with any of the rare earth element, such as Erbium (Er), Ytterbium (Yb), Neodymium (Nd) or Praseodymium (Pr), Thulium (Tm). The host fiber material can be either standard silica, a fluoride based glass or a multicomponent glass. The operating regions of these devices depend on the host material and the doping elements. Fluorozirconate glasses doped with Pr or Nd are used for operation in the 1300nm window, since neither of the ions can amplify 1300nm signals when embedded in silica glass. The next popular material for long haul telecommunication applications is a silica fiber doped with Thulium, which is known as Thulium Doped Fiber Amplifier (TDFA). In some cases as Yb is added to increase the pumping efficiency and the amplifier gain. The TDFA are used in S-band (1460-1530nm). The energy state diagram of Tm^{3+} is shown in figure 7 [Aozasa et.al.2008].

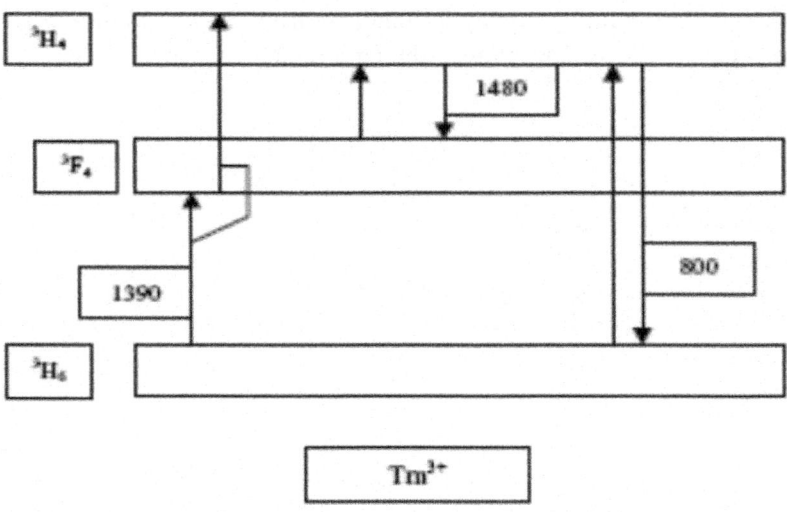

Figure 7: Energy Level Diagram of Tm^{3+}.

Tm^{+3} has three energy levels that are considered with respect to the Tm^{+3} populations. TDFA uses upconversion pumping method. The upconversion pumping consists of the twostep excitation of $^3H_6$ to $^3F_4$ and $^3F_4$ to $^3H_4$ with the same pump wavelength and this makes it possible to form a population inversion state between $^3F_4$ and $^3H_4$. The gain and loss of TDFA in the 1460-1530nm wavelength region are determined not only by the excited state absorption (ESA) ($^3F_4$ to $3H_4$) and stimulated emission(SE) ($^3H_4$ to $^3F_4$) but also by the ground state absorption (GSA) ($^3H_6$ to $^3F_4$) The absorption, emission and ground state cross section emission of Tm^{3+} is shown in figure 8 [Aozasa et.al.2008,2002].

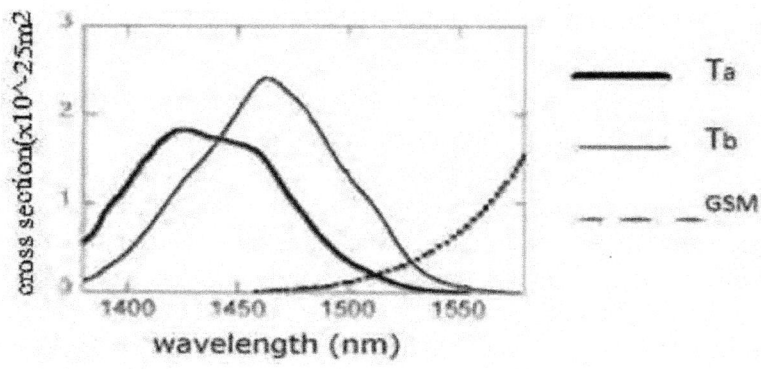

Figure 8: Cross Sections of Tm3+.

The gain spectrum of TDFA is shown in figure 9 [Aozasa et.al.2008, 2002]. A gain of 22dB (approximate) is obtained from 1460-1485nm wavelength range. After this wavelength range the gain reduces sharply.

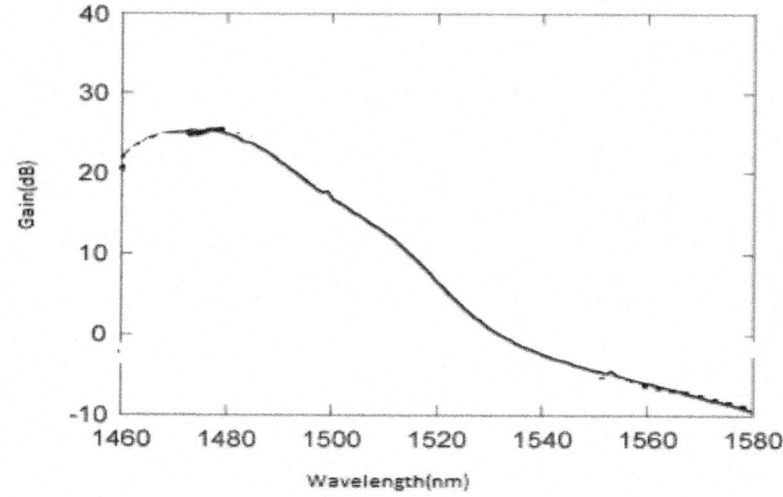

Figure 9: Gain Spectrum of TDFA.

So, to utilize the S-band, TDFA is proposed to be used by various authors. The S-band has attracted attentions because it has low fiber loss, low dispersion and also high gain and efficiency. The summary of the work done on EDFA, TDFA and TDFA-EDFA amplifiers is given in Table 3.

Table 3: Summary of work done in [Qiao & Vella2007], [Aozasa et.al.2008], [Sakamoto et.al.2006]

Type of Amplifier/ Parameters	EDFA in [Qiao & Vella2007]	TDFA [Aozasa et.al.2008]	TDFA-EDFA [Sakamoto et.al.2006]
Number of Stages	Two	Single	Four
ASE and its Correction Function	Considered with signal wavelength 1546nm	Considered	Not considered
Peak Gain /Range of Gain	0-37dB	22.6dB	20dB
Gain Excursion	0.35dB	0.35dB	2dB
Range of Input Power	-5 to 5 dBm	-32 to -2 dBm	-20 to -10dBm
Signal Gain Band	1525-1565nm	1479-1507nm	1460-1537nm
Noise Figure	Not considered	< 6.5dB	<7dB
No. of DWDM Channels Considered	8	8	4

HYBRID AMPLIFIERS

There is one more method of utilizing fiber amplifiers for optimum utilization of available fiber bandwidth i.e. by way of using various combinations of optical amplifiers in different wavelength ranges. The amplifiers can be connected either in parallel or in series. This configuration is termed as Hybrid Amplifier which is highly viable for the above discussed cause. In parallel configuration, the DWDM signals are first demultiplexed into several wavelength-band groups with a coupler, then they are amplified by amplifiers that have gains in the corresponding wavelength band and then they are multiplexed again with a coupler. The parallel configuration is very simple and applicable to all amplifiers. However, it has disadvantages also e.g. an unusable wavelength region exists between each gain band originated from the guard band of the coupler. Also, the noise figure degrades due to the loss of the coupler located in front of each amplifier. On the contrary, the amplifiers connected in series have relatively wide gain band, because they do not require couplers. Hybrid configurations can be made by combination of the following:

- **EDFAs and FRAs:** It has been observed that the gain spectrum of FRAs can be tailored by adjusting the pump powers and pump wavelengths. So this property is used to increase the amplification bandwidth of EDFA [Thyagarajan& Kakkar 2004; Oliveira et.al.2007, Kaur &Gupta 2008].

- **TDFAs and FRAs**: Combining FRAs with TDFAs is very effective approach, because FRAs can provide any gain bandwidth by selecting the appropriate pump wavelengths. However, a drawback with FRAs is that double Rayleigh scattering (DRS) degrades the amplified signals [Percival & Williams 1994; Komukai et.al.1995,2001; Royet.al.2002and Aozasa et.al.2002].

• **TDFAs and EDFAs**: Hybrid amplifiers consisting of all rare –earth –doped- fiber amplifiers are easier to utilize than those incorporating FRAs, because these are free from DRS. These hybrid amplifiers are relatively simple in gain spectra control [Sakamoto et.al.2006 and Kaur & Gupta 2010]. Hybrid doped fiber amplifiers with different gain bandwidths have attracted a large interest for increasing the transmission capacity of long haul wavelength multiplexed optical communication systems in Cband and L-band.

Out of these, one of the-state-of-art hybrid amplifiers is TDFA -EDFA configuration. It is observed that for TDFA-EDFA configuration, the total gain of hybrid amplifier is given as product of gain of TDFA and gain of EDFA. The gain bandwidth is extended by cascading EDFA with TDFA. When EDFA is cascaded with TDFA in series, the total gain of hybrid amplifier is given by product of individual gains of each amplifier.The gain of TDFA is given as

$$G_{T(\lambda)} = \exp\left[\left(\sigma_{T(1480)}NT_{2-\sigma T(1390)}NT_{1-\sigma T(800)}NT_0\right)\right](\eta_T\ L_T) \tag{8}$$

The gain of EDFA is given as:

$$G_{E(\lambda)} = \exp\left[\left(\sigma_{E(1530)}NE_{2-\sigma E(1480)}NE_1\right)\left(\eta_E\ L_E\right)\right] \tag{9}$$

This means the total gain of hybrid amplifier is given as:

$$G_{(\lambda)} = G_{T(\lambda)} X G_{E(\lambda)} =$$
$$= \exp\left[\left(\sigma_{T(1480)}NT_{2-\sigma T(1390)}NT_{1-\sigma T(800)}NT_0\right)(\eta_T\ L_T)\right] X \exp\left[\left(\sigma_{E(1530)}\ NE_{2-\sigma E(1480)}NE_1(\eta_E\ L_E)\right)\right] \tag{10}$$

In the above equations, σ_T (1480), σ_T (1390) and σ_E (980) denotes cross-sections of excited state absorption, stimulated emission and ground state emission of TDFA. Similarly, σ_E (1530), σ_E (1480) represents the respective cross sections of EDFA. η_T and η_E represents the confinement factors of TDFA and EDFA respectively.

The above stated mathematical equations clearly illustrate the fact that the gain of hybrid amplifier broadens from 1460 nm to 1530 nm wavelength range. Further there is a noticeable reduction in the noise figure correspondingly in the hybrid amplifier. This affects in the gradual increase in the number of transmission channels of DWDM system, thereby increasing the overall transmission capacity of the optical communication system. The statistical analysis of TDFA-EDFA hybrid amplifier and EDFA-TDFA hybrid amplifier is done. The configuration TDFA-EDFA means the hybrid amplifier in which TDFA is in first stage and EDFA is in second stage, whereas configuration EDFA-TDFA means EDFA is in first stage and TDFA is in second stage. For this analysis, it is assumed that both fibers have step refractive index homogeneously broadened spectrum of thulium and erbium ions. We consider three levels of TDFA i.e. $^3H^6$, $^3H^4$ and $^3F^4$, the other two levels of TDFA i.e. $^3H^5$ and $^3F^{2,3}$ are ignored as the rate of their non radiation (τnr) to the corresponding levels are very high. Similarly, in case of EDFA, two levels $_4I^{15/2}$, $_4I^{13/2}$ is considered and level $_4I^{11/2}$ is ignored for the same reasons. From the absorption and emission spectra, it is clear that the absorption and emission peaks of EDFA coincides at 1530nm, while the absorption peak of TDFA lies at 1430nm and emission peak of TDFA lies at 1460nm. Form the gain spectrum of EDFA, from which it is clear that EDFA has peak gain at 1530nm, beyond which the gain reduces slightly and remains flat almost until 1550nm. After that gain reduces sharply. This gain can be flattened by cascading TDFA with EDFA. Fig. 10 shows the flattened gain spectrum of Hybrid amplifier by cascading TDFA and EDFA. The thulium doped fiber (TDF) in first stage was forward pumped with a 1390nm pump laser diode (LD) and erbium doped fiber (EDF) in second stage is pumped with a 980nm LD. For efficient amplification the concentration of TDF $^{+3}$ ions was kept very high (approx. 7500 ppm) [Percival & Williams 1994; Komukai et.al.1995]. Table 4 shows the different characteristics of both configurations.

Table 4: Features of TDFA-EDFA & EDFA-TDFA Amplifiers

Feature	TDFA-EDFA	EDFA-TDFA
Gain	25 dB for 1456nm-1556nm range	20 dB for 1485nm-1550nm range
Noise Figure	<6Db	<7dB

F-ratio is calculated for the parameters mentioned in above Table 4. The calculated and tabulated value of F-ratio is shown in Table 5.

Table 5: F-Measure Results

Source of variation	D.F	SS	MS	F-ratio	Tabulated F-ratio (1,2)
Between samples	1	1.5	1.5	.05	18.51
Within Samples	2	39.6	29.9		
Total	3	41.1			

The table 5 shows that the F-ratio is significant of 5% level which means that both hybrid configurations work differently. F-ratio is used to judge whether the difference among several sample means is significant or just a matter of sampling fluctuations. MS residual is always due to fluctuations of sampling and so serves as the basis for the significance test. The F-ratio is compared with its corresponding table value for the given degree of freedom at a specified level of significance. The table 5 shows that both the F-ratios are significant of 5% level which means that TDFA-EDFA amplifier work differently as compared with EDFA-TDFA amplifier. The gain spectrum of TDFA-EDFA is more widened as compared to that of EDFA-TDFA configuration. Since there is a large difference between the calculated and the table value of F. So, the null hypothesis is rejected. For WDM systems, TDFA-EDFA has a great impact as a hybrid amplifier as compared with EDFA-TDFA amplifier[Kaur & Gupta 2009]. As studied from the existing schemes, the amplification of DWDM signals using TDFA-EDFA hybrid amplifiers have major problems and short comings which are as listed below:

- The use of increasing number of channels in the present day DWDM optical networks requires a flat gain spectrum across the whole usable bandwidth. The unflattened gain spectrum of hybrid amplifiers implies that different channels of a DWDM system are amplified by different amounts. Hence a need is felt to broaden as well as flatten the gain spectrum of hybrid amplifier.

- It is observed that amplified spontaneous emission and its correction function for hybrid amplifiers have not been carried out leading to lesser gain and more noise of the signal. So, there is a need to analyze different parameters e.g. gain, noise figure, amplified spontaneous emission of hybrid amplifier and its correction function.

- No scheme or algorithm has been designed to allow the hybrid amplifier to maintain a constant output signal power, independent of the optical wavelength and input power level. There are many occasions when constant optical signal power, independent of input power, is more desirable e.g. in an optical preamplifier in an optical receiver and automatic power control cannot guarantee constant signal output power.

- Hybrid amplifiers proposed till dates are using four or more than four amplifiers to achieve desirable gain, leading to higher complexity, noise. Hence there is a dire need to minimize number of amplifiers in hybrid configuration.

- DWDM system till date are upto thirty two (32) channels but with lesser gain and high noise figure. The systems having adequate gain have been designed only upto eight (8) channels. Hence, there is a drastic need to increase the number of channels in DWDM system.

Although there are many improvements in gain spectrum of EDFA, but still the improved configurations are unable to provide enough bandwidth for emerging high quality parameters like gain, noise figure, amplified spontaneous emission etc. Therefore, there is a need to search for a new and versatile approach that enables an effective system with adequate bandwidth to accommodate large number of DWDM channels. One approach that can do the job is use of hybrid amplifier consisting of TDFA and EDFA in cascaded series combination. This hybrid amplifier is proven effective in DWDM systems. Several challenging points of research are realization and development of hybrid amplifiers, which can increase the bandwidth for S-band, C-band and L-band. The biggest challenge with hybrid amplifier is to maintain and offer high bandwidth in case of higher number of channels.

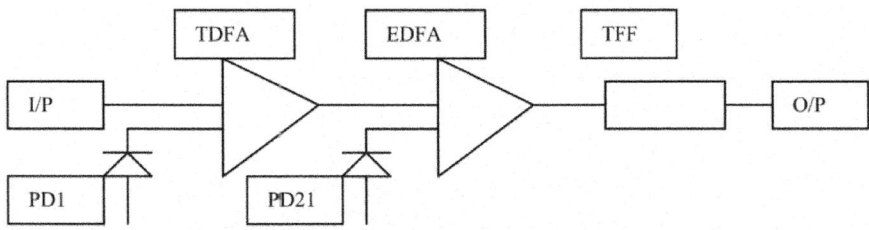

Figure 10: Schematic Diagram of Cascaded TDFA-EDFA Hybrid amplifier using TFF.

It is clear with the configuration as shown in figure 10 that a wide bandwidth spectrum of nearly 100nm i.e. from 1460nm to 1560nm wavelength range is obtained. This also includes the 1510nm-1520nm range where EDFA as well as TDFA has no large gain for themselves. This is also observed that this gain is unflattened mainly from 1520nm to 1540nm region. This whole wavelength range is flattened by using a seven layer interference filter (TFF).A seven layer optical thin film filters consists of a stack of seven dielectric thin a film is used along with a cascaded TDFA and EDFA [Kaur & Gupta 2010]. There are so many ways to flatten the gain bandwidth of EFDA such as gain equalizers based on Mach-Zehnder optical filters, interference filters or long period grating and fluoride or tellurite based EDFA. The figure (11) shows a schematic diagram of a seven layer dielectric interference filter for gain flattening of hybrid amplifier consisting of cascaded TDFA with EDFA. A seven layer dielectric film is proposed as a gain equalizer. In figure (11) the shaded layer as high index layer having refractive index 2.4. The unshaded layer is a low index layer having refractive index as 1.46. The refractive index of fiber is assumed as 1.46. The third and sixth layer of this filter is half wavelength thick and all other layers have one fourth wavelength thickness. The filter is so designed that transmission loss occurs around the maximum gain of hybrid amplifier i.e. at 1531nm. The transmission loss is about 9dB. The flattened gain bandwidth of hybrid amplifier with the help seven layer deictic filter is shown in figure (12). The gain variation is less than ± 2.5% in the wavelength region of 1460-1560 nm. A [2X2] square matrix of a dielectric filter for TM mode is given as

$$[\text{Matrix}]= [M] = \prod_{x=1} b \begin{bmatrix} cos\beta x & \frac{j}{q}sin\beta x \\ jqsin\beta x & cos\beta x \end{bmatrix}$$

(11)

The transmission of a TFF is given as

$$T= \left[1 + \frac{4R}{(1-R)2} sin^2(\frac{\Phi}{2})\right]^{-1}$$

(12)

From equations (11) & (12), we get following equation for transmission

$$T= \left[\frac{2\, n\, fiber}{\left(m_{11}+m_{12}n_{fiber}\right)n_{fiber} +\left(m_{21}+m_{22}n_{fiber}\right)}\right]^2$$

(13)

Where m_{ij} are the components of the matrix [M]. Here we assume n_{fiber}=1.46.Since the gain peak of hybrid amplifier occurs at 1530nm. So here we designed TFF such that the maximum transmission loss occurs at around the gain peak at wavelength 1530nm, which is observed as 9dB. In case of TE mode all parameters remain same except q_x is replaced by p_x. It is clear from formula given in equation (13) that designing of interference filters with desired wavelength spectrum and transmittance is possible by selecting proper of layer of dielectric films, their thickness and refractive indices of core and cladding of the optical fiber.

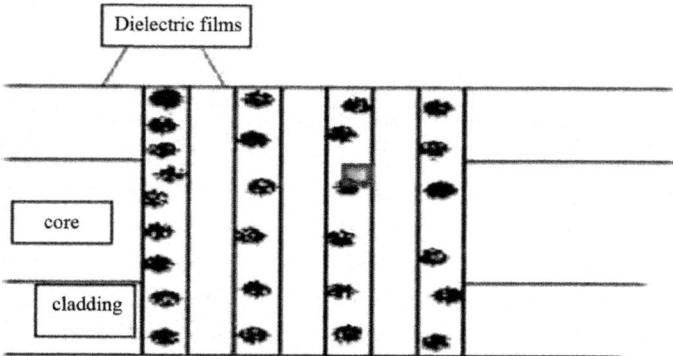

Figure 11: Schematic Diagram of the dielectric multi-layer Interference Filter (TFF).

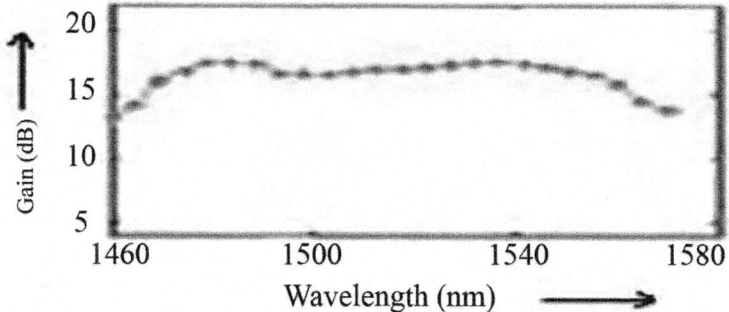

Figure 12: Broadened and Flattened Gain Spectrum of Hybrid Amplifier.

CONCLUSION

Wavelength multiplexing (WDM) technology along with optical amplifiers is used for optical communication systems in S-band, C-band and L-band. To improve the overall system performance Hybrid amplifiers consisting of cascaded TDFA and EDFA with different gain bandwidths are preferred for long haul wavelength multiplexed optical communication systems. It has found that calculated value of F ratio is very much different from the tabulated value, so the difference between parameters is considered as significant and we reject the null hypothesis. Here, we are able to conclude that for WDM systems, TDFA-EDFA hybrid fiber doped amplifier has higher gain and lower noise figure. So, this configuration gives better performance in WDM systems as compared with the EDFA-TDFA hybrid configuration.

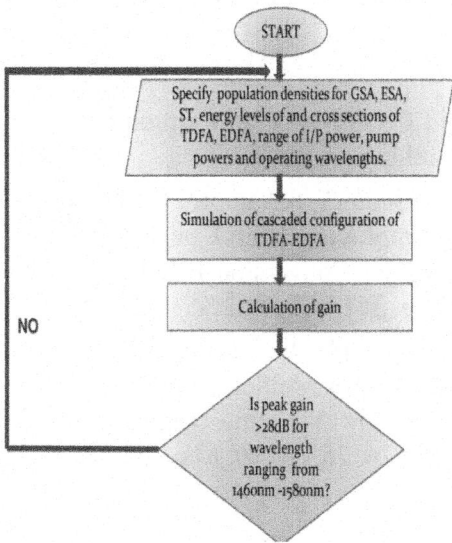

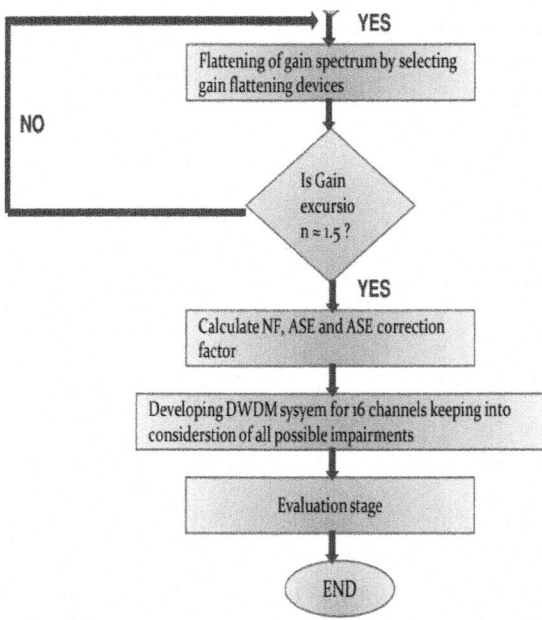

Figure 13: Flowchart Showing Simulation Process.

With this design it has also been found that when TDFA with EDFA are cascaded in series then gain spectrum is broadened. The gain variation is less than ± 2.5% in the wavelength region of 1460-1560 nm. The TF filter is so designed that transmission loss occurs around the maximum gain of hybrid amplifier i.e. at 1531nm. The transmission loss is about 9dB.The simulation process can be represented by a flowchart as shown in figure 13.

REFERENCES

1. Agarwal Govid P., "Fiber Optic Communication Systems", John Wiley & sons, Inc. Publication, 2003.

2. Aozasa S., H.Masuda, T. Sakamoto, K.Shikano and M.Shimizu, "Gain-Shifted TDFA Employing High Concentration Doping Technique with High Internal Power Conversion Efficiency of 70%,"Electron. Letters, Vol. 38, No. 8, pp. 361-363,2002.

3. Aozasa S, Hiroji Masuda, Makoto Shimizu and Makoto Yamada, "Novel Gain Spectrum Control Method Employing Gain Clamping and Pump Power Adjustment in Thulium- Doped Fiber Amplifier" Journal of Lightwave Tech. Vol. 26, No.10, May 2008.

4. Arbore .M.A., Y. Zhou, G.Keaton and T.Kane, "34 dB Gain at 1500nm in S- Band with Distributed ASE Suppression," presented at Eur. Conf. Opt. Commun., Denmark, Sept. 2002.

5. Bakshi B., M.Vaa, E.A.Golovchenko, H.Li and G.T. Harvey, " Impact of Gain Flattening Filter Ripple in long Haul WDM Systems", Proc. 27th Eur. Conference on Optical Communication 2001 Vol. 3, pp. 448-449, Sept. 2001.

6. Becker P.C, Olsson N.A. and. Simpson J.R, "Erbium Doped Fiber Amplifiers". New York: Academic, 1999, Chapter 7.

7. Desurvire Emmanuel, "Erbium Doped Fiber Amplifiers- Principles and Application", Hoboken, NJ: John Wiley & Sons, Inc. ISBN 0-471-58977-2,2002 Chapter 5.

8. Hwang B. C. et al., "Cooperative Up- Conversion and Energy Transfer of New High Er+3 and Er+3 – Yb+3 Doped Phosphate Glasses," Journal of Optical Society of America, Vol. 17, No. 5, pp. 833-839, 2000.

9. Kasamatsu T., Yano Y. and Ono T., "Gain-Shifted Dual –Wavelength-Pumped ThuliumDoped Fiber Amplifier for WDM Signals in the 1.48-1.51-μm Wavelength Region," IEEE Photonics Technology Letters, Vol. 13, No. 1, pp.31-33,2001.

10. Kaur Inderpreet, Gupta Neena, "Statistical Analysis of Different Configurations of Hybrid Doped FiberAmplifiers", International Journal of Electrical and Electronics Engineering 3:8 2009 pp 515-520.

11. Kaur Inderpreet, Gupta Neena, "Enhancing the Performance of WDM Systems By Using TFF in Hybrid Amplifiers", 2010 IEEE 2nd International Advance Computing Conference, Patiala, India, pp 106-109, Feb 2010.

12. Kaur Inderpreet, Gupta Neena,, "Increasing the Amplification Bandwidth of Erbium Doped Fiber Amplifiers by Using a Cascaded Raman-EDFA Configuration", Photonics 2008, P.284, Dec. 2008 at IIT, Delhi.

13. Kawai Shingo, Masuda Hiroji, Suzuki Ken Ichi, Aida Kazuo, "Wide Bandwidth and Long Distance WDM Transmission Using Highly Gain Flattened Hybrid Amplifier, IEEE Photonics Technology Letters, Vol. 11, pp. 886-888, July 1999.

14. Keiser Gerd, "Optical Fiber Communications", 4th Edition, Tata McGraw-Hill Education Pvt. Ltd., New Delhi, Inc. 2009, ISBN-13: 978-0-07-064810-4.

15. Kemtchou J., M. l, Chatton F. and Lecoy T. G., "Comparision of Temperature Dependences of Absorption and Emission Cross – Section

in Different Hosts of EDFAS," Proc. Opt. Amplifiers Appl., 1996, pp. 126-129, Paper FD2.

16. Kim Hyo Sang, Yun Seok Hyun, Hyang Kyun, Kim Namkyoo Park and Kim Byoung Yoon, " Actively Gain-Flattened EDFA Over 35nm by Using All-Fiber Acousto-Optic Tunable Filters", IEEE Photonics Technology Letters, Vol. 10, pp. 790-792, June1998.

17. Komukai T., Yamamoto T., Sugawa T andMiyajima Y, "Upconversion Pumped ThuliumDoped Fluoride Fiber Amplifier and Laser Operating at 1.47µm," Journal of Quantum Electronics, Vol. 31, pp. 1880-1889, Nov. 1995.

18. Lee Y.W, Nilsson J, Hwang S.T. and Kim S.J., "Experimental Characterization of a Dynamically Gain –Flattened EDFA", IEEE Photonics Technology Letters, Vol. 8, No. 12, pp. 1612-1614, Dec.1996.

19. Lobo Audrey Elisa, Besley James A. and C.Martijin De Sterke, " Gain Flattening Filter Design Using Rotationally Symmetric Crossed Gratings", Journal of Lightwave Tech. Vol. 21, No. 9, pp. 2084-2088, 2003.

20. Lu Yi Bin and Chu P.L., "Gain Flattening by Using Dual-Core Fiber in Erbium Doped Fiber Amplifier",IEEE Photonics Technology Letters, Vol. 12, No. 12, pp. 1616-1617, Dec. 2000.

21. Mynbaev D.K, L.L.Schiner' "Fiber Optics Communications Technology", Pearson Education, Delhi, Inc.,2003, ISBN 81-7808-317-5.

22. Oliveira J.C. .Silva R.F., Rossi S.M, Rosolem J.B. and .Bordonalli A.C, "An EDFA Hybrid Gain Control Technique for Extended Input Power and Dynamic Gain Ranges with Suppressed Transients", IMOC 2007, pp. 683-687, 2007 SBMO/IEEE MTT-S.

23. Olsson N.A., "Lightwave Systems with Optical Amplifiers", Journal of Lightwave Tech. Vol. 7, July 1989.

24. Ono Hirotaka, Yamada Makoto and Ohishi Yasutake, "Broadband and Gain Flattened Amplifier Composed of a 1.55µm- Band and a 1.58 µm- Band Er3+- Doped Fiber Amplifier in a Parallel Configuration", Electronics Letters, Vol. 33, No. 8, pp. 710-711, May 1997.

25. Park Seo Yeon, Kim Hyang Kyun, Lyu Gap Yeol, Sun Mo Kang and Sang Yung Shin, " Dynamic Gain and Output Power Control in a Gain-Flattened EDFA", IEEE Photonics Technology Letters, Vol. 10, No. 6, pp. 787-789, June1998.

26. Pasquale F. D. and Federighi M., "Modeling of Uniform and Pair – Induced Up Conversion Mechanisms in High-Concentration Erbium

Doped Silica Waveguides," Journal of Lightwave Technology, Vol.13, pp. 1858-1864, 1995.

27. Percival R.M. and Williams J.R, "Highly Efficient 1.064µm Upconversion Pumped 1.47µm Thulium Doped Fluoride Fiber Amplifier," Electron Letters, Vol. 30, No. 20, pp. 1684-1685, June 1994.

28. Qiao Lijie and Vella Paul J., "ASE Analysis and Correction for EDFA Automatic Control," Journal of Lightwave Tech. Vol. 25, No.3, May 2007.

29. Roy F., Sauze A., Baniel P. and Bayart D., "0.8-µm +1.4µm Pumping for Gain Shifted TDFA With Power Conversion Efficiency Exceeding 50%," Presented at the Opt. Amplifiers Appl. Vancouver, BC, Canada, Jul.2002, PD4.

30. Sakamoto Tadashi, Aozasa Shin- ichi, Yamada Makoto and Shimizu Makoto, "Hybrid Fiber Amplifier Consisting Of Cascaded TDFA and EDFA for WDM Signals", Journal of Lightwave Tech. Vol. 24 No.6 , June 2006.

31. Sohn Ik-Bu, Baek Jang Gi, Lee Nam Kwon, Kwon Hyung Woo and Song Jae Won, "Gain Flattened and Improved EDFA Using Microbending Long-Period Fiber Gratings", Electronics Letters, Vol.38, No. 22, pp. 1324-1325, Oct. 2002.

32. Sun .Y, Judkins J.B., Srivastava A.K., Garrett L.,J.L.Zyskind, Sulhoff J.W, Wolf C., Derosier R.M., Gnauck A.H., R.W.Tkach, J.Zhou, Espindola R.P., Vengsarkar A.M and Chraplvy A.R., "EDFA Transmission Response to Channel Loss in WDM Transmission System", IEEE Photonics Technology Letters, Vol. 9,No. 3, pp. 386-388, March 1997.

33. Y. Sun, Zyskind J.L. and Srivastava A. K., "Average Inversion Level, Modeling and Physics of Erbium Doped Fiber Amplifiers," Journal of IEEE Sel. Topics Quantum Electronics., Vol. 3, no. 4, pp. 991-10007, Aug.1997.

34. Thyagarajan K., Kakkar Charu, "Novel Fiber Design for Flat Gain Raman Amplification Using Single Pump and Dispersion Compensation in S-Band", IEEE Photonics Technology Letters, Vol.22, pp. 2279-2286, Oct. 2004.

35. Yun Seok Hyun, Lee Bong Wan Lee, Kim Hyang Kyun and Kim Byoung Yoon, "Dynamic Erbium Doped Fiber Amplifier Based on Active Gain Flattening with Fiber Acousto-Optic Tunable Filters", IEEE Photonics Technology Letters, Vol. 11, No. 10, pp. 1229-1231, Oct. 1999.

Chapter 4

THE DOHERTY POWER AMPLIFIER

Paolo Colantonio, Franco Giannini, Rocco Giofrè and Luca Piazzon

University of Roma Tor Vergata Italy

INTRODUCTION

The Doherty Power Amplifier (DPA) was invented in the far 1936 by W. H. Doherty, at the Bell Telephone Laboratories of Whippany, New Jersey (Doherty, 1936). It was the results of research activities devoted to find a solution to increase the efficiency of the first broadcasting transmitters, based on vacuum tubes. The latter, as it happens in current transistors, deliver maximum efficiency when they achieve their saturation, i.e. when the maximum voltage swing is achieved at their output terminals. Therefore, when the signal to be transmitted is amplitude modulated, the typical single ended power amplifiers achieve their saturation only during modulation peaks, keeping their average efficiency very low. The solution to this issue, proposed by Doherty, was to devise a technique able to increase the output power, while increasing the input power envelope, by simultaneously maintaining a constant saturation level of the tube, and thus a high efficiency. The first DPA realization was based on two tube amplifiers, both biased in Class B and able to deliver tens of kilowatts.

Nowadays, wireless systems are based on solid state technologies and also the required power level, as well as the adopted modulation schemes, are completely different with respect to the first broadcasting transmitters. However, in spite of more than 70th years from its introduction, the DPA actually seems to be the best candidate to realize power amplifier (PA) stage for current and future generations of wireless systems. In fact, the increasing complexity of modulation schemes, used to achieve higher and higher data rate transfer, is requiring PAs able to manage signals with a large time-varying envelope. The resulting peak-to-average power ratio (PAPR) of the involved signals critically affects the achievable average efficiency with traditional PAs. For instance, in the European UMTS standard with W-CDMA modulation, a PAPR of 5-10 dB is typical registered. As schematically reported in Fig. 1, such high values of

PAPR imply a great back-off operating condition, dramatically reducing the average efficiency levels attained by using traditional PA solutions.

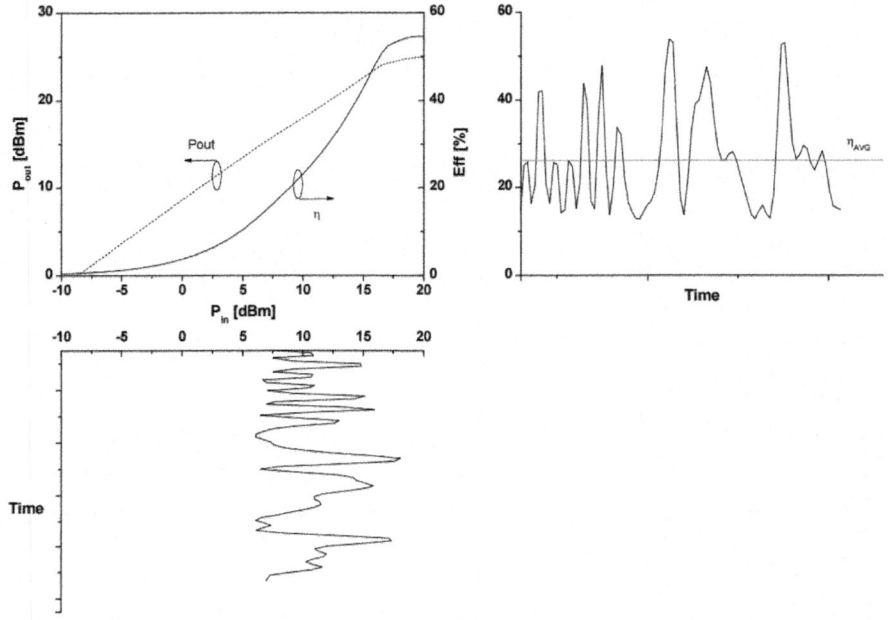

Figure 1: Average efficiency using traditional PA.

To stress this effect, it is helpful to refer to an ideal Class B PA, which delivers an efficiency of 78.6% at its maximum output power, whereas it becomes only 25% at 10dB back-off. Therefore, when dealing with amplitude modulation signal, it is more useful to refer to the average efficiency, which is defined as the ratio of the average output power ($P_{out,AVG}$) to the average supply DC power ($P_{DC,AVG}$) (Raab, 1986):

$$\eta_{AVG} = \frac{P_{out,AVG}}{P_{DC,AVG}}$$

(1)

Clearly, the average efficiency depends on both the PA instantaneous efficiency and the probability density function (PDF), i.e. the relative amount of time spent by the input signal envelope at different amplitudes. Therefore, to obtain high average efficiency when timevarying envelope signals are used, the PA should work at the highest efficiency level in a wide range of its output (i.e. input) power. This requirement represents the main feature of the DPA architecture, as shown in Fig. 2, where its theoretical efficiency behavior is reported.

The region with almost constant efficiency identifies the DPA Output Back-Off (OBO) range, and it is fixed according to the PAPR of the signal to be amplified. As will be later detailed, the OBO value represents the first parameter to be chose in the design process.

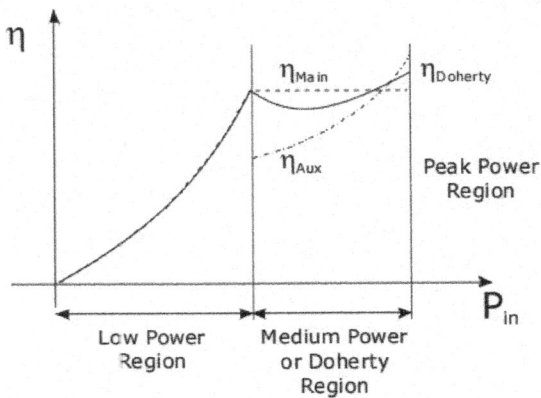

Figure 2: Typical DPA efficiency behavior versus input power.

Due to this attractive characteristic and the relative simple implementation scheme, the DPA is being the preferred architecture for new communication systems.

The Doherty technique is usually adopted to design PA for wireless systems and, in particular, in base stations, working in L-S-C Band with time-varying envelope signals such as WiMax, WLAN, Cellular network etc. In this field, a lot of experimental results have been published using different active device technologies such as Si LDMOS, GaN HEMT, GaAs PHEMT and GaAs HBT. Typically, these DPAs are realised in hybrid form and they work around 2.14 GHz with W-CDMA input signals. Drain efficiencies up to 70% have been demonstrated for output powers between 5W and 10W (Kim et al., 2008 – Lee et al., 2008 – Markos et al., 2007 – Kim et al., 2005), whereas 50% of drain efficiency has been demonstrated for 250W output power (Steinbeiser et al., 2008). Also for high frequency applications the DPA has been successfully implemented using GaAs MMIC technologies (McCarroll et al., 2000 – Campbell, 1999 – Tsai & Huang, 2007). For instance, in (Tsai & Huang, 2007) it has been reported a fully integrated DPA at millimeter-wave frequency band with 22dBm and 25% of output power and efficiency peak, respectively. Also DPA realizations based on CMOS technology was proposed (Kang et al., 2006 – Elmala et al., 2006 – Wongkomet et al., 2006). However, in this case, due to the high losses related to the realization of required transmission lines, the achieved performances are quite low (peak efficiency lower than 15%).

In this chapter the theory and the design guidelines of the DPA will be reviewed in deep detail with the aim to show to the reader the proper way to design a DPA.

THE DOHERTY OPERATING PRINCIPLE

The DPA operating principle is based on the idea to modulate the load of the active device, namely Main (or Carrier) typically biased in Class AB, exploiting the active load pull concept (Cripps, 2002), by using a second active device, namely Auxiliary (or Peaking), usually biased in Class C.

In order to understand the active load-pull concept, it is possible to consider the schematic reported in Fig. 3, where two current sources are shunt connected to an impedance Z_L.

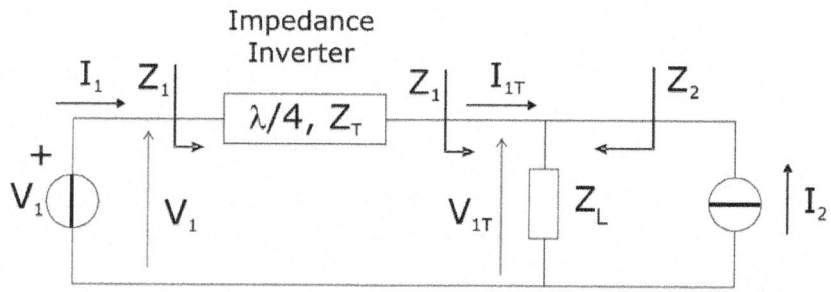

Figure 3: Schematic of the active load-pull.

Appling Kirchhoff law, the voltage across the generic loading impedance Z_L is given by:

$$V_L = Z_L \left(I_1 + I_2 \right)$$

(2)

Where I_1 and I_2 are the currents supplied by source 1 and 2, respectively. Therefore, if both currents are different from zero, the load seen by each current source is given by:

$$Z_1 = Z_L \cdot \left(1 + \frac{I_2}{I_1} \right)$$

(3)

$$Z_2 = Z_L \cdot \left(1 + \frac{I_1}{I_2} \right)$$

(4)

Thus, the actual impedance seen by one current source is dependent from the current supplied by the other one.

In particular, if I_2 is in phase with I_1, Z_L will be transformed in a higher impedance Z_1 at the source 1 terminals. Conversely, if I_2 is opposite in phase with I_1, Z_L will be transformed in a lower impedance Z_1. However, in both cases also the voltage across Z_L changes becoming higher in the former and lower in the latter situation.

Replacing the current sources with two equivalent transconductance sources, representing two separate RF transistors (Main and Auxiliary respectively), it is easy to understand that to maximize the efficiency of one device (i.e. Main) while its output load is changing (by the current supplied by the Auxiliary device), the voltage swing across it has to be maintained constant. In order to guarantee such constrain, it is necessary to interpose an Impedance Inverting Network (IIN) between the load (Z_L) and the Main source, as reported in Fig. 4. In this way, the constant voltage value V_1 at the Main terminals will be transformed in a constant current value I_{1T} at the other IIN terminals, independently from the value of Z_L.

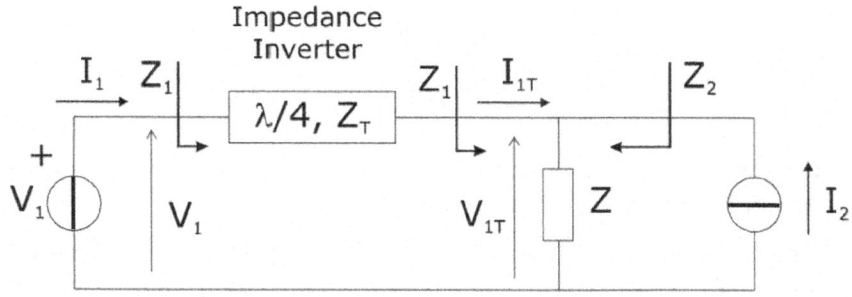

Figure 4: Simplified schema of the DPA.

For the IIN implementations, several design solutions could be adopted (Cripps, 2002). The most typical implementation is through a lambda quarter transmission line (l/4 TL), which ABCD matrix is given by:

$$\begin{pmatrix} V_1 \\ I_1 \end{pmatrix} = \begin{pmatrix} 0 & j \cdot Z_0 \\ \dfrac{j}{Z_0} & 0 \end{pmatrix} \cdot \begin{pmatrix} V_2 \\ I_2 \end{pmatrix}$$

(5)

being Z_0 the characteristic impedance of the line.

From (5) it is evident that the voltage at one side (V_1) is dependent only on the current at the other side (I_2) through Z_0, but it is independent from the output load (Z_L) in which the current I_2 is flowing.

Thus, actual DPAs are implemented following the scheme reported in Fig. 5, which is composed by two active devices, one IIN connected at the output of the Main branch, one Phase Compensation Network (PCN) connected at the input of the Auxiliary device and by an input power splitter besides the output load (R_L). The role of the PCN is to allow the in phase sum on R_L of the signals arising from the two active devices, while the splitter is required to divide in a proper way the input signal to the device gates.

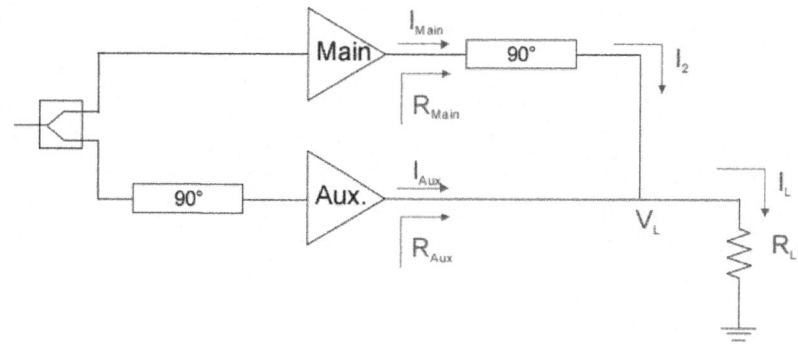

Figure 5: Typical DPA structure.

In order to easy understand the DPA behavior, the following operating regions can be recognized (Raab, 1987).

For low input power level (i.e. Low Power Region, see Fig. 2), the DPA acts as a typical PA, since the Main device is conducting while the Auxiliary is OFF due to its Class C bias condition.

Increasing the input power level, the current supplied by the Main device to R_L increases reaching the device saturation ($I_{critical}$), thus the maximum efficiency condition. The corresponding input power level reaches a "break point" condition, while the expected load curve of both active devices are indicated in Fig. 6 with the letter A. For higher input power level ($P_{in\_DPA} > P_{in\_DPA}$(break point)), the Auxiliary device will automatically turned on, injecting current into the output load R_L. Consequently, the impedance (Z_1) seen by the Main device is modulated and, thanks to the 1/4 TL, its value becomes lower with respect to the one at the break point (load curve "A" in the Fig. 6). In this way, the efficiency of the Main device remains constant, due to the constant level of saturation, while the efficiency of the Auxiliary device starts to increase (see Fig. 2). As a result, the overall DPA efficiency shows the typical behavior reported in Fig. 2.

At the end of the DPA dynamic, i.e. for the peak envelope value, both devices achieve their saturation corresponding to the load curves "C" in Fig. 6.

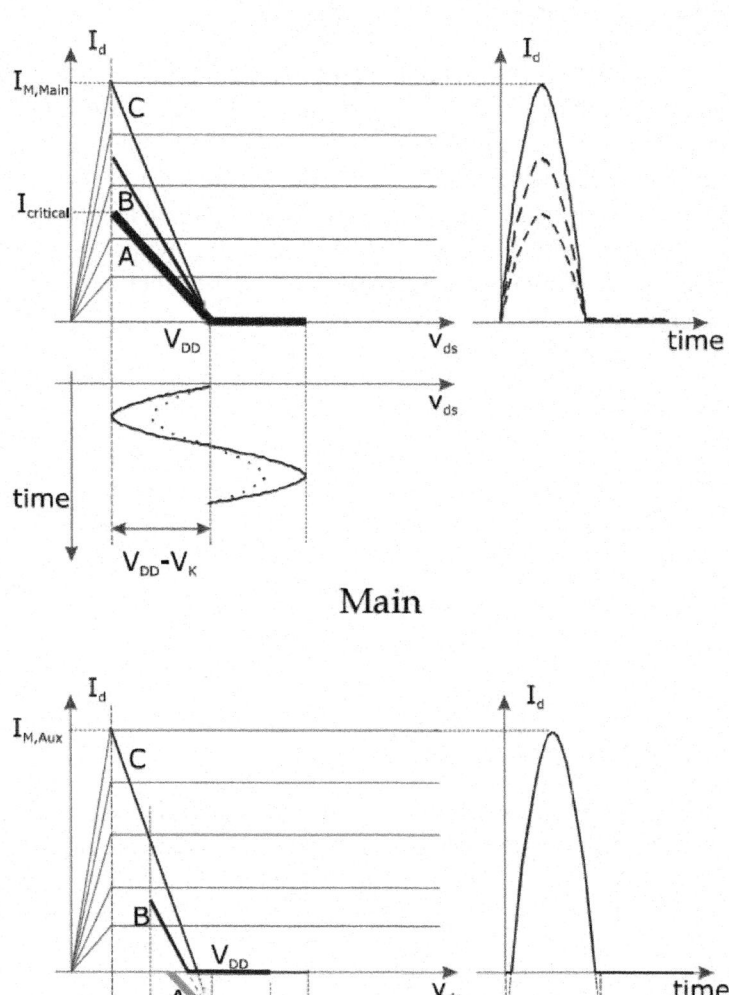

Figure 6: Evolution of the load curves for both DPA active devices: Main (left) and Auxiliary (right) amplifiers.

THE DOHERTY DESIGN GUIDELINES

In order to infer useful design relationships and guidelines, simplified models are assumed for the elements which are included in the DPA architecture. In particular, the passive components (l/4 TLs and power splitting) are assumed to be ideally lossless, while for the active device (in the following assumed as a FET device) an equivalent linearised model is assumed, as shown in Fig. 7. It is represented by a voltage-controlled current source, while for simplicity any parasitic feedback elements are neglected and all the other ones are embedded in the matching networks.

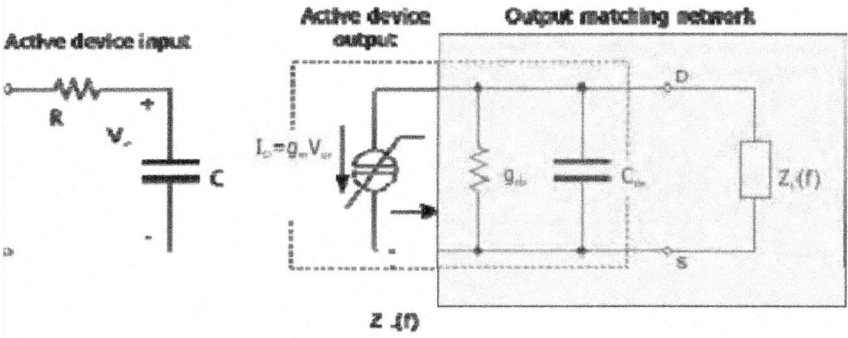

Figure 7: Simplified model assumed for the active device.

The device output current source is described by a constant transconductance (g_m) in the saturation region, while a constant ON resistance (R_{ON}) is assumed for the ohmic region, resulting in the output I-V linearised characteristics depicted in Fig. 8.

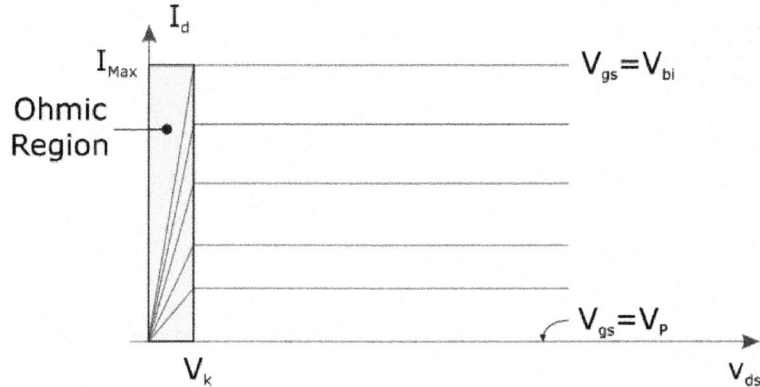

Figure 8: I-V output characteristics of the simplified model assumed for the active device.

The main parameter taken into account to represent the simplified I-V characteristics are the maximum achievable output current (I_{Max}), the constant knee voltage (V_k) and the pinch-off voltage (V_p).

As it commonly happens in the amplifiers design, some parameters are assumed as starting requirements, thus imposed by the designer, while other ones are consequently derived. Obviously, the following guidelines outline only one of the possible design flows. The design starts by fixing the OBO level, required to the DPA, accounting for the peculiar PAPR of the application which the DPA is oriented for. The OBO can be defined by the following equation:

$$OBO = \frac{P_{out,DPA(x=x_{break})}}{P_{out,DPA(x=1)}} = \frac{P_{out,Main(x=x_{break})}}{P_{out,Main(x=1)} + P_{out,Aux(x=1)}}$$

(6)

where the subscripts are used to refer to the entire DPA or to the single amplifiers (Main and Auxiliary respectively). Moreover a parameter x ($0 \leq x \leq 1$) is used to identify the dynamic point in which those quantities are considered. In particular x=0 identifies the quiescent state, i.e. when no RF signal is applied to the input, while x=1 identifies the saturation condition, i.e. when the DPA reaches its maximum output power level. Similarly, x=x$_{break}$ identifies the break point condition, i.e. when the Auxiliary amplifier is turned on.

Clearly, eqn. (6) is based on the assumption that only the Main amplifier delivers output power until the break point condition is reached, and the output network is assumed lossless.

In order to understand how the selected OBO affects the design, it is useful to investigate the expected DLLs of the Main and Auxiliary amplifiers for x=x$_{break}$ (load curves "A" in Fig. 6) and x=1 (load curves "C" in Fig. 6). It is to remark that the shape of the DLLs is due, for sake of simplicity, to the assumption of a Tuned Load configuration (Colantonio et al., 2002) both for Main and Auxiliary amplifiers.

Assuming a bias voltage V_{DD}, the drain voltage amplitude of the Main device is equal to V_{DD}-V_k both for x=x$_{break}$ and x=1 The same amplitude value is reached by the drain voltage of the Auxiliary device for x=1, as shown by the load curve "C" in Fig. 6. Consequently the output powers delivered by the Main and Auxiliary amplifiers in such peculiar conditions become:

$$P_{out,Main(x=x_{break})} = \frac{1}{2} \cdot (V_{DD} - V_k) \cdot I_{1,Main(x=x_{break})}$$

(7)

$$P_{out,Main(x=1)} = \frac{1}{2} \cdot (V_{DD} - V_k) \cdot I_{1,Main(x=1)}$$

(8)

$$P_{out,Aux(x=1)} = \frac{1}{2} \cdot (V_{DD} - V_k) \cdot I_{1,Aux(x=1)}$$

(9)

where the subscript "1" is added to the current in order to refer to its fundamental component.

Referring to Fig. 5, the power balance at the two ports of the l/4 both for $x=x_{break}$ and $x=1$ is given by:

$$\frac{1}{2} \cdot (V_{DD} - V_k) \cdot I_{1,Main(x=x_{break})} = \frac{1}{2} \cdot V_{L(x=x_{break})} \cdot I_{2(x=x_{break})}$$

(10)

$$\frac{1}{2} \cdot (V_{DD} - V_k) \cdot I_{1,Main(x=1)} = \frac{1}{2} \cdot (V_{DD} - V_k) \cdot I_{2(x=1)}$$

(11)

being I_2 the current flowing into the load R_L from the Main branch. From (11) it follows:

$$I_{1,Main(x=1)} = I_{2(x=1)}$$

(12)

Moreover, remembering that the current of one side of the l/4 is function only of the voltage of the other side, it is possible to write

$$I_{2(x=x_{break})} = I_{2(x=1)}$$

(13)

since the voltage at the other side is assumed constant to $V_{DD}-V_k$ in all medium power region, i.e. both for $x=x_{break}$ and $x=1$.

Consequently, taking into account (11), the output voltage for $x=x_{break}$ is given by:

$$V_{L(x=x_{break})} = (V_{DD} - V_k) \cdot \frac{I_{1,Main(x=x_{break})}}{I_{1,Main(x=1)}} = \alpha \cdot (V_{DD} - V_k)$$

(14)

where a defines the ratio between the currents of the Main amplifier at $x=x_{break}$ and $x=1$:

$$\alpha = \frac{I_{1,Main(x=x_{break})}}{I_{1,Main(x=1)}} \tag{15}$$

Regarding the output resistance (R_L), its value has to satisfy two conditions, imposed by the voltage and current ratios at $x=x_{break}$ and $x=1$ respectively:

$$R_L = \frac{V_{L(x=x_{break})}}{I_{2(x=x_{break})}} = \frac{\alpha \cdot (V_{DD} - V_k)}{I_{1,Main(x=1)}} \tag{16}$$

$$R_L = \frac{V_{L(x=1)}}{I_{2(x=1)} + I_{1,Aux(x=1)}} = \frac{(V_{DD} - V_k)}{I_{1,Main(x=1)} + I_{1,Aux(x=1)}} \tag{17}$$

Therefore, from the previous equations it follows:

$$I_{1,Aux(x=1)} = \frac{1-\alpha}{\alpha} \cdot I_{1,Main(x=1)} \tag{18}$$

Consequently, substituting (7)-(9) (9) in (6) and taking into account for (18), the following relationship can be derived:

$$OBO = \alpha^2 \tag{19}$$

which demonstrates that, selecting the desired OBO, the ratio between the Main amplifier currents for $x=x_{break}$ and $x=1$ is fixed also.

Since the maximum output power value is usually fixed by the application requirement, it represents another constraints to be selected by the designer. Such maximum output power is reached for $x=1$ and it can be estimated by the following relationship:

$$P_{out,DPA(x=1)} = P_{out,Main(x=1)} + P_{out,Aux(x=1)} = \frac{1}{\alpha} \cdot \frac{1}{2} \cdot (V_{DD} - V_k) \cdot I_{1,Main(x=1)} \tag{20}$$

which can be used to derive the maximum value of fundamental current of Main amplifier ($I_{1,Main(x=1)}$), once its drain bias voltage (V_{DD}) and the device knee voltage (V_k) are selected. Knowing the maximum current at fundamental, it is possible to compute the values of R_L by (16)(16) and the required characteristic impedance of the output 1/4 TL (Z_0) by using:

$$Z_0 = \frac{(V_{DD} - V_k)}{I_{1,Main(x=1)}}$$

(21)

which is derived assuming that the output voltage (V_L) reaches the value V_{DD}-V_k for x=1. Clearly the maximum value $I_{1,Main(x=1)}$ depends on the Main device maximum allowable output current I_{Max} and its selected bias point.

Referring to Fig. 9, where it is reported for clearness a simplified current waveform, assuming a generic Class AB bias condition, the bias condition can be easily identified defining the following parameter

$$\xi = \frac{I_{DC,Main}}{I_{Max,Main}}$$

(22)

being $I_{DC,Main}$ the quiescent (i.e. bias) current of the Main device.

Consequently, x=0.5 and x=0 refer to a Class A and Class B bias conditions respectively, while 0<ξ

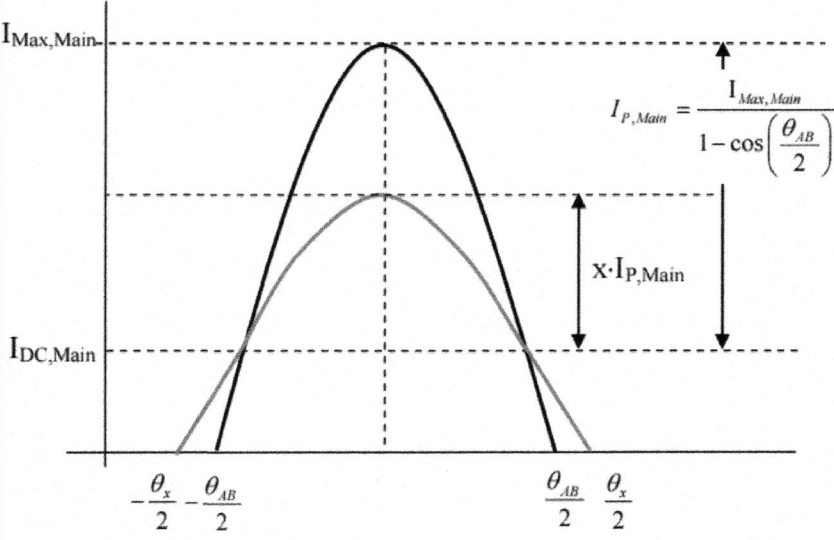

Figure 9: Current waveform in time domain of the Main amplifier.

The current waveform of Fig. 9 can be analytically described by the following expression:

$$i_{D,Main} = I_{DC,Main} + x \cdot \frac{I_{Max,Main}}{1 - \cos\left(\dfrac{\theta_{AB}}{2}\right)} \cdot \cos(\theta)$$

(23)

whose fundamental component can be written as following:

$$I_{1,Main(x=1)} = \frac{I_{Max,Main}}{2\pi} \cdot \frac{\theta_{AB} - \sin(\theta_{AB})}{1 - \cos\left(\dfrac{\theta_{AB}}{2}\right)}$$

(24)

being q_{AB} the current conduction angle (CCA) of the Main output current, achieved for x=1. The bias point x and the CCA q_{AB} can be easily related by the following relationship:

$$\theta_{AB} = 2\pi - 2\arccos\left(\frac{\xi}{1-\xi}\right)$$

(25)

Manipulating (24), the value of $I_{Max,Main}$, required to reach the desired maximum power, can be estimated, once the bias point x of the Main amplifier has been selected (the last parameter should be fixed by the designer).

As made with Main amplifier, the value of the Auxiliary maximum current can be obtained by using the equation of the first order coefficient of the Furier series, since the value of $I_{1,Aux,(x=1)}$ should fulfill (18).

Consequently, it follows:

$$I_{1,Aux(x=1)} = \frac{I_{Max,Aux}}{2\pi} \cdot \frac{\theta_C - \sin(\theta_C)}{1 - \cos\left(\dfrac{\theta_C}{2}\right)}$$

(26)

being q_C the CCA of the Auxiliary device output current for x=1.

Referring to Fig 10, where it is reported the current waveform of the Auxiliary amplifier, assuming a virtual negative bias point, the Auxiliary device current can be written similarly to (23), thus:

$$i_{D,Aux} = I_{DC,Aux} + x \cdot \frac{I_{Max,Aux}}{1 - \cos\left(\dfrac{\theta_C}{2}\right)} \cdot \cos(\theta)$$

(27)

Moreover, for a proper behavior of the Auxiliary amplifier, the peak of the current has to reach zero for x=x$_{break}$, as highlighted in Fig10. Consequently the following condition has to be taken into account.

$$x_{break} \cdot \frac{I_{Max,Aux}}{1 - \cos\left(\dfrac{\theta_C}{2}\right)} = -I_{DC,Aux}$$

(28)

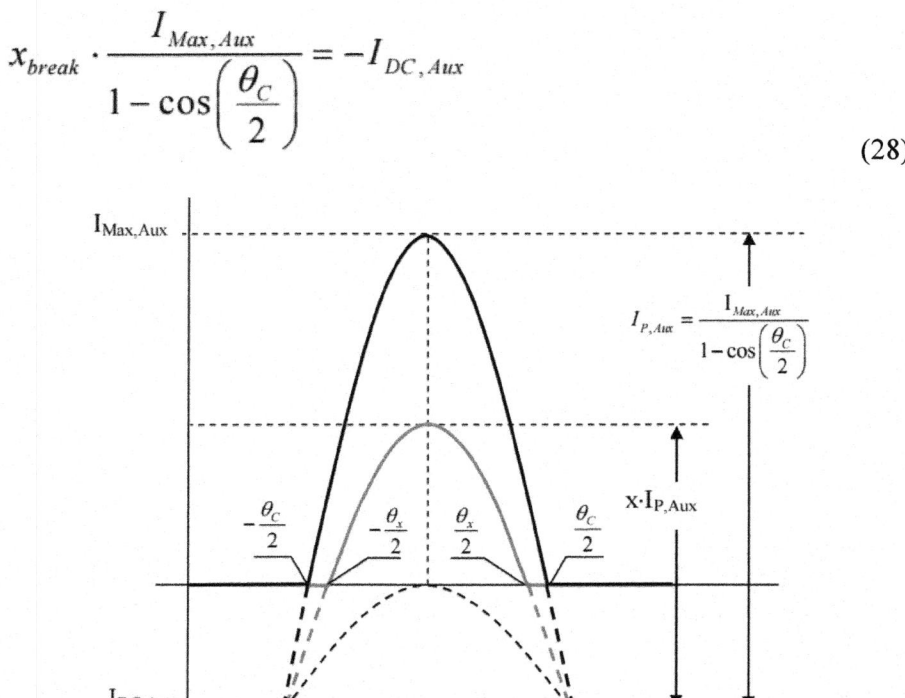

Figure 10: Current waveform in time domain of the Auxiliary amplifier for x=x$_{break}$ and x=1.

Substituting (28) in (27), it is possible to refer the value of q$_C$ directly to x$_{break}$:

$$\theta_C = 2 \cdot \arccos\left(x_{break}\right)$$

(29)

Now, from (15) and replacing the respective Fourier expressions, it follows:

$$x_{break} \cdot \left[\theta_{Main(x=x_{break})} - \sin\left(\theta_{Main(x=x_{break})}\right)\right] = \alpha \cdot \left(\theta_{AB} - \sin\left(\theta_{AB}\right)\right)$$

(30)

where from (23) it can be inferred:

$$\theta_{Main(x=x_{break})} = 2\pi - 2\arccos\left(\frac{\xi}{x_{break} \cdot \left(1 - \xi\right)}\right)$$

(31)

The value of x_{break} has to be numerically obtained solving (30), having fixed the OBO (i.e. α) and the Main device bias point (i.e. x). Once the value of $I_{Max,Aux}$ is obtained, the one of $I_{DC,Aux}$ is immediately estimable manipulating (28):

$$I_{DC,Aux} = -I_{Max,Aux} \cdot \frac{x_{break}}{1 - x_{break}}$$

(32)

At this point, an interesting consideration can be done about the ratio between the maximum currents required by the devices. Fig. 11 reports this ratio as function of OBO and x. As it is possible to note, the dependence on x can be practically neglected, while the one by the OBO is very high. Moreover, the same amount of maximum current is required from both devices in case of nearly 5dB as OBO, while an higher current has to be provided by the Auxiliary device for greater OBO.

From the designer point of view, the maximum currents ratio can be used as an useful information to choice the proper device periphery. In fact, supposing for the used technology a linear relationship between maximum current and drain periphery, Fig. 11 gives the possibility to directly derive the drain periphery of the Auxiliary device, once the Main one has been selected in order to respect the maximum output power constraint.

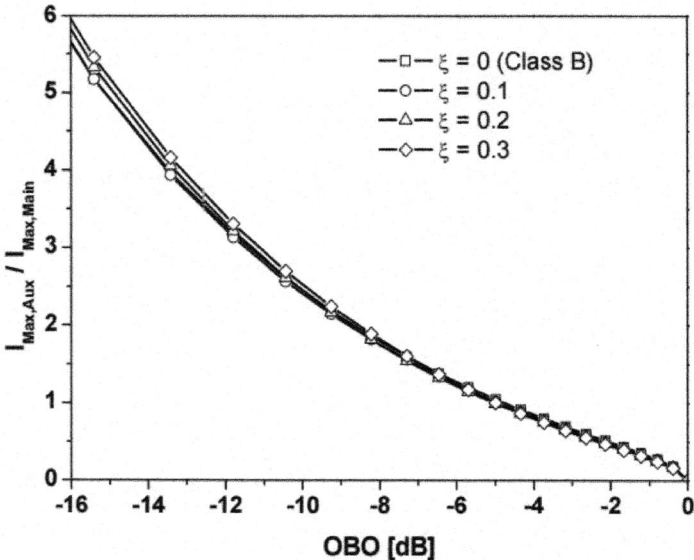

Figure 11: Ratio between Auxiliary and Main maximum currents as function of OBO and x.

Power Splitter Dimensioning

In this subsection the dimensioning of the input power splitter is discussed, highlighting its critical role in the DPA architecture.

Following the simplified analysis based on an active device with constant transconductance (g_m), the amplitude of the gate voltage for x=1, for Main and Auxiliary devices respectively, can be written as

$$V_{gs,Main(x=1)} = \frac{I_{Max,Main} - I_{DC,Main}}{g_{m,Aux}} = \frac{I_{Max,Main} \cdot (1 - \xi)}{g_{m,Main}} \tag{33}$$

$$V_{gs,Aux(x=1)} = \frac{I_{Max,Aux} - I_{DC,Aux}}{g_{m,Aux}} = \frac{I_{Max,Aux}}{g_{m,Aux}} \cdot \frac{1}{1 - x_{break}} \tag{34}$$

Using the previous equations, it is possible to derive the powers at the input of the devices by using the following relationships:

$$P_{in,Main(x=1)} = \frac{1}{2} \cdot \frac{\left(V_{gs,Main(x=1)}\right)^2}{R_{in,Main}} = \frac{1}{2} \cdot \frac{\left(I_{Max,Main} \cdot (1 - \xi)\right)^2}{R_{in,Main} \cdot \left(g_{m,Main}\right)^2} \tag{35}$$

$$P_{in,Aux(x=1)} = \frac{1}{2} \cdot \frac{\left(V_{gs,Aux(x=1)}\right)^2}{R_{in,Aux}} = \frac{1}{2} \cdot \frac{\left(I_{Max,Aux}\right)^2}{R_{in,Aux} \cdot \left(g_{m,Aux} \cdot (1 - x_{break})\right)^2} \tag{36}$$

where $R_{in,Main}$ and $R_{in,Aux}$ are the input resistances respectively of Main and Auxiliary devices.

Therefore, it is possible to compute the power splitting factor, i.e. the amount of power delivered to the Auxiliary device with respect to the total input power, by using:

$$\Lambda_{Aux} = \frac{P_{in,Aux(x=1)}}{P_{in,Main(x=1)} + P_{in,Aux(x=1)}} = \frac{1}{\left(\frac{I_{Max,Main}}{I_{Max,Aux}} \cdot \frac{1 - \xi}{1 - x_{break}} \cdot \frac{g_{m,Aux}}{g_{m,Main}}\right)^2 \cdot \frac{R_{in,Aux}}{R_{in,Main}} + 1} \tag{37}$$

and consequently for the Main device:

$$\Lambda_{Main} = 1 - \Lambda_{Aux} \tag{38}$$

In Fig. 12 is reported the computed values for L_{Aux}, as function of OBO and x parameters, assuming for both devices the same values for gm and R_{in}.

Fig. 12 highlights that large amount of input power has to be sent to the Auxiliary device, requiring an uneven power splitting. For example, considering a DPA with 6dB as OBO and a Class B bias condition (i.e x=0) for the Main amplifier, 87% of input power has to be provided to Auxiliary device, while only the remaining 13% is used to drive the Main amplifier. This aspect dramatically affects in a detrimental way the overall gain of the DPA, which becomes 5-6 dB lower if compared to the gain achievable by using a single amplifier only.

Nevertheless, it has to remark that this largely unbalanced splitting factor has been inferred assuming a constant transconductance (g_m) for both devices. Such approximation issufficiently accurate in the saturation region (x=1), while becomes unsatisfactory for low power operation. In this case, the actual transconductance behavior can be very different depending on the technology and bias point of the selected active device. In general, it is possible to state that the transconductance value of actual devices, in low power region, is lower than the average one, when the chosen bias point is close to the Class B. Thus, if the bias point of Main amplifier x is selected roughly lower than 0.2, the predicted gain in low power region is higher than the experimentally resulting one, being the former affected by the higher value assumed for the transconductance in the theoretical analysis.

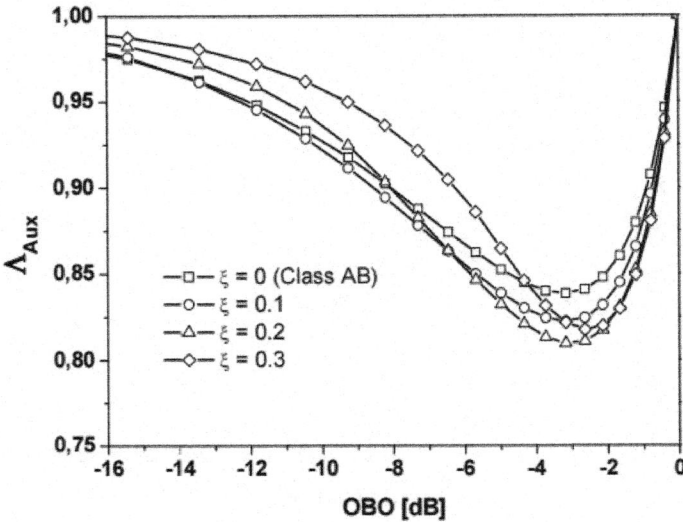

Figure 12: LAux behavior as a function of OBO and x, assuming for both devices the same values for gm and R_{in}.

From a practical point of view, if the theoretical splitting factor is assumed in actual design, usually the Auxiliary amplifier turns on before the Main amplifier reaches its saturation (i.e. its maximum of efficiency). Consequently a reduction of the unbalancing in the power splitter is usually required in actual DPA design with respect to the theoretical value, in order to compensate the non constant transconductance behavior and, thus, to switch on the Auxiliary amplifier at the proper dynamic point.

Performance Behavior

Once the DPA design parameters have been dimensioned, closed form equations for the estimation of the achievable performances can be obtained. Since the approach is based on the electronic basic laws, it will be here neglected, in order to avoid that this chapter dull reading and to focus the attention on the analysis of the performance behavior in terms of output power, gain, efficiency and AM/AM distortion. The complete relationships can be found in (Colantonio et al., 2009 - a).

The theoretical performance of a DPA designed to fulfill 7dB of OBO and 6W as maximum output power, are shown in Fig. 12. Moreover, the same physical parameters have been assumed for both Main and Auxiliary devices: $V_k=0V$, $g_m=0.22S$ and $R_{in}=50W$. Finally the drain bias voltage and the Main amplifier quiescent point have been fixed to $V_{DD}=10V$ and x=0.1 respectively.

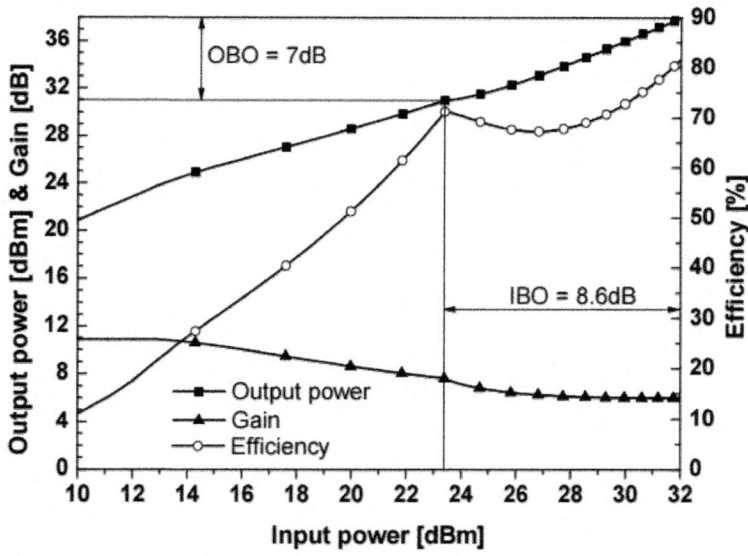

Figure 12: Theoretical performances of a DPA with 7dB OBO and 6W as maximum output power.

As it appears looking at Fig. 13, the efficiency value at the saturation is higher than the one at the break point. The latter, in fact, is the one of the Main device, which is a Class AB amplifier. The efficiency at the saturation, instead, is increased by the one of the Auxiliary device, which has a Class C bias point, with a consequent greater efficiency value.

It is possible to note as the gain behaves linearly until 13dBm of input power, while becomes a monotonic decreasing function up to about 23.5dBm. Along this dynamic region, the Main amplifier only is working and the variation of the gain behavior is due to the pinch-off limitation in the output current.

In particular, until 13dBm, the Main device operates as a Class A amplifier, since its DLL did not reach yet the pinch-off physical limitation. Then, the Main device becomes a Class AB amplifier, coming up to the near Class B increasing the input power, with a consequent decreasing of the gain. However this evident effect of class (and gain) changing is due to the assumption of a constant transconductance model for the active device. In actual devices, in fact, the value of the transconductance is lower than the average one, when the selected bias point is close to the Class B, as it has been discussed in section 3.1. Consequently, in practical DPA design, the gain, for small input power levels, is lower than the theoretical one estimated by the average gm value, thus reducing the effect highlighted in Fig. 12.

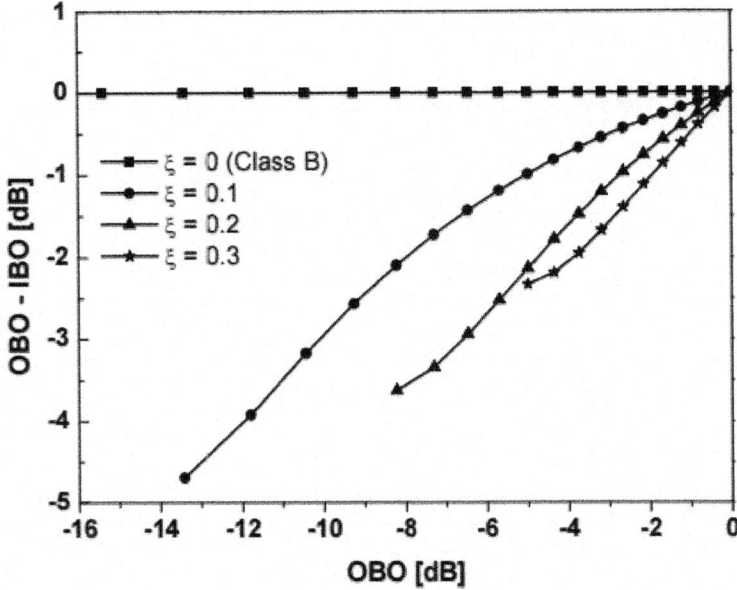

Figure 13: Theoretical difference between OBO and IBO for several values of x.

In the Doherty region, from 23.5dBm up to 32dBm of input power, the gain changes its behavior again. The latter change is due to the combination of the gain decreasing of the Main amplifier, whose output resistance is diminishing, and the gain increasing of the Auxiliary amplifier, which passes from the switched off condition to the proper operative Class C.

The non constant gain behavior is further highlighted in Fig. 12 by the difference between the resulting OBO and input back-off (IBO), resulting in an AM/AM distortion in the overall DPA. In order to deeply analyze this effect, Fig. 13 reports the difference between OBO and IBO for several values of x.

In order to proper select the Main device bias point x to reduce AM/AM distortion, it is useful to introduce another parameter, the Linear Factor (LF), defined as:

$$LF = \frac{1}{1 - x_{break}} \cdot \int_{x_{break}}^{1} \left[P_{out,DPA}(x) - \left(x^2 \cdot P_{out,DPA(x=1)} \right) \right] dx$$

(39)

The Linear Factor represents the variation in the Doherty region of the DPA output power, with respect to a linear PA having the same maximum output power and represented in (39)(39) by $x^2 \cdot P_{out,DPA(x=1)}$. Thus LF gives the simplified estimation of the average AM/AM distortion in the Doherty region.

Consequently, the optimum bias condition should be assumed to assure LF=0. Obviously this condition, if it exists, can be obtained only for one x, once the OBO has been selected.

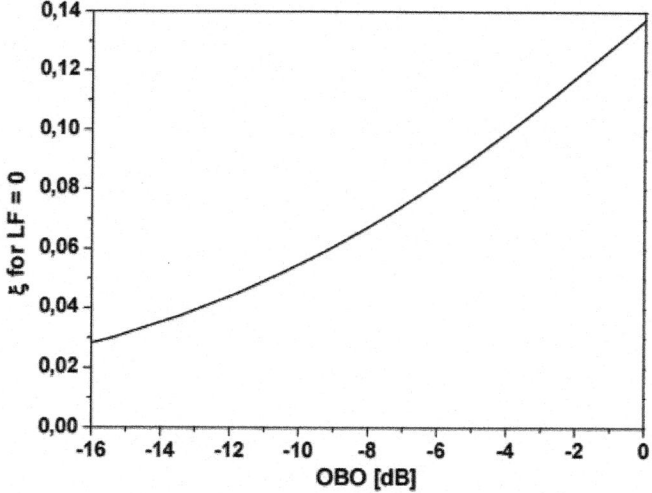

Figure 14: Values of x assuring LF=0, as function of the OBO.

Fig. 14 shows the values of x, which theoretically assures LF=0, as function of the selected OBO. This design chart provides a guideline to select the proper bias point of the Main amplifier (x), having fixed the desired OBO of the DPA. In order to further clarify the DPA behavior, Fig. 15 shows the fundamental drain currents and voltages for both Main and Auxiliary devices. These behaviors can be used in the design flow to verify if the DPA operates in a proper way. In particular, the attention has to be focused on the Main voltage, which has to reach, at the break point (x_{break}), the maximum achievable amplitude (10V in this example) in order to maximize the efficiency. Moreover the Auxiliary current can be used to verify that the device is turned on in the proper dynamic instant. Finally, the designer has to pay attention if the Auxiliary current reaches the expected value at the saturation (x=1), in order to perform the desired modulation of the Main resistance. This aspect can be evaluated also observing the behavior of Main and Auxiliary resistances, as reported in Fig. 17.

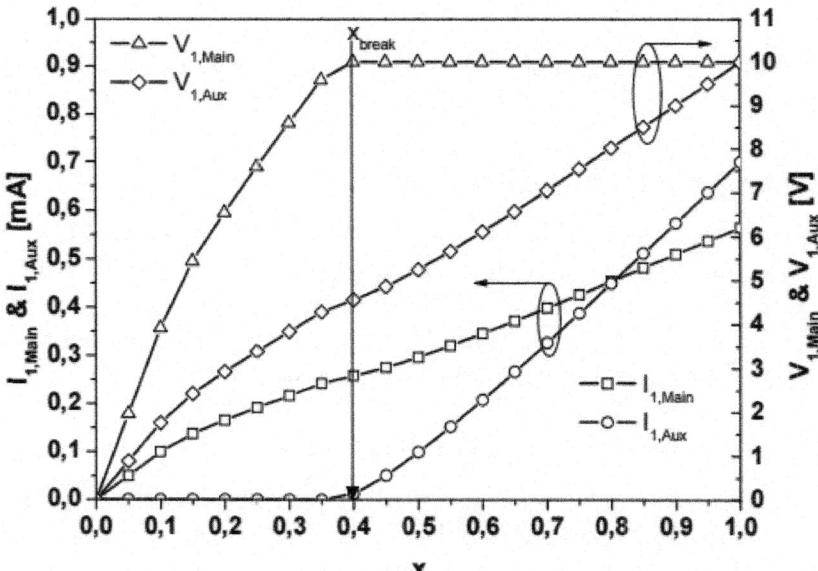

Figure 15: Fundamental current and voltage components of Main and Auxiliary amplifiers, as function of the dynamic variable x.

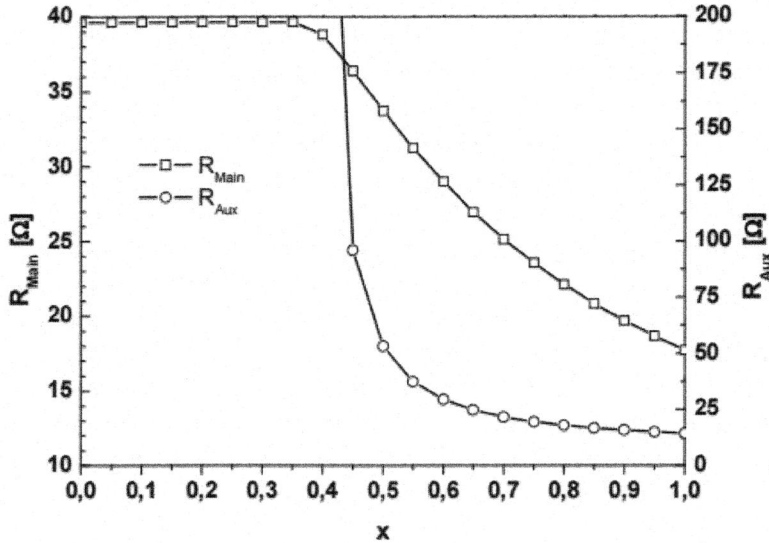

Figure 17: Drain resistance at fundamental frequency of Main and Auxiliary amplifiers, as function of the dynamic variable x.

ADVANCED DPA DESIGN

In the previous paragraphs the classical Doherty scheme based on Tuned Load configuration for both Main and Auxiliary amplifiers has been analyzed. Obviously, other solutions are available, still based on the load modulation principle, but developed with the aim to further improve the features of the DPA, by using additional some free design parameters.

DPA Design by using different Bias Voltage

For instance, the adoption of different drain bias voltage for the two amplifiers (Main and Auxiliary) could be useful to increase the gain of the overall DPA. In fact, in the DPA topology the voltage at the output common node, V_L in Fig. 5, at saturation is imposed by the Auxiliary drain bias voltage ($V_{DD,Aux}$) in order to fulfill the condition $V_L = V_{DD,Aux} - V_{k,Aux}$. Thus, assuming a different bias, i.e. $V_{DD,Main}$ and $V_{DD,Aux}$ for the Main and Auxiliary devices respectively, and defining the parameter

$$\beta = \frac{V_{DD,Main} - V_{k,Main}}{V_{DD,Aux} - V_{k,Aux}}$$

(40)

then the design relationships previously inferred have to be tailored accounting for such different supplying voltages.

Therefore, the DPA elements R_L and Z_0 becomes:

$$R_L = \frac{\alpha^2}{\beta^2} \cdot R_{Main}\left(x_{break}\right)$$

(41)

$$Z_0 = \frac{V_{DD,Aux} - V_k}{I_{1,Main}\left(\theta_{AB}\right)}$$

(42)

Where

$$R_{Main}\left(x_{break}\right) = 2\frac{V_{DD,Main} - V_{k,Main}}{I_{M,Main}}\frac{\pi}{\alpha} \cdot \frac{1 - \cos\left(\dfrac{\theta_{AB}}{2}\right)}{\theta_{AB} - \sin\left(\theta_{AB}\right)}$$

(43)

Moreover, the Auxiliary and Main devices maximum output currents are now related through the following relationship:

$$I_{M,Aux} = \beta \cdot I_{M,Main} \cdot \frac{1-\alpha}{\alpha} \cdot \frac{1 - \cos\left(\dfrac{\theta_C}{2}\right)}{\theta_C - \sin\left(\theta_C\right)} \cdot \frac{\theta_{AB} - \sin\left(\theta_{AB}\right)}{1 - \cos\left(\dfrac{\theta_{AB}}{2}\right)}$$

(44)

Which, clearly, highlights that a suitable selection of the Auxiliary device supply voltage, i.e. $\beta<1$, could imply a lower maximum output current required from the Auxiliary device. Conversely, the saturated output power of the Doherty (for x=1) is still related to both the Main device supply voltage and its maximum output current, i.e.:

$$P_{out,DPA(x=1)} = \frac{1}{2}\left(V_{DD,Main} - V_k\right) \cdot \frac{I_{1,Main}\left(\theta_{AB}\right)}{\alpha} = \frac{1}{\alpha} \cdot P_{out,Main,Max} = \frac{1}{\alpha^2} \cdot P_{out,Main,break}$$

(45)

Thus being not affected by the different drain supply voltage adopted for the Auxiliary device.

DPA Design by using Harmonic Tuning strategies

To further improve the overall efficiency in a Doherty configuration, high efficiency design strategies can be adopted in the synthesis of both Main and

Auxiliary amplifiers. For this purpose, harmonic tuning strategies have been proposed (Colantonio et al., 2002).

However, due to the Class C bias condition for the Auxiliary device, thus implying a wrong phase relationships between current (and voltage) harmonic components, the optimum solution for such amplifier is the classical Tuned Load one.

Conversely, for the Main amplifier, which is normally operating in a Class AB bias, the efficiency can be improved by using for instance Class F strategy (Raab, 2001). Such design strategy implies that the second harmonic current component I_2 should be short circuited, while the fundamental (I_1) and the third one (I_3) should be terminated on impedance R_1 and R_3, respectively, to obtain a proper voltage harmonic component ratio (Colantonio et al., 2002):

$$k_3 = \frac{V_3}{V_1} = \frac{R_3 \cdot I_3}{R_1 \cdot I_1} = 0.167$$

(46)

Thus, in a Class F Doherty amplifier (i.e. with Main amplifier in Class F configuration), the proper output harmonic loading conditions have to be fulfilled across the Main device, accounting for the load modulation effect in the medium power region.

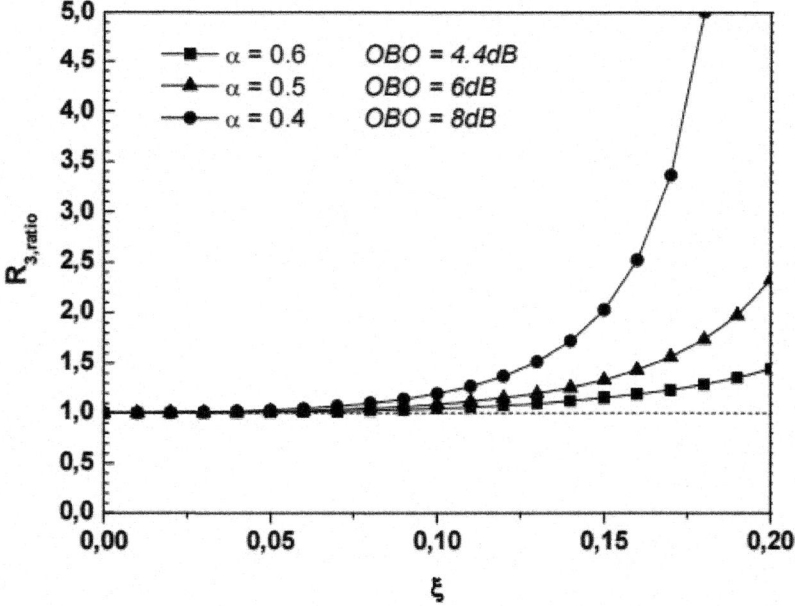

Figure 16: $R_{3,ratio}$ as function of x for different OBO (a) values.

In particular, it is possible to compute the theoretical load modulation required for the third harmonic, in order to fulfill (46) accounting for the modulation of R_1. In Fig. 16 is reported the ratio between the values required for R_3 at the end of the low power region ($x=x_{break}$) and at the end of the Medium (or Doherty) power region, i.e. at saturation ($x=1$), as a function of the Main device bias point (x) and the selected OBO.

As it can be noted, the $R_{3,ratio}$ (i.e. the degree of modulation required for the third harmonic loading condition) increases with the bias point (x) and OBO values (a). Nevertheless, the modulation of R_3 through the output 1/4 line and the Auxiliary current, critically complicate the design and can be usually neglected if the Main device bias point is chosen nearly Class B condition, i.e. x<0.1, being $R_{3,ratio} \approx 1$.

Under such assumption, it is possible to compute the Class F DPA design parameter as compared to the Tuned Load case. It can be inferred, referring to Fig. 3, that the output load R_L and the characteristic impedance of the output 1/4 TL become

$$R_{L,F} \simeq \frac{R_L}{1.15} \qquad\qquad Z_{0,F} = Z_{0,TL}$$

(47)

being R_L given equivalently by (16) or (17) and Z_0 by (21). Finally, regarding the power splitter dimensioning, it is required a different power splitting ratio, resulting in:

$$\Lambda_{Aux,F} = \frac{1.322 \cdot \Lambda_{Aux}}{1+0.322 \cdot \Lambda_{Aux}} > \Lambda_{Aux} \quad \Lambda_{Main,F} = 1 - \Lambda_{Aux,F} < \Lambda_{AB}$$

(48)

The expected behavior of the output current and voltage fundamental components for the Main and the Auxiliary devices are reported in Fig. 19, assuming x=0.082 and OBO=6dB (i.e. α=0.5).

Similarly, in Fig. 17 are reported the comparisons in terms of output power and efficiency of Class F DPA with respect to Tuned Load DPA, normalized as functions of the input signal x.

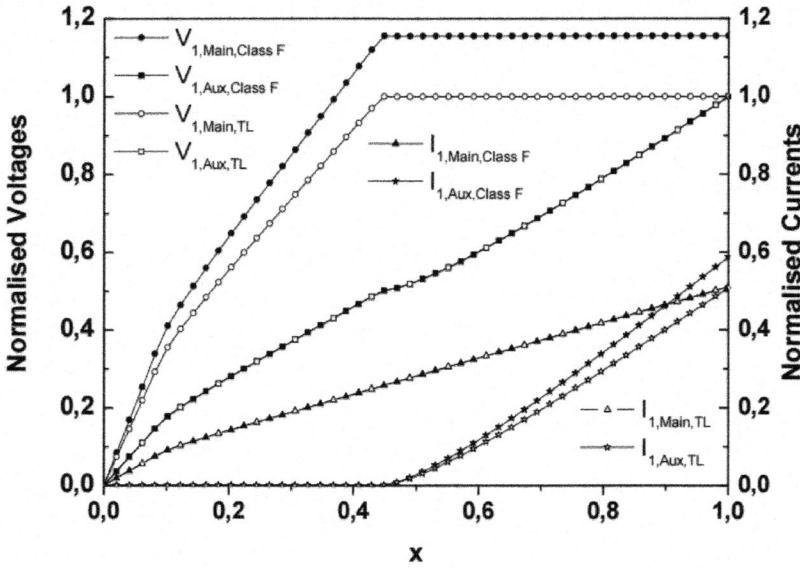

Figure 19: Theoretical behavior of drain current and voltage fundamental components for Main and Auxiliary devices, assuming Class F or Tuned Load design strategies for Main amplifier.

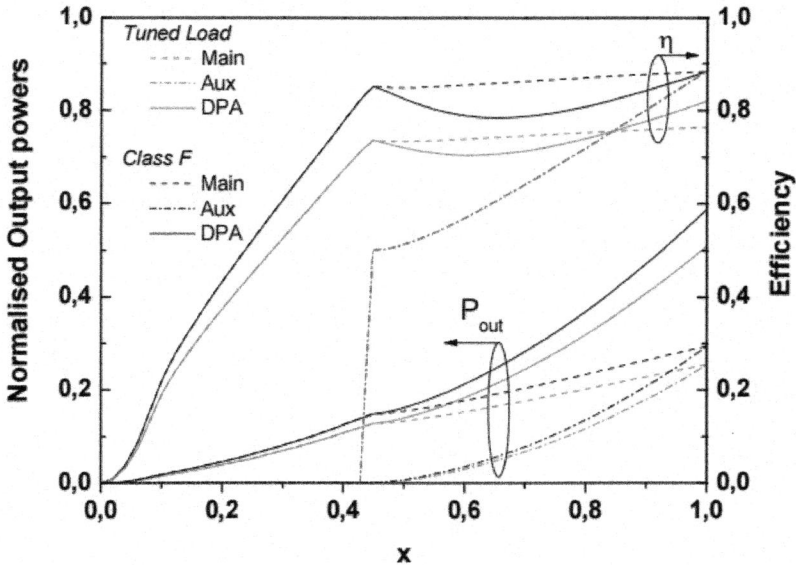

Figure 17: Expected drain efficiency and output power behaviors for Class F and Tuned Load Doherty amplifiers.

Multi-way Doherty Amplifiers

In order to reduce the Auxiliary device size, while still providing the required current for the load modulation, a different solution is based on the so called Multi-Way Doherty configuration, usually referred as N-Way Doherty amplifier also (Yang et al., 2003 – Kim et al., 2006 – Cho et al., 2007). It is realized by paralleling one Main amplifier and N-1 Auxiliary amplifiers, aimed to acquire an N-1 times larger-sized Auxiliary amplifier, as schematically shown in Fig. 18.

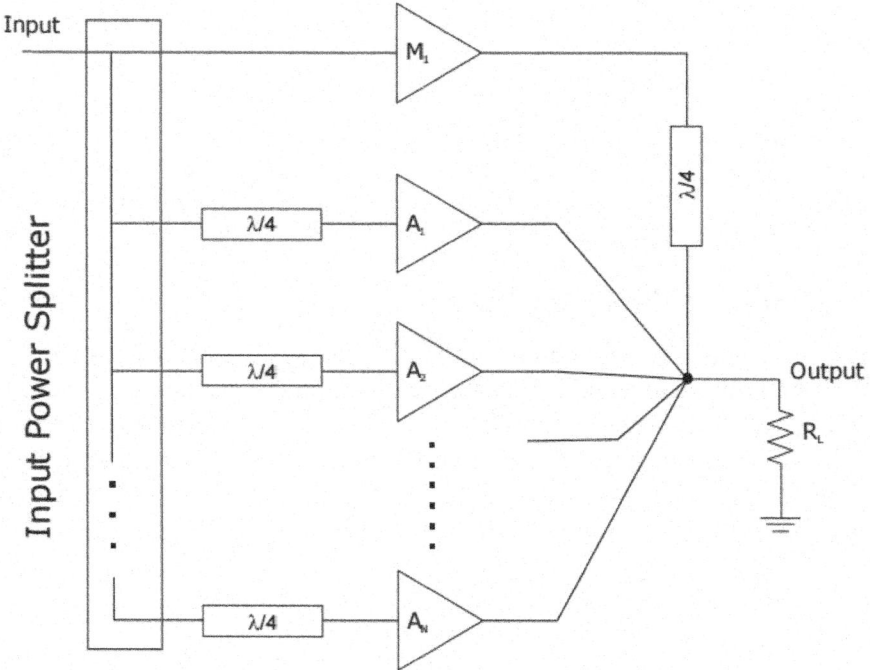

Figure 18: Schematic diagram of the N-way Doherty amplifier.

With the proposed device combination, it becomes possible to implement larger OBO using smaller devices, resulting in the theoretical efficiency performance shown in Fig. 19.

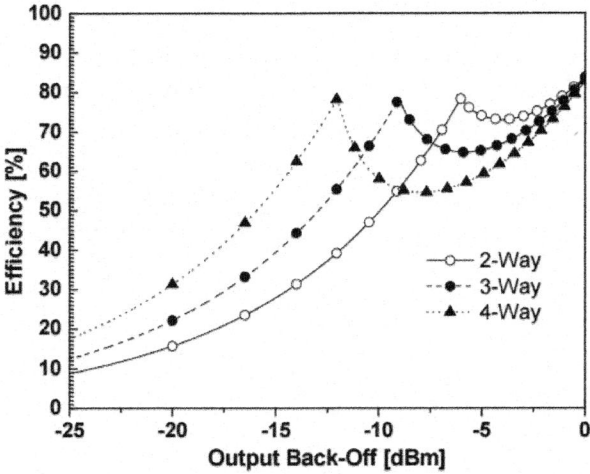

Figure 19: Theoretical efficiency of the N-Way Doherty amplifier.

Multi-Stage Doherty Amplifiers

The Multi-Stage Doherty amplifier is conceptually different from the Multi-Way configuration, since it is based on a subsequent turning on condition of several Auxiliary devices, with the aim to assure a multiple Doherty region in a cascade configuration, overcoming the reduction of the average value due to the increased drop-down phenomenon in efficiency, especially when larger OBO are required (Neo et al., 2007 – Pelk et al., 2008 – Srirattana et al., 2005).

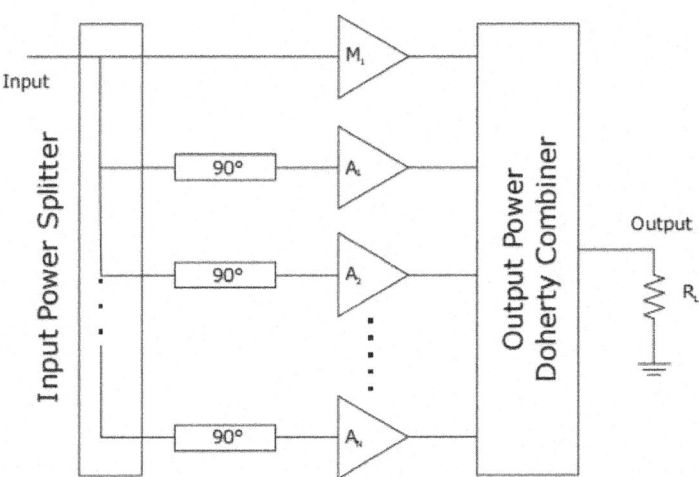

Figure 20: Theoretical diagram of a Multi-Stage Doherty amplifier.

For this purpose, referring to the theoretical diagram shown in Fig. 23, amplifiers M_1 and A_1 have to be designed to act as Main and Auxiliary amplifiers in a standard Dohertyconfiguration. Then, when both amplifiers are approaching their saturation, amplifier A_2 is turned on operating as another Auxiliary amplifier, thus modulating the load seen by the previous M_1-A_1 pair, that must be considered, from now onward, as a single amplifier. Such concept is then iterated inserting N Auxiliary amplifiers, each introducing a new breakpoint, resulting in a theoretical efficiency behavior as depicted in Fig. 21.

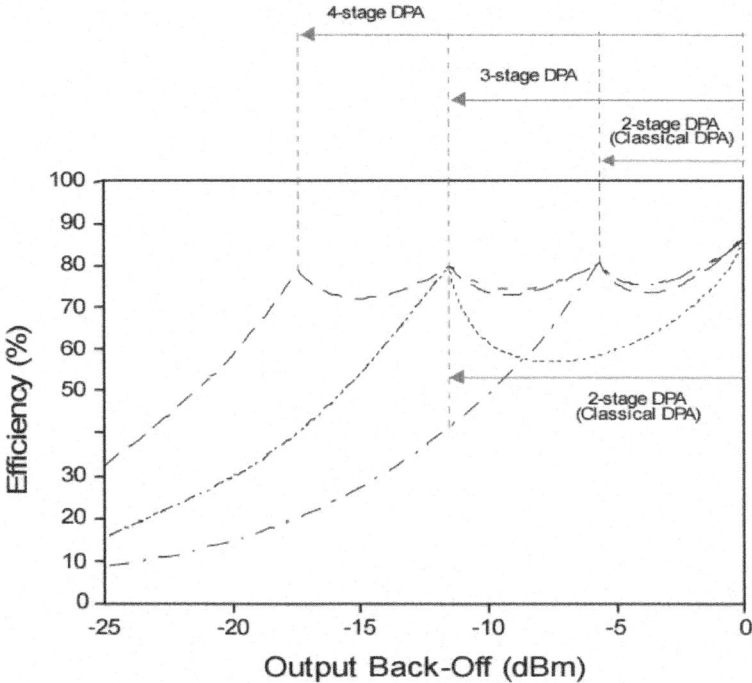

Figure 21: Theoretical behavior of the efficiency for a Multi-Stage Doherty amplifier.

From the design issues, it is easy to note that the most critical one resides in the practical implementation of the output power combining network, required to properly exploit the load modulation concept for all the cascaded stages. A proposed solution is reported in (Neo et al., 2007 – Pelk et al., 2008), based on the scheme depicted in Fig. 22, where the relationships to design the output l/4 transmission lines adopted are given by

$$Z_{0,i} = R_L \cdot \prod_{j=1}^{i} \frac{1}{\alpha_j}$$

$$\prod_{j=k}^{\frac{i+k}{2}} \alpha_{2j-k} = 10^{-\frac{OBO_i}{20}}$$

(49)

where i=1,2,...N, k=1 (for odd i) or k=2 (for even i), N is the total number of Auxiliary amplifiers, OBO_i is the back-off level from the maximum output power of the system at which the efficiency will peak (i.e. the turning on condition of the Auxiliary A_i). The R_L value is determined by the optimum loading condition of the last Auxiliary stage, according to the following relationship:

$$R_L = (1 - \alpha_1) \cdot R_{opt, Aux_N}$$

(50)

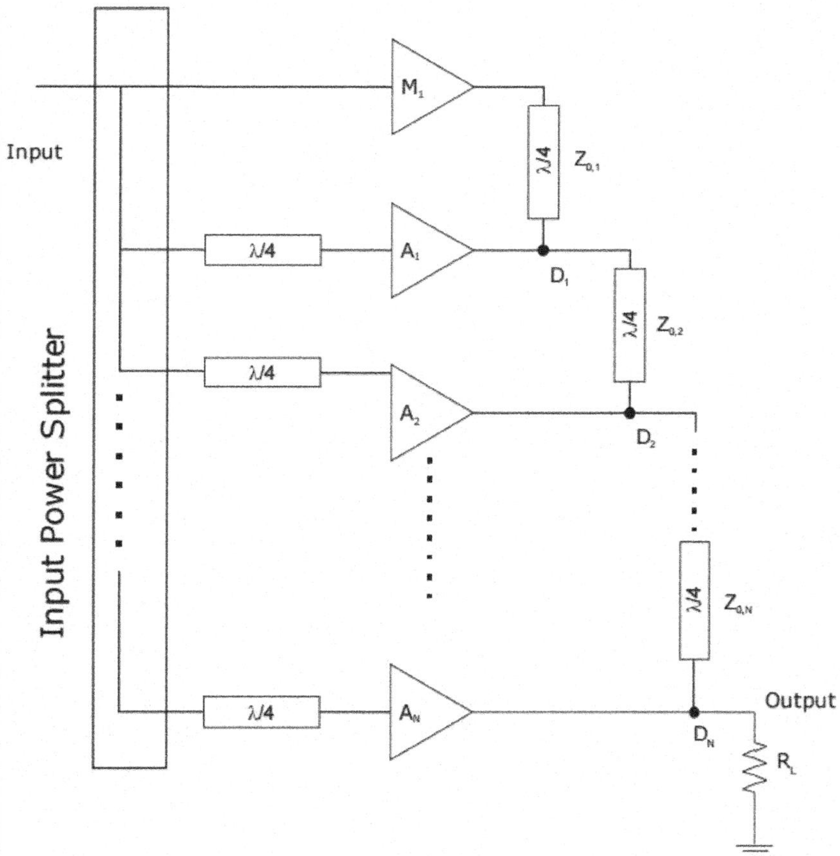

Figure 22: Proposed schematic diagram for a multi-stage Doherty amplifier.

However, some practical drawbacks arise from the scheme depicted in Fig. 22. In fact, the Auxiliary device A_1 is turned on to increase the load at D_1 node and consequently, due to the 1/4 line impedance $Z_{0,1}$, to properly decrease the load seen by M_1. However, when A_2 is turned on, its output current contributes to increase the load impedance seen at D2 node. Such increase, while it is reflected in a suitable decreasing load condition for A_1 (at D_1 node), it also results in an unwanted increased load condition for M_1, still due to the 1/4 line transformer. As a consequence, such device results to be overdriven, therefore saturating the overall amplifier and introducing a strong non linearity phenomencn in such device. To overcome such a drawback, it is mandatory to change the operating conditions, by turning on, for instance, the corresponding Auxiliary device before the Main device M_1 has reached its maximum efficiency, or similarly, changing the input signal amplitudes to each device (Pelk et al., 2008).

Different solutions cculd be adopted for the output power combiner in order to properly exploit the Doherty idea and perform the correct load modulation, and a optimized solution has been identified as the one in (Colantonio et al., 2009 - a).

REFERENCES

1. Campbell, C. F. (1999). A Fully Integrated Ku-Band Doherty Amplifier MMIC, IEEE Microwave and Guided Wave Letters, Vol. 9, No. 3, March 1999, pp. 114-116.

2. Cho, K. J.; Kim, W. J.; Stapleton, S. P.; Kim, J. H.; Lee, B.; Choi, J. J.; Kim, J. Y. & Lee, J. C. (2007). Design of N-way distributed Doherty amplifier for WCDMA and OFDM applications, Electronics Letters, Vol. 43, No. 10, May 2007, pp. 577-578.

3. Colantonio, P.; Giannini, F.; Leuzzi, F. & Limiti, E. (2002). Harmonic tuned PAs design criteria, IEEE MTT-S International Microwave Symposium Digest, Vol. 3, June 2002, pp. 1639–1642.

4. Colantonio, P.; Giannini, F.; Giofrè, R. & Piazzon, L. (2009 - a). AMPLIFICATORE DI TIPO DOHERTY, Italian Patent, No. RM2008A000480, 2009.

5. Colantonio, P.; Giannini, F.; Giofrè, R. & Piazzon, L. (2009 - b). The AB-C Doherty power amplifier. Part I: Theory, International Journal of RF and Microwave Computer-Aided Engineering, Vol. 19, Is. 3, May 2009, pp. 293–306.

6. Cripps, S. C. (2002). Advanced Techniques in RF Power Amplifiers Design, Artech House, Norwood (Massachusetts).

7. Doherty, W. H. (1936). A New High Efficiency Power Amplifier for Modulated Waves, Proceedings of Institute of Radio Engineers, pp. 1163-1182, September 1936.

8. Elmala, M.; Paramesh, J. & Soumyanath, K. (2006). A 90-nm CMOS Doherty power amplifier with minimum AM-PM distortion, IEEE Journal of Solid-State Circuits, Vol. 41, No. 6, June 2006, pp. 1323–1332.

9. Kang, J.; Yu, D.; Min, K. & Kim, B. (2006). A Ultra-High PAE Doherty Amplifier Based on 0.13-mm CMOS Process, IEEE Microwave and Wireless Components Letters, Vol. 16, No. 9, September 2006, pp. 505–507.

10. Kim, J.; Cha, J.; Kim, I. & Kim, B. (2005). Optimum Operation of Asymmetrical-Cells-Based Linear Doherty Power Amplifier-Uneven Power Drive and Power Matching, IEEE Transaction on Microwaves Theory and Techniques, Vol. 53, No. 5, May 2005, pp. 1802- 1809.

11. Kim, I.; Cha, J.; Hong, S.; Kim, J.; Woo, Y. Y.; Park, C. S. & Kim, B. (2006). Highly Linear Three-Way Doherty Amplifier With Uneven Power Drive for Repeater System, IEEE Microwave and Wireless Components Letters, Vol. 16, No. 4, April 2006, pp. 176- 178.

12. Kim, J.; Moon, J.; Woo, Y. Y.; Hong, S.; Kim, I.; Kim, J. & Kim, B. (2008). Analysis of a Fully Matched Saturated Doherty Amplifier With Excellent Efficiency, IEEE Transaction on Microwaves Theory and Techniques, Vol. 56, No. 2, February 2008, pp. 328-338.

13. Lee, Y.; Lee, M. & Jeong, Y. (2008). Unequal-Cells-Based GaN HEMT Doherty Amplifier With an Extended Efficiency Range, IEEE Microwave and Wireless Components Letters, Vol. 18, No. 8, August 2008, pp. 536–538.

14. Markos, Z.; Colantonio, P.; Giannini, F.; Giofrè, R.; Imbimbo, M. & Kompa, G. (2007). A 6W Uneven Doherty Power Amplifier in GaN Technology, Proceedings of 37th European Microwave Conference, pp. 1097-1100, Germany, October 2007, IEEE, Munich.

15. McCarroll, C.P.; Alley, G.D.; Yates, S. & Matreci, R. (2000). A 20 GHz Doherty power amplifier MMIC with high efficiency and low distortion designed for broad band digital communication systems, IEEE MTT-S International Microwave Symposium Digest, Vol. 1, June 2000, pp. 537–540.

16. Neo, W. C. E.; Qureshi, J.; Pelk, M. J.; Gajadharsing, J. R. & de Vreede, L. C. N. (2007). A Mixed-Signal Approach Towards Linear and Efficient N-Way Doherty Amplifiers, IEEE Transaction on Microwaves Theory and Techniques, Vol. 55, No. 5, May 2007, pp. 866-879.

17. Pelk, M. J.; Neo, W. C. E.; Gajadharsing, J. R.; Pengelly, R. S. & de Vreede, L. C. N. (2008). A High-Efficiency 100-W GaN Three-Way Doherty Amplifier for Base-Station Applications, IEEE Transaction on Microwaves Theory and Techniques, Vol. 56, No. 7, July 2008, pp. 1582-1591.

18. Raab, F. H. (1987). Efficiency of Doherty RF power-amplifier systems, IEEE Transaction on Broadcasting, Vol. BC-33, No. 3, September 1987, pp. 77–83.

19. Raab, F. H. (2001). Class-E, Class-C and Class-F power amplifiers based upon a finite number of harmonics, IEEE Transaction on Microwaves Theory and Techniques, Vol. 49, No. 8, August 2001, pp. 1462-1468.

20. Srirattana, N.; Raghavan, A.; Heo, D.; Allen, P. E. & Laskar, J. (2005). Analysis and Design of a High-Efficiency Multistage Doherty Power Amplifier for Wireless Communications, IEEE Transaction on Microwaves Theory and Techniques, Vol. 53, No. 3, March 2005, pp. 852-860.

21. Steinbeiser, C.; Landon, T.; Suckling, C.; Nelson, J.; Delaney, J.; Hitt, J.; Witkowski, L.; Burgin, G.; Hajji, R. & Krutko, O. (2008). 250W HVHBT Doherty With 57% WCDMA Efficiency Linearized to -55 dBc for 2c11 6.5 dB PAR, IEEE Journal of SolidState Circuits, Vol. 43, No. 10, October 2008, pp. 2218–2228.

22. Tsai, J. & Huang, T. (2007). A 38–46 GHz MMIC Doherty Power Amplifier Using PostDistortion Linearization, IEEE Microwave and Wireless Components Letters, Vol. 17, No. 5, May 2007, pp. 388–390.

23. Wongkomet, N.; Tee, L. & Gray, P. R. (2006). A 31.5 dBm CMOS RF Doherty Power Amplifier for Wireless Communications, IEEE Journal of Solid-State Circuits, Vol. 41, No. 12, December 2006, pp. 2852–2859.

24. Yang, Y.; Cha, J.; Shin, B. & Kim, B. (2003). A Fully Matched N-Way Doherty Amplifier With Optimized Linearity, IEEE Transaction on Microwaves Theory and Techniques, Vol. 51, No. 3, March 2003, pp. 986-993.

Chapter 5

DESIGN AND PERFORMANCE INVESTIGATION OF A NEW DISTRIBUTED AMPLIFIER ARCHITECTURE FOR 40 AND 100 GB/S OPTICAL RECEIVERS

Essra E. Al-Bayati[1], R. S. Fyath[2]

[1]Department of Electronic and Communications Engineering, Al-Nahrain University, Baghdad, Iraq

[2]Department of Computer Engineering, Al-Nahrain University, Baghdad, Iraq

ABSTRACT

The design of distributed amplifiers (DAs) is one of the challenging aspects in emerging ultra high bit rate optical communication systems. This is especially important when implementation in submicron silicon complementary metal oxide semiconductor (CMOS) process is considered. This work presents a novel design scheme for DAs suitable for frontend amplification in 40 and 100 Gb/s optical receivers. The goal is to achieve high flat gain and low noise figure (NF) over the ultra wideband operating bandwidth (BW). The design scheme combines shifted second tire (SST) matrix configuration with cascode amplification cell configuration and uses m-derived technique. Performance investigation of the proposed DA architecture is carried out and the results are compared with that of other DA architectures reported in the literature. The investigation covers the gain and NF spectra when the DAs are implemented in 180, 130, 90, 65 and 45 CMOS standards.The simulation results reveal that the proposed DA architecture offers the highest gain with highest degree of flatness and low NF when compared with other DA configurations. Gain-BW products of 42772 and 21137 GHz are achieved when the amplifier is designed for 40 and 100 Gb/s operation, respectively, using 45 nm CMOS standard. The simulation is performed using AWR Microwave Office (version 10).

INTRODUCTION

Everybody wants to benefit from the evaluation in the field of communication especially through internet. Due to the expanding demand of communication services, the volume of data exchanged in the communication systems has increased. This leads to increase data rate of the global communication systems from tens of Gb/s to Tb/s. BW requirements will increase by more than 100 times and applications such as virtual reality require data rate that are 10,000 times higher than currently available. To transport such data rate, a media with low loss and high BW is required [1, 2]. Among the available medium to transfer the data, optical fibers have the best performance. Optical fibers are very common these days to transport very high rate digital data. Such high speed data rates can be transported over kilometers of optical fiber and without significant loss. Normally loss is very low when the signal is transmitted using light rather than electrical signal. These fibers also have the advantage of being low cost in addition to improvement of performance. Fiber optic devices and systems are evidently employed to realize very high data rates. Fiber optic communication is a solution because high data rates can be transmitted through this high capacity cable with high performance [3].

The exponential growth of Internet traffic is fueling the research and development of wavelength division multiplexed passive optical network technology in the access network segment [4]. Driven by the continues increase in BW demand and number of subscribers, future access networks will require 40 Gb/s high speed service per wavelength channel [5, 6] on a 50 GHz dense wavelength grid. The realization of high-speed analog-to-digital conversation and digital signal processing have enabled a bit rate of 100 Gb/s in long-haul coherent optical communication system [7]. However, for short-reach 100 Gb/s applications, solutions that use intensity modulation and direction detection are seen as more practical [8]. With the advances in semiconductor technologies, integrated circuits operating at 40 Gb/s have been realized in standard CMOS process [9]. Among all kinds of high speed circuits, the broadband amplifier is a key building block at both the transmitting and receiving ends, see Fig. 1. In fact there is demand for wideband CMOS amplifiers in the frontend section of the optical receivers. Distributed amplification is one of the well-known methods to provide such performance by absorbing the parasitic capacitances of parallel gain distributed cell into an artificial transmission line, which in return, guarantees the gain uniformity and input/output matching within the BW of operation. However, design DA for 40 Gb/s (and above) optical receivers needs careful consideration related to the gain, frequency spectrum and NF. This paper addresses the design issues and performance investigation of CMOS DA for the front-end amplification stage in 40 and 100 Gb/s optical receivers.

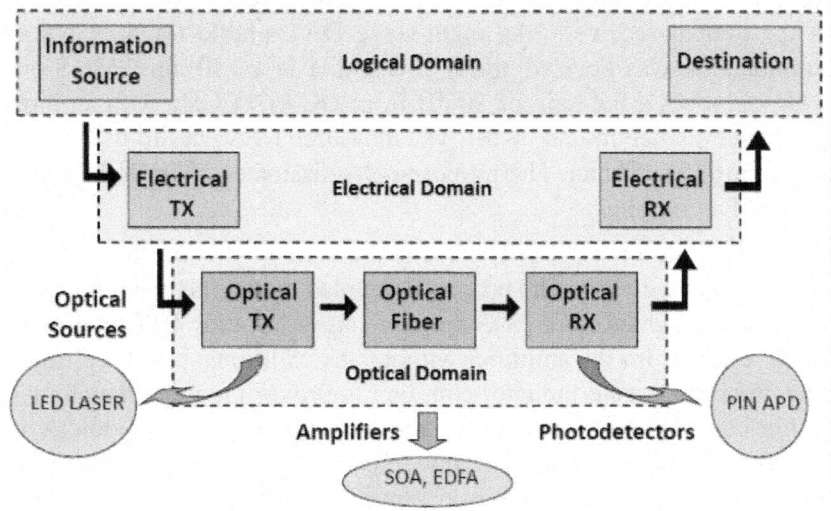

Figure 1: Optical communication system [10].

RELATED WORK

In 2005, Wolf et al [11] demonstrated an eight-stage DA with 12.5 dB ± 0.45 dB gain and 50 GHz BW in a commercially available 0.1 μm metamorphic GaAs HEMT technology. The amplifier has a minimum NF lower than 2.5 dB in the BW. The group delay variation from 9 to 40 GHz is ± 7.5 ps and circuit power consumption is 0.4 W. Such amplifier packaged with a high responsively PD into a fiber pig-tailed module. Eye diagrams measurements demonstrate the successful highspeed operation of the photoreceiver.

In 2007, Chien and Lu [12] presented a novel circuit topology for high-gain 40 Gb/s DAs. Based on the conventional distributed architecture, the gain cells were realized by cascading cascode stages for gain enhancement. In addition, the stagger-tuning technique was extensively utilized in the design of the cascode stages as well as the cascaded stages, leading to significant improvement in terms of the operating BW and the gain flatness. With the proposed circuit architecture, two amplifiers were implemented in a standard 180 nm CMOS technology. The amplifier with a 3 x3 configuration exhibits a gain of 16.2 dB and a 3 dB BW of 33.4 GHz, while the one in a form of 2x4 demonstrates a gain of20 dB and a BW of 39.4 GHz. Consuming a DC power of 260 mW from a 2.8-V supply voltage, both circuits provide clear eye-opening with a pseudorandom bit sequence at 40 Gb/s.

In 2009, Moez and Elmasry [13] presented a circuit technique to compensate for the metal and substrate loss of the onchip transmission lines, and, consequently, to improve the gain flatness and BW of CMOS DAs for optical receivers. An eight-stage DA suitable for 40 Gb/s optical communication was devised and implemented in a 130 nm CMOS process. The DA achieves a flat gain of 10 dB from DC to 44 GHz with an input and output matching better than −8 dB. The measured NF varies from 2.5 to 7.5 dB with the amplifier's band. The proposed DA dissipates 103mW from two 1-V and 1.2-V DC supplies.

In 2009, Entesari et al [14] presented a state-of-the-art DA with coupled inductors in the gate line. The proposed coupled inductors, in conjunction with series-peaking inductors in cascode gain stages, provide BW extension with flat gain response for the amplifier without any additional power consumption. On the other hand, gate-inductor coupling improves the input matching of the amplifier considerably. The detailed analysis and design methodology for the proposed DA were presented. The new four-stage DA, fabricated using an IBM 0.18- m CMOS process, achieves a power gain of around 10 dB, input and output return losses better than 16 and 18 dB, respectively, a NF of 3.6–4.9 dB, and a power consumption of 21mW over a 16-GHz flat 1-dB BW. is between 0.1 and 3.75 dBm across the entire band.

In 2010, Ghadiri and Moez [15] presented a new high-gain structure for DA. Negative capacitance cells were exploited to ameliorate the loading effects of parasitic capacitors of gain cells in order to improve the gain of the DA while keeping the desired BW. In addition, the negative capacitance circuit creates a negative resistance that can be used to increase the amplifier BW. Implemented in 130 nm IBM's CMRF8SF CMOS, the proposed six-stage DA presents an average gain of 13.2 dB over a BW of 29.4 GHz. The measured input return loss is less than 9 dB and the output return loss is less than 9.5 dB over the entire BW. With a chip area of 1.5 mm X 0.8 mm, the amplifier consumes 136 mW from a 1.5-V DC power supply.

In 2010, Chien et al [16] presented a transimpedance amplifier (TIA) with a tunable BW for optical communications. The proposed TIA is composed of two cascaded stages in which an input network with inductive peaking elements is employed in the first stage for broadband operations while a modified DA is utilized as the second stage for enhanced transimpedance gain. In addition, a feedback loop is incorporated as the bandwidth-tuning mechanism. By tuning the BW of the TIA, optimum circuit operation with lowest bit error rate (BER) can be achieved in the receiver front end for highspeed data transmission. The proposed circuit was implemented in a 180 nm CMOS process. Consuming a DC power of 33.3 mW from a 1.8-V supply, the fabricated TIA exhibits a transimpedance gain of 47.8 dB and a variable 3-dB BW from 6.2 to 10.5 GHz. Providing a 2^{11}-1 pseudorandombit sequence at 9-15 Gb/s, a BER less than $10^{-12}$was demonstrated experimentally by the TIA with the BW tuning mechanism.

In 2011, KimandBuckwalter [17] demonstrated a low-power cascode DA in a 45 nm silicon-on-insulator (SOI) CMOS process. The amplifier achieves a 3 dB BW of 92 GHz. The peak gain is 9 dB with a gain-ripple of less 1.5 dB over the BW. The group-delay variation is under ± 4.7ps over the 3 dB BW. The amplifier consumes 73.5mW from a 1.2V supply and results in a GBW efficiency figure of merit of 3.53 GHz/mW. The chip occupies an area of 0.45 mm² including the pads.

In 2012, Jahanian and Heydari [18] presented a CMOS DA with distributed active input balun that achieves a GBW product of 818 GHz, while improving linearity. Each cell within the DA employs dual-output two-stage topology that improves gain and linearity without adversely affecting BW and power. Comprehensive analysis and simulations were carried out to investigate gain, BW, linearity, noise, and stability of the proposed cell, and compare them with conventional cells. Fabricated in a 65-nm low-power CMOS process, the 0.9-mm DA achieves 22 dB of gain and a P1dB of 10 dBm, while consuming DC power of 97 mW from a 1.3-V supply. A distributed balun, designed and fabricated in the same process, using the same topology achieves a BW larger than 70 GHz and a gain of 4 dB with 19.5-mW power consumption from 1.3-V supply.

In 2013, Feng et al [19] realized compact self-biased wideband low noise amplifier (LNA) in Global Foundries 65 nm CMOS technology. Wideband input matching characteristic is achieved by placing a series gate inductor and a parallel tuning capacitor in the resistive-feedback network. Combined with the inductive-series peaking technique which further extends the BW, the proposed cascaded three-stage resistive-feedback amplifier obtains a large operating BW which is comparable with the DA. Measurement shows that the proposed amplifier achieves a power gain of with input and output return losses better than 8 dB and NF ranging from 4.5 to 6.8 dB between 2.1–39 GHz. The fabricated low LNA occupies a silicon area of 0.16 including all testing pads and draws 17 mA from a 1.5 V power supply.

In 2013, Kao et al. [20] proposed a new DA topology which is a combination of the conventional DA and the cascaded single-stage DA. This DA topology can provide wide BW with considerations of the gain, NF, and output power simultaneously, and requires reasonable DC power consumption. Two termination methods of this combination were investigated. From the measurements, the first DA has a small-signal gain of 20.5 dB, a 3-dB BW of 35 GHz, and a GBW product of 371 GHz. The maximum output power at 1-dB output compression point is 8.6 dBm and the NF is between 6.8– 8 dB at frequencies lower than 18 GHz. The chip size, including testing pads, is only 0.78 mm, and the ratio of the GBW to chip size is 476GHz/mm. The second DA has a small-signal gain of 24dB, a 3-dB BW of 33 GHz, and a GBW product of 523 GHz. The maximum output power is 9 dBm and the NF is between 6.5-7.5 dB at frequencies lower than 18 GHz. The chip size including testing pads is only 0.83 mm, and the ratio of the GBW to chip size is 630 GHz mm.

In 2013, Cho et al. [21] proposed a wideband switchless bi-directional DA in a commercial 130 nm CMOS technology, which realizes multi-octave BW with high gain and low NF using DA technique and cascode amplifier pair. The measured gain is over 10 dB and measured NF is 3.2-6.5 dB. The input and output return losses are better than 9 dB at 3–20 GHz. The measured output power 1dB and output input power (OIP3) is larger than 8 dBm and 17 dBm at 4–15 GHz. The chip sizeis 0.96 x 0.85 including pads. The proposed switchless bi-directional amplifier has almost the same chip size compared to the conventional uni-directional DA.

In 2014, Kim and Nguyen [22] presented a new tri-band power amplifier on a 180 nm SiGeBiCMOS process, operating concurrently in Ku/K/Ka and -band, is presented. The concurrent tri-band PA design is based on the DA structure with capacitive coupling to enable large device size, while maintaining wide BW, gain cells with the enhanced-gain peaking inductor, and negative-resistance active notch filters for improved tri-band gain response. The concurrent tri-band PA exhibits measured small-signal gain around 15.4, 14.7 and 12.3 dB in the low band (10–19 GHz), midband (23–29 GHz), and high band (33–40 GHz), respectively.

It is clear from the above survey that the reported designs of DAs suitable for high bit rate optical receiver don't achieves simultaneously wide BW, high flat gain and low NF. This issue is addressed in this thesis where a modified DA topology is introduced to reach these goals.

PROPOSED DA ARCITECTURE AND DESIGN CONCEPTS

In the literature, different DA architectures have been discussed. Each one has its own design topology and uses different techniques to enhance one of the design requirements: wide BW, high flat gain and low NF. In this section, a new DA architecture is proposed to achieve ultrawide band operation under high amplification gain and low NF conditions. The proposed architecture collects the main features behind different topologies and techniques adopted in previous DA designs such as shifted second stage (SST) topology, matrix configuration, cascode cell amplifier configuration, and mderived matching technique. Design issues based on deep submicron CMOS technology are discussed toward achieving efficient front-end amplification in 40 Gb/s and 100 Gb/s optical receivers.

Architecture of the Proposed Distributed Amplifier

The main idea behind the proposed DA is to combine high-feature topologies and techniques adopted by other DA designs. The task in this work is primarily to design CMOS-based DA with ultrawide BW, high flat gain and low NF suitable for high bit rate optical receivers (>40 Gb/s). The proposed DA uses the following techniques and topologies

m-Derived Technique

m-derived techniques is used at the input and output of constant-k filter sections in order to obtain flat BW and constant impedance matching.

Shifted-Second Tire Matrix Configuration

This type of DA uses additive and multiplicative technique to achieve high flat gain. The configuration uses M stages and N distributed cells (N tires). Simulation results show that design the DA with two stages and four distributed cells is sufficient to get the require design target. Using M>2 and N>4 will increase the complexity of the design without offering reasonable performance improvement.

Cascode Amplification Cell Configuration

Cascode DA is obtained by adding common-gate amplifier section to the drain line of common-source amplifier stage. Using cascode amplification cells in DA configuration will decrease the variation of the capacitance seen by the transmission line and offer a nearly flat gain over wider BW as compared with the simple common-source amplifier.

The cascode amplifier can be considered as a two-stage amplifier composed of a transconductance amplifier followed by a current buffer [23]. Compared to a single amplifier stage, this combination may have one or more of the following characteristics: higher input-output isolation, higher input impedance, higher output impedance, higher gain or higher BW. Thus cascade configuration can be designed to improve input-output isolation (or reverse transmission) as there is no direct coupling from the output to the input.

Fig. 2a shows a MOSFET cascode amplifier. If one make a small-signal analysis for this amplifier using the simplified small-signal equivalent circuit shown in Fig. 2b, the following results are obtained [23]. The voltage gain

$$Av = \frac{V_{out}}{V_{in}} = -(g_{m1}r_{o1} + 1)\, g_{m2}r_{o2}$$

(1.a)

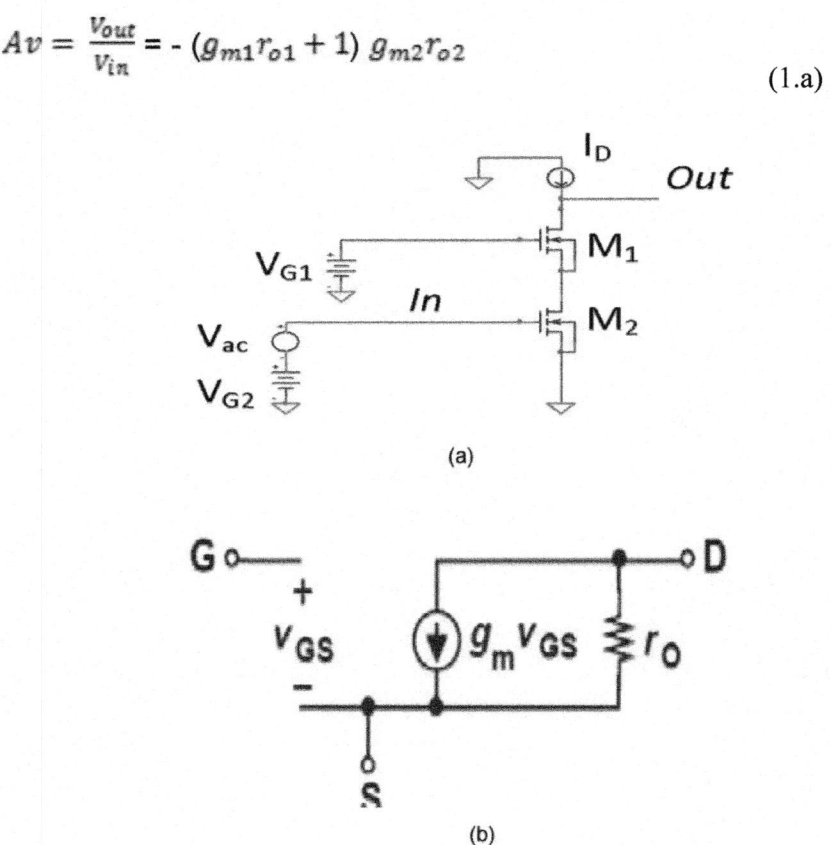

(a)

(b)

Figure 2: a- Cascode MOSFET amplifier b- Simplified small-signal equivalent circuit of MOSFET [23].

$$Av \approx g_{m1}g_{m2}r_{o1}\, r_{o2} \qquad \text{when } g_{m1}r_{o1} >> 1$$

(1.b)

Note that the cascode amplifier behaves as a common-source amplifier with higher gain. If identical transistor is used, the $Av \approx g_m^2 r_o^2$. The input resistance of the cascode amplifier tends to infinitely since the input signal source is applied to the high impedance gate terminal.

The output resistance of the cascode amplifier is given by

$$R_{out} = (r_{o1} + r_{o2})(1 + g_{m1}\,(r_{o1}//r_{o2}))$$

(2.a)

For identical transistors

$$R_{out} = 2r_o\left(1 + g_m\frac{r_o}{2}\right)$$

(2.b)

$$\approx g_m^2\, r_o^2 \qquad \text{when } g_m r_o >> 2.$$

This yields higher output impedance compared with conventional common-source amplifier.

Fig. 3 shows a block diagram for the proposed DA configuration. The structure contains two amplification stages each with four amplification cells. The corresponding circuit-level description of the proposed DA is illustrated in Fig. 4.

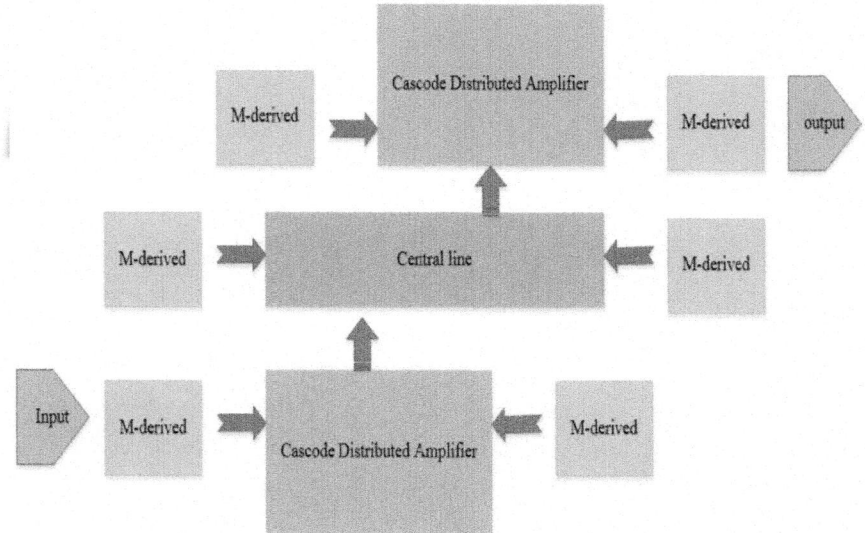

Figure 3: Block diagram of the proposed shifted second tier with m-derived DA.

Design Concepts

The gate and drain transmission lines are designed with equal inductance and capacitance to ensure equal wave propagation velocities along both lines. The cutoff frequency f_c and characteristic impedance Z_0 of the transmission lines are given by

$$f_c = \frac{1}{\pi\sqrt{L_g C_g}} = \frac{1}{\pi\sqrt{L_d C_d}} = \frac{1}{\pi\sqrt{L_l C_c}}$$

(3.a)

$$Z_o = \sqrt{\frac{L_g}{C_g}} = \sqrt{\frac{L_d}{C_d}} = \sqrt{\frac{L_c}{C_c}}$$

(3.b)

where L and C stand, respectively, to the inductance and capacitance of the transmission line, while the subscripts g and d denote gate and drain, respectively. For given values of f_c and z_0, the transmission line elements can be calculated as follows

$$C_g = C_d = \frac{1}{\pi Z_o f_c}$$

(4.a)

$$L_g = L_d = \frac{Z_0}{\pi f_c}$$

(4.b)

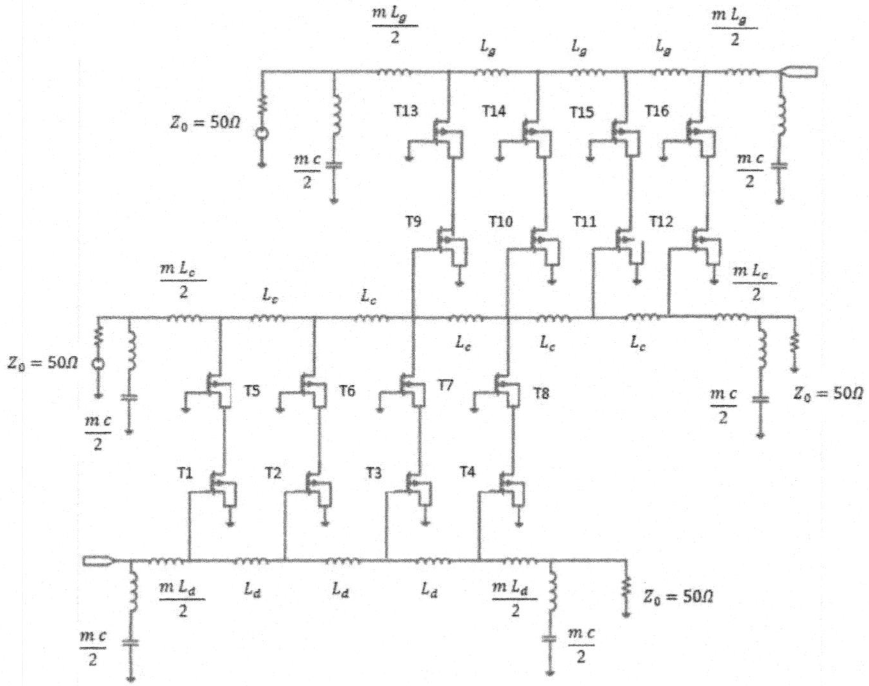

Figure 4: Circuit-level description of the proposed DA having two stages each with four amplifications cells with m-drive techniques.

Note that both capacitance and inductance are inversely proportional to the cutoff frequency f_c. This is illustrated in Figs. 5 a and b which show respectively, the dependence of $C_g = C_d$ and $L_g = L_d$ on the transmission line cutoff frequency f_c when $Z_0 = 50\ \Omega$. At $f_c = 10$ GHz, $C_g = C_d = 0.63$ pF and $L_g = L_d = 1.57$ nH. These values are to be compared with 0.08 pF and 0.19 nH for $f_c = 80$ GHz.

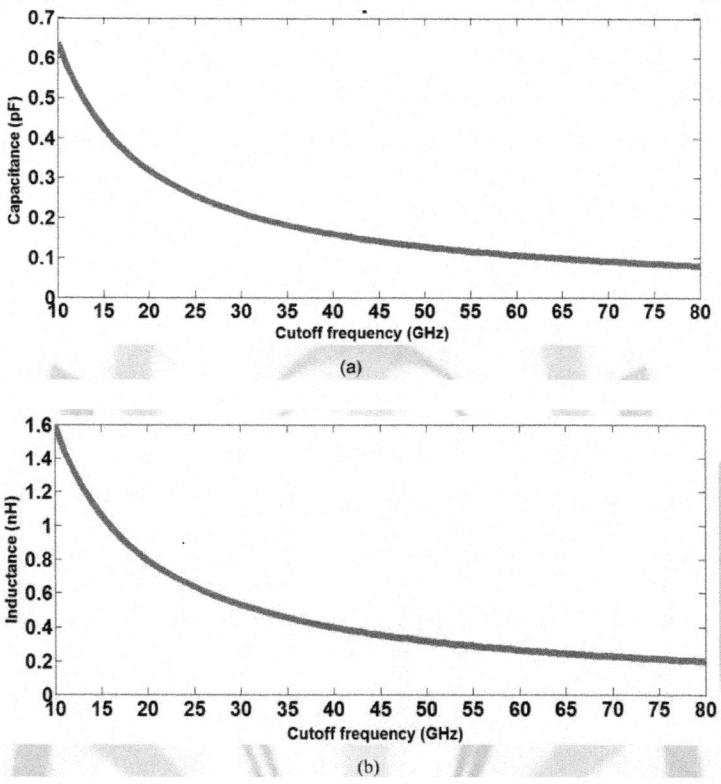

(a)

(b)

Figure 5: Dependence of the transmission line capacitance (a) and inductance (b) on the cutoff frequency.

The next step is to determine transistor gate width for different submicron CMOS standards. The gate length (channel length) L is set equal to the used CMOS standard [24].The gate width is estimated using the following relation.

$$C_g = \frac{2}{3} W\, L C_{ox}$$

(5)

$$W = \frac{3C_g}{2\, L\, C_{ox}}$$

(6.a)

$$W = \frac{3}{2\pi\, L\, C_{ox}\, Z_o f_c}$$

(6.b)

where is the oxide capacitance per unit area. Unfortunately, values of C_{ox} for submicron CMOS standards used in this thesis<180 nm are not reported in the literature. For example, Ref. [25] gives C_{ox} for CMOS standards 800, 500, 250 and 180 nm (see Table 1). These data are curve fitted to the flowing equation

$$C_{ox} = a_2 x^2 + a_1 x + a_0$$

(7)

Where x is the CMOS standard in nm. The values of the fitting coefficients are

$a_2 = 0.017188\ fF/mm^2.n^2\ m^2$

$a_1 = -25.876\ fF/mm^2.n\ m^2$

$a_0 = 12084\ fF/mm^2$

Table 1: Oxide Capacitance C_{ox} for various CMOS standards

CMOSStandard (nm)	OxideCapacitance	Referenc
800	2300	[25]
500	3800	[25]
250	5800	[25]
180	8600	[25]
130	9001	This work
90	9089	This work
65	10005	This work
45	11000	This work

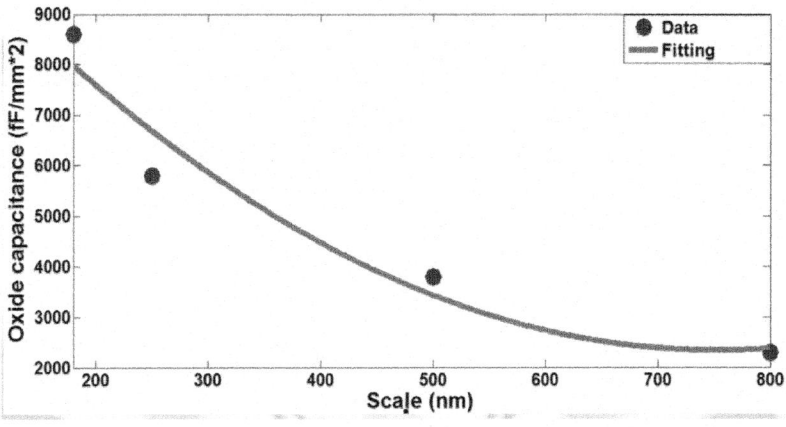

Figure 6: Variation of oxide capacitance with CMOS standard (>180 nm).

Fig. 6 shows the variation of C_{ox} with CMOS standard. The marks are the data taken for Ref. [25] while the solid line denotes curve fitting. The curve fitting is used to extract C_{ox} for CMOS standards 130, 90, 65 and 45 nm (see Table 1 and Fig. 7).

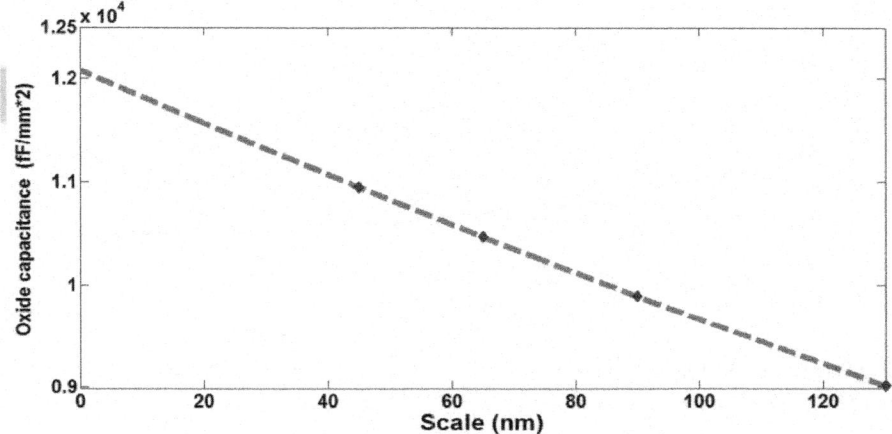

Figure 7: Extracted values of oxide capacitance for 130, 90, 65 and 45 nm standards.

It is clear from Table 1 and Figs. 6 and 7 that C_{ox} increases as CMOS standard increases. This result can be explained by treating the gate capacitance simply as a two-parallel plate capacitance arises from the dioxide layer.

$$C_g = \frac{A\,\varepsilon_{SiO2}}{t_{ox}}$$

$$\text{(8.a)}$$

$$C_{ox} = \frac{C_g}{A} = \frac{\varepsilon_{ox}}{t_{ox}}$$

$$\text{(8.b)}$$

$$\varepsilon_{ox} = 3.9\,\varepsilon_0 = 3.45 \times 10^{-11} F/m$$

Where A is the gate area and e_{ox} is the permittivity of the silicon oxide which has t_{ox} thickness. As CMOS standard decreases, t_{ox} thickness decreases too and leading to higher values for C_{ox}.

The dependence of the gate width on the cutoff frequency of the transmission line is reflected in Equ. 6b. For a given CMOS standard, the gate width is inversely proportional to the cutoff frequency. This relation is illustrated graphically in Fig. 8 where the gate width is plotted versus cutoff frequency for different values of CMOS standards. According to Equ. 6b, W Increases as CMOS standard decreases.

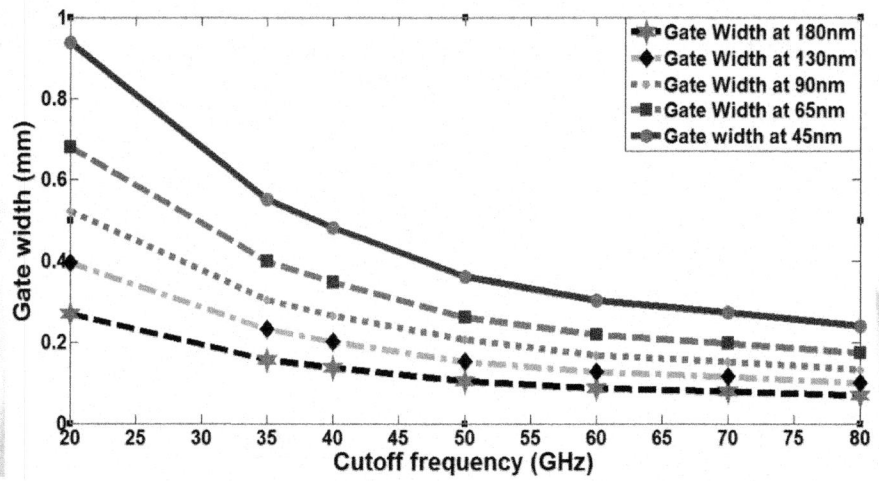

Figure 8: Dependence of gate width on transmission line cutoff frequency for different CMOS standards.

The next design step is to calculate the transistor transconductance g_m as a function of transmission line cutoff frequency f_c for different values of CMOS standards. Recall that $g_m = (W/L)m_n C_{ox} V_{ov}$ when the transistor operates in the saturation region. Here m_n is the electron mobility in the channel (assuming NMOS structure) and V_{ov} is the over drive voltage which is set to 0.2V in the simulation. With the aid of Equ. 6b one can arrive to the following expression for the transconductance

$$g_m = \frac{3\,\mu_n V_{ov}}{2\,\pi L^2 Z_o f_c}$$

(9)

Investigating Equ. 9 reveals the following findings

1. g_m is independent of oxide capacitance C_{ox} and gate width W.
2. g_m is inversely proportion to both cutoff frequency f_c and the square of the gate length.

The electron mobility for NMOS transistors fabricated with standards 180 nm and above are reported in Ref. [25] and listed in Table 2. These data is curve fitted to the following polynomial in order to extract the values of m_n for standards below 180 nm (see Fig. 9)

$$\mu_n = b_2 x^2 + b_1 x + b_0$$

(10)

where μ_n in $cm^2/V.s$

$b2 = 1.7265 \times 10^{-5} cm^2/V.s.n^2m^2$

$b1 = 0.14487 \ cm^2/V.s.nm$

$b0 = 423.09 \ cm^2/V.s$

The extracted values of m_n for standards 130, 90, 65 and 45 nm are listed in Table 2.

Table 2: Electron mobility for various CMOS standards

CMOS Standard (nm)	Electron Mobility ($cm^2/V.sec$)	Reference
800	550	[25]
500	500	[25]
250	460	[25]
180	450	[25]
130	442	This work
90	436	This work
65	433	This work
45	430	This work

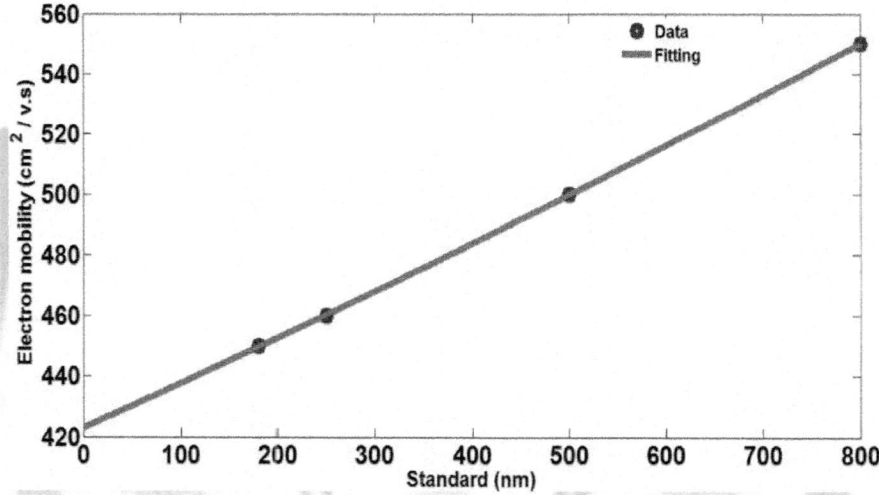

Figure 9: Variation of electron mobility with CMOS standard.

Figs. 10 a-e show the variation of transconductance with transmission line cutoff frequency for different CMOS standards 180, 130, 90, 65 and 45 nm, respectively. Note that g_m increases rapidly as the fabrication process goes to a smaller standard. This result is expected since g_m is inversely proportional to L^2 and the gate length L decreases as one dopted deeper submicron standards. For example, the values of g_m at f_c = 40 GHz are 0.053S, 0.124S, 0.256S, 0.488S and 1.012S for 180, 130, 90, 65 and 45 nm standards, respectively.

Table 3 lists the required transistor design parameters (L, W and g_m) for NMOS transistor fabricated using deep submicron standards and operates at a specific value of cutoff frequency f_c.

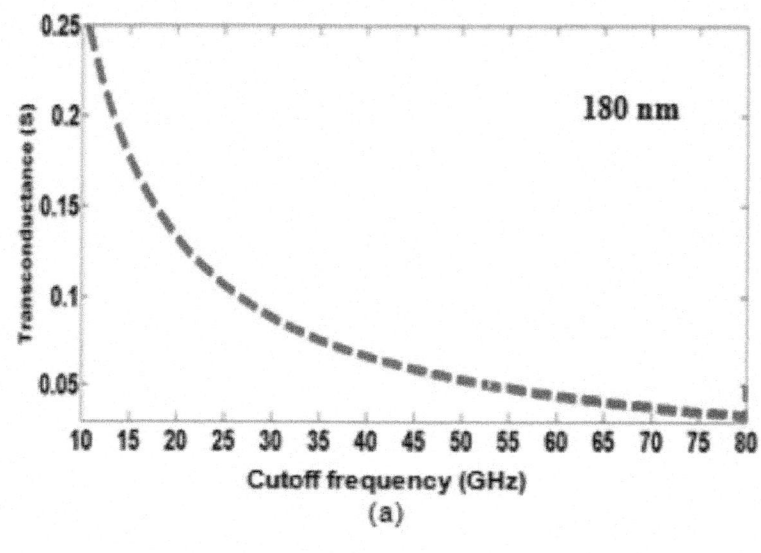
(a)

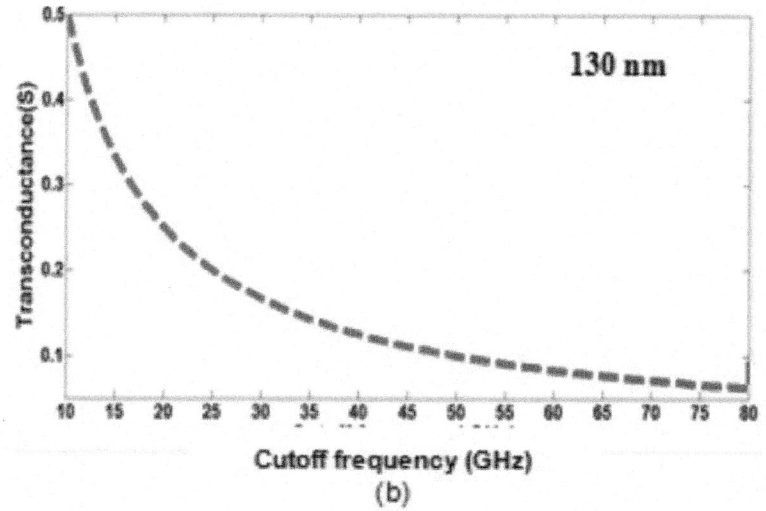
(b)

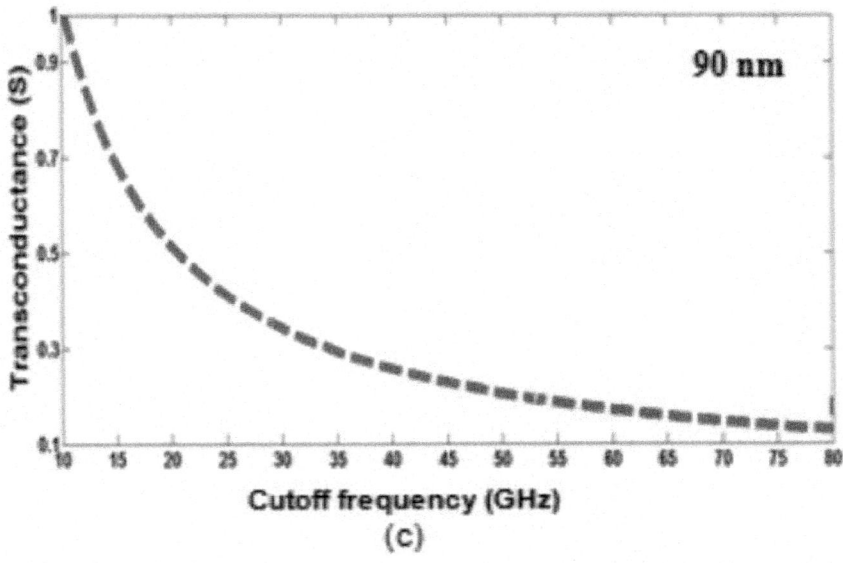

Cutoff frequency (GHz)

(c)

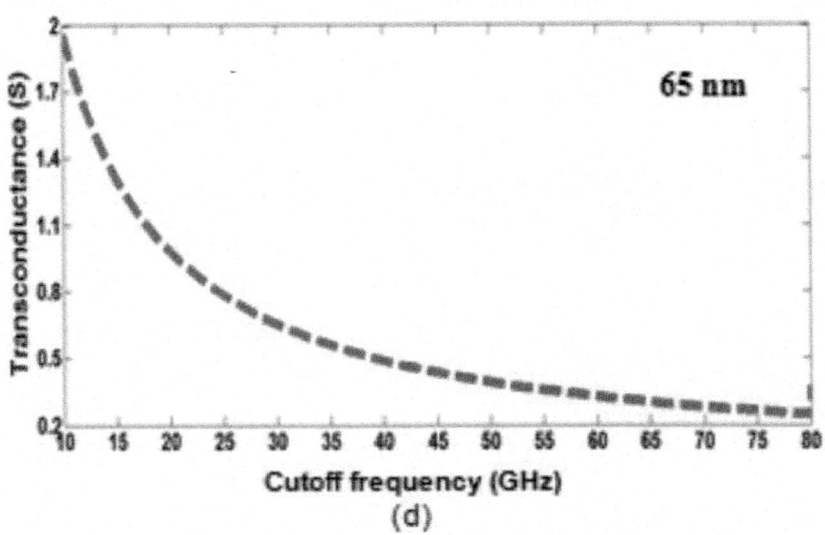

Cutoff frequency (GHz)

(d)

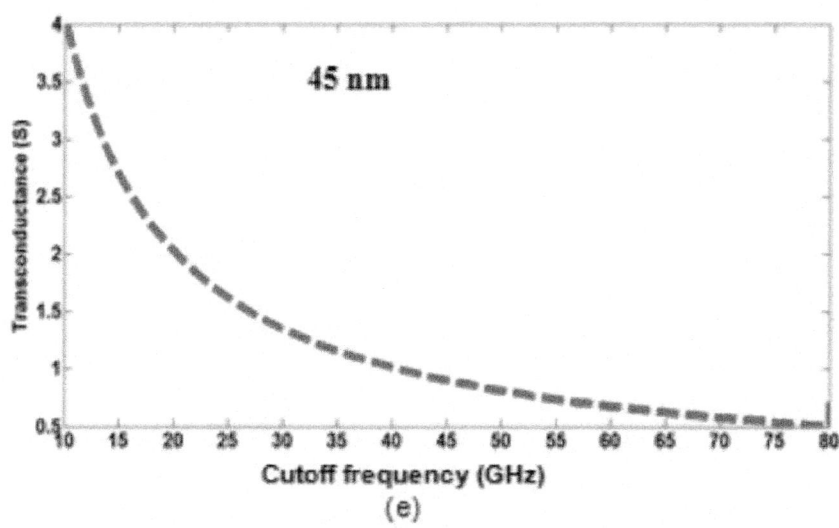

Cutoff frequency (GHz)

(e)

Figure 10: Dependence of NMOS transconductance with transmission line cutoff frequency for different standards. (a) 180 nm, (b) 130 nm, (c) 90 nm, (d) 65 nm, (e) 45 nm.

Table 3: NMOS transistor parameters for different values of cutoff frequency and fabrication standards

Standard (nm)	Transistor Parameter	Cutoff frequency (GHz)			
		20	40	60	80
180	L (nm)	180	180	180	180
	W (μm)	270	138	87	69
	g_m (S)	0.104	0.053	0.033	0.026
130	L (nm)	130	130	130	130
	W (μm)	396	203	128	101
	g_m (S)	0.243	0.124	0.078	0.062
90	L (nm)	90	90	90	90
	W (μm)	522	267	168	134
	g_m (S)	0.500	0.256	0.161	0.128
65	L (nm)	65	65	65	65
	W (μm)	681	349	219	174
	g_m (S)	0.953	0.488	0.307	0.244
45	L (nm)	45	45	45	45
	W (μm)	939	481	303	241
	g_m (S)	1.974	1.012	0.637	0.507

PERFORMANCE SIMULATION RESULTS

Introduction

This section presents simulation results characterizing the gain, NF and BW of the proposed distributed amplifier for 40 and 100 Gb/s operation. The results are compared with the performance of five DA architectures, namely

1.   Conventional DA which uses common-source amplification cells.

2.   Cascode DA which uses cascode configuration for the amplification.

3.   m-derived DA which is a conventional DA supported with both input and output m-derived stages.

4.   Matrix DA where the amplification cells are arranged in matrix-form topology.

5.   Shifted second tier (SST) DA.

All the DAs considered here are designed with four amplification cells per stage. The proposed DA, matrix and SST distributed amplifiers have two stages.

Simulation results related to DA characteristic are obtained using AWR Microwave Office (version 10). The results are then used to design the proposed DA for 40 and 100 Gb/s optical receivers.

Gain and Noise Figure Spectra

The aim of this section is to investigate the gain and NF spectra when the proposed DA is designed using different nano scale CMOS technologies. The results are to be compared with other DA architectures to assess the main features behind the proposed DA.

Figs. 11 a-e illustrates the variation of the DA gains with frequency when the amplifiers are designed with 35 GHz cutoff frequency. This value of cutoff frequency is found to be suitable for operation around 40 Gb/s. Parts a-e of this figure are related to 180, 130, 90, 65 and 45 nm standards, respectively. Investigating these figures highlights the following finding. The proposed DA offers the highest gain with almost flat characteristic in the passband region compared with other DAs. Tables 4 and 5 list, respectively, the DC gain and BW of various DAs under investigation and taking the CMOS standard as independent parameter.

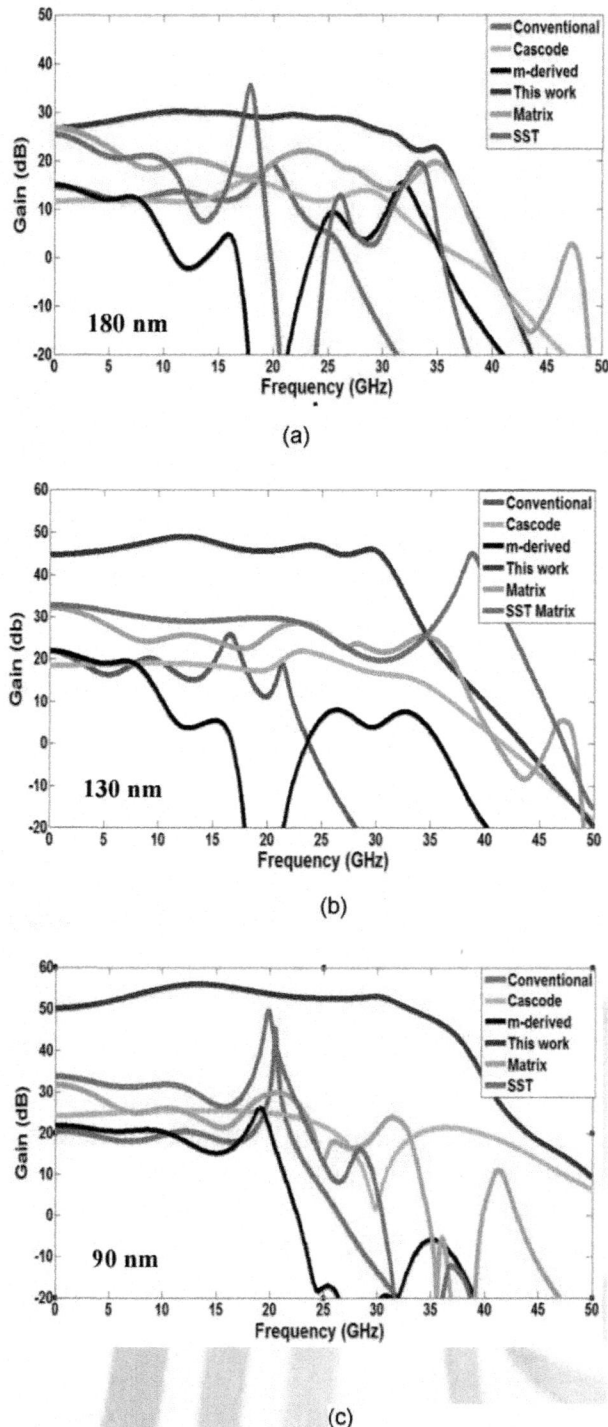

(a)

(b)

(c)

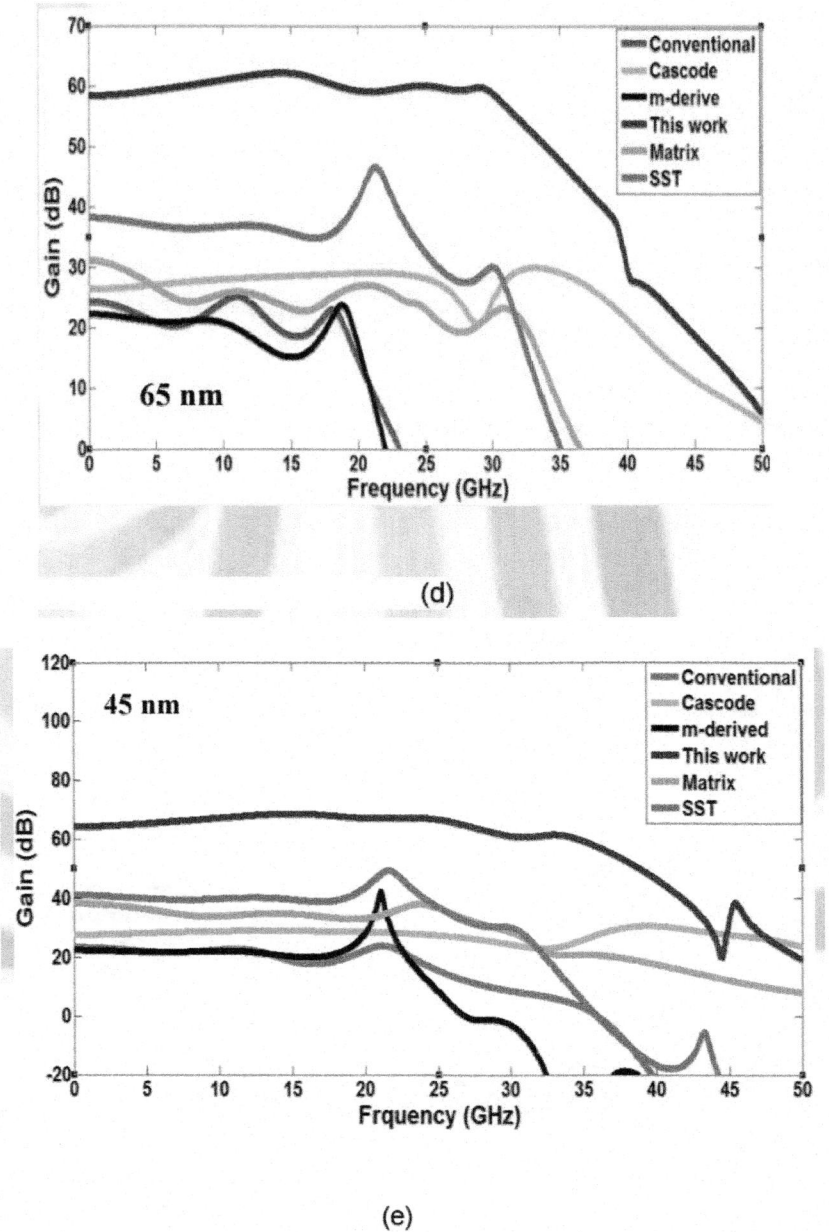

(d)

(e)

Figure 11: Gain spectra of various DA configurations designed with 35 GHz cutoff frequency using (a) 180 nm standard, (b) 130 nm standard, (c) 90 nm standard, (d) 65 nm standard, (e) 45 nm standard.

Table 4: DC gain of various DAs designed with 35 GHz cutoff frequency

Distributed Amplifier	DC Gain (dB)				
	180 nm	130 nm	90 nm	65 nm	45 nm
Conventional	13 dB	19 dB	17 dB	20 dB	20 dB
Cascode	9 dB	15 dB	23 dB	23 dB	24 dB
m-derived	12 dB	19 dB	17 dB	19 dB	20 dB
Matrix	22 dB	27 dB	29 dB	29 dB	35 dB
Shifted second tier	21 dB	29 dB	33 dB	33 dB	39 dB
This work	24 dB	42 dB	47 dB	56 dB	62 dB

Table 5: Bandwidth of various DAs designed with 35 GHz cutoff frequency

Distributed Amplifier	Bandwidth (GHz)				
	180 nm	130 nm	90 nm	65 nm	45 nm
Conventional	1.2 GHz	3 GHz	6 GHz	5 GHz	6 GHz
Cascode	30 GHz	32 GHz	26 GHz	28 GHz	29 GHz
m-derived	7 GHz	8 GHz	11 GHz	10 GHz	19 GHz
Matrix	3.7 GHz	4 GHz	5 GHz	5.5 GHz	7 GHz
Shifted second tier	9.7 GHz	20 GHz	13 GHz	16 GHz	18 GHz
This work	31 GHz	31 GHz	34 GHz	32 GHz	34 GHz

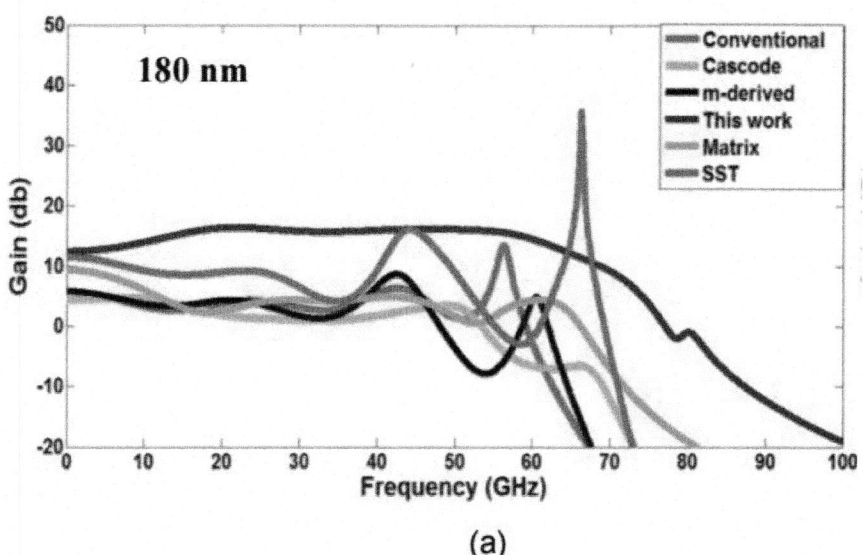

(a)

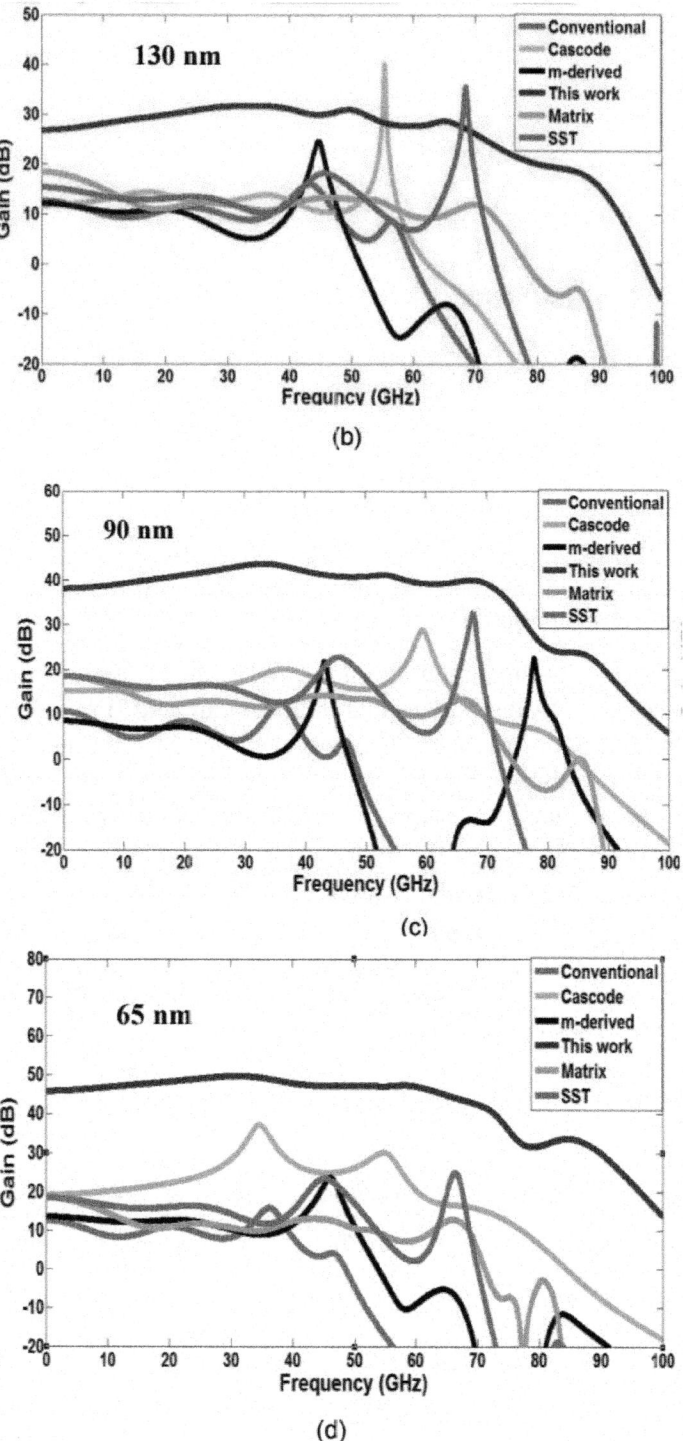

(b)

(c)

(d)

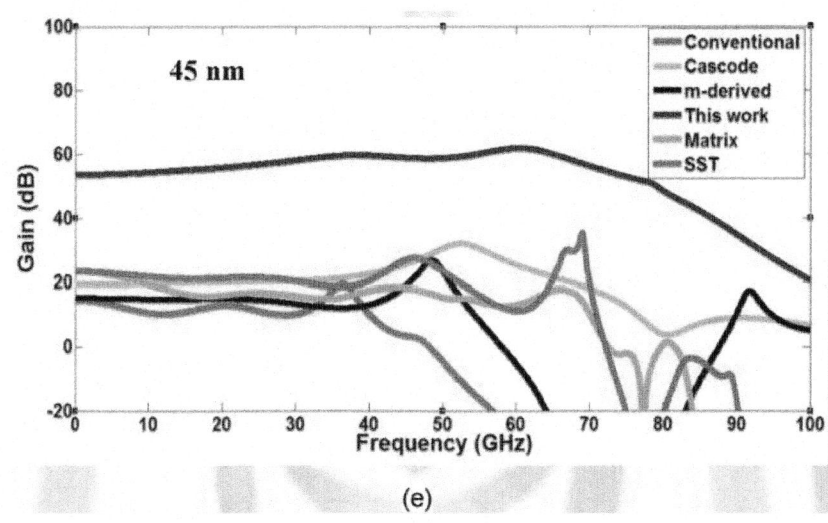

(e)

Figure12: Gain spectra of various DA configurations designed with 80 GHz cut off frequency using (a) 180 nm standard, (b) 130 nm standard, (c) 90 nm standard, (d) 65 nm standard, (e) 45 nm standard.

From the results illustrated in Fig. 11 and Tables 4 and 5 one can find out that the proposed DA the best results (highest gain and BW) when compared with other DAs. Conventional DA gives medium gain but not flat with low BW. Cascode DA gives low gain with BW range close to this work. M-derived DA gives flat medium gain with low BW. Matrix DA gives high gain but not flat with low BW. SST gives flat high gain with low BW.

The simulation is repeated to investigate the characteristics of 80 GHz-cut off frequency DAs suitable for around 100 Gb/s operation. The results are presented in Fig. 12 to highlight the gain spectrum and summarised in Tables 6 and 7 to assess the DC gain and BW, respectively, of various DAs synthesized with nanoscale CMOS standards.

Table 6: DC gain of various DAs designed with 80 GHz cutoff frequency

Distributed Amplifier	DC Gain (dB)				
	180 nm	130 nm	90 nm	65 nm	45 nm
Conventional	3 dB	9 dB	9 dB	9 dB	11 dB
Cascode	2 dB	8 dB	12 dB	15 dB	17 dB
m-derived	3 dB	8 dB	8 dB	10 dB	12 dB
Matrix	6 dB	15 dB	15 dB	15 dB	19 dB
Shifted second tier	9 dB	12 dB	14 dB	15 dB	20 dB
This work	10 dB	23 dB	34 dB	42 dB	49 dB

Table 7: Bandwidth of various DAs designed with 80 GHz cutoff frequency

Distributed Amplifier	Bandwidth (GHz)				
	130 nm	130 nm	90 nm	65 nm	45 nm
Conventional	19 GHz	20 GHz	16 GHz	17 GHz	12 GHz
Cascode	50 GHz	52 GHz	59 GHz	63 GHz	68 GHz
m-derived	27 GHz	28 GHz	25 GHz	29 GHz	35 GHz
Matrix	10 GHz	10 GHz	9 GHz	9 GHz	13 GHz
Shifted second tier	27 GHz	31 GHz	32 GHz	30 GHz	58 GHz
This work	70 GHz	75 GHz	74 GHz	71 GHz	75 GHz

Investigating the results in Fig 12 and Tables 6 and 7 reveals that the same conclusions drawn from the 35 GHz cutoff frequency DAs are also applied.

The results also show that when the cutoff frequency increases, the gain decreases for all DAs. However, the proposed DA gives the highest gain. One can apply these results to all standards (180, 130, 90, 65 and 45 nm).

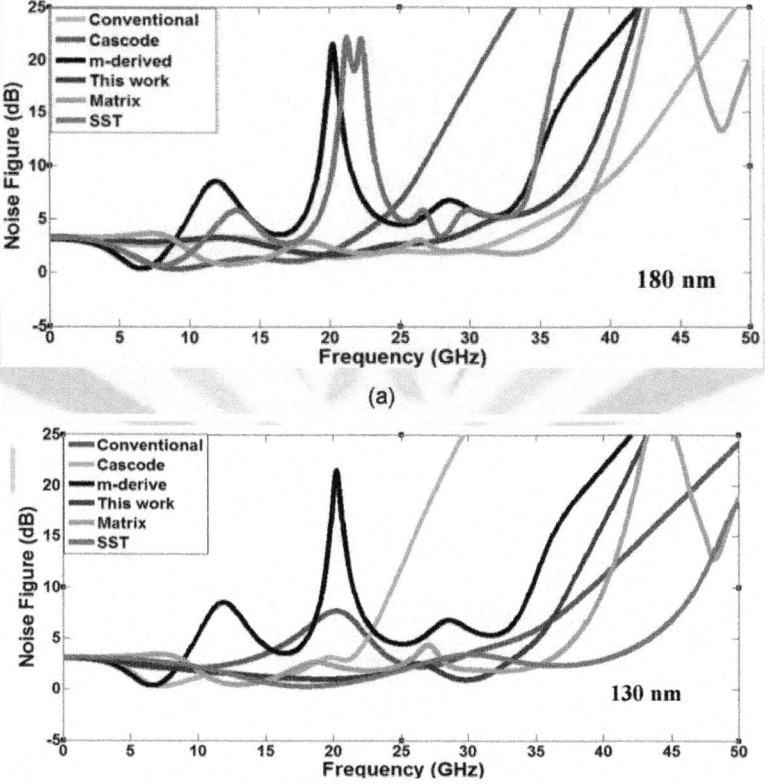

(a)

(b)

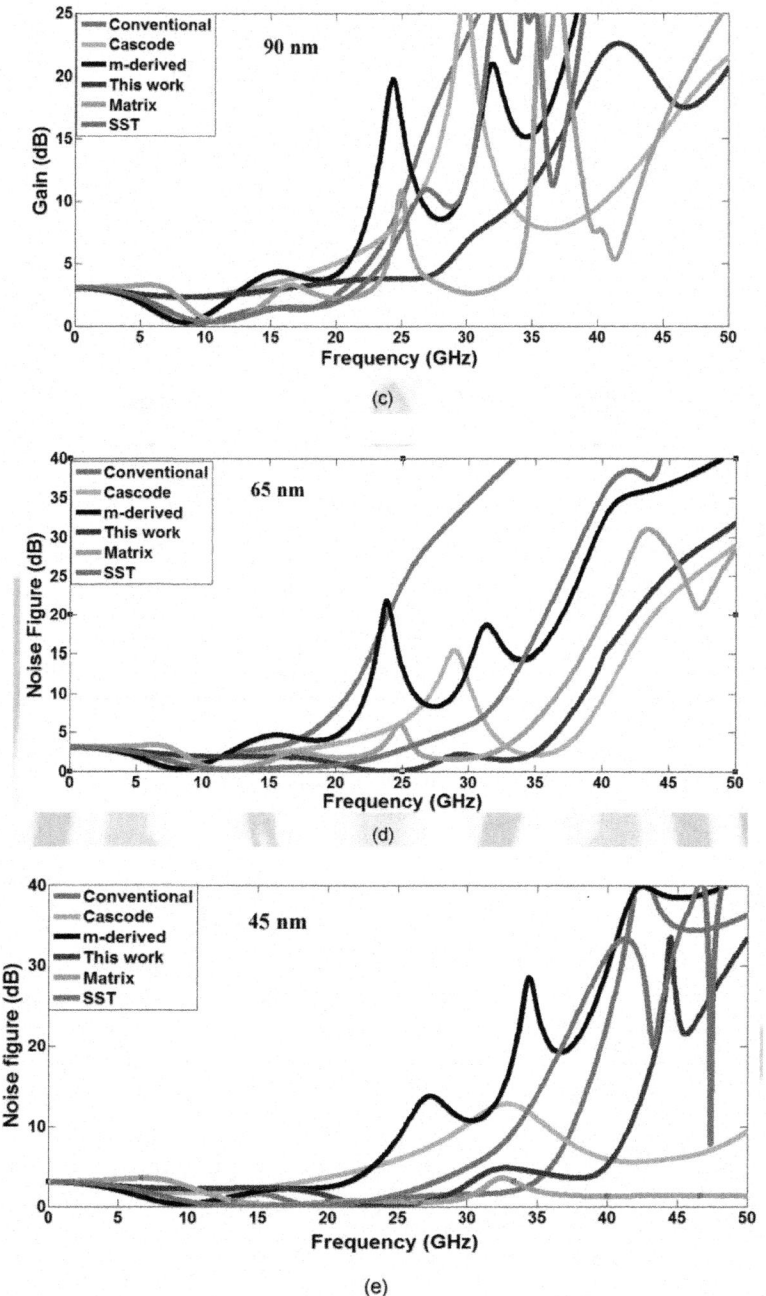

Figure13: Noise figure spectra of various DA configurations designed with 35 GHz cutoff frequency using (a) 180 nm standard, (b) 130 nm standard, (c) 90 nm standard, (d) 65 nm standard, (e) 45 nm standard.

The simulation is carried further to assess the NF spectrum of 35 GHz – and 80 GHz cut off frequency DAs and the results are displayed in Figs. 13 and 4.4, respectively. Tables 8 and 9 summarises the dependants of low frequency NF on CMOS standard for various DA configurations.

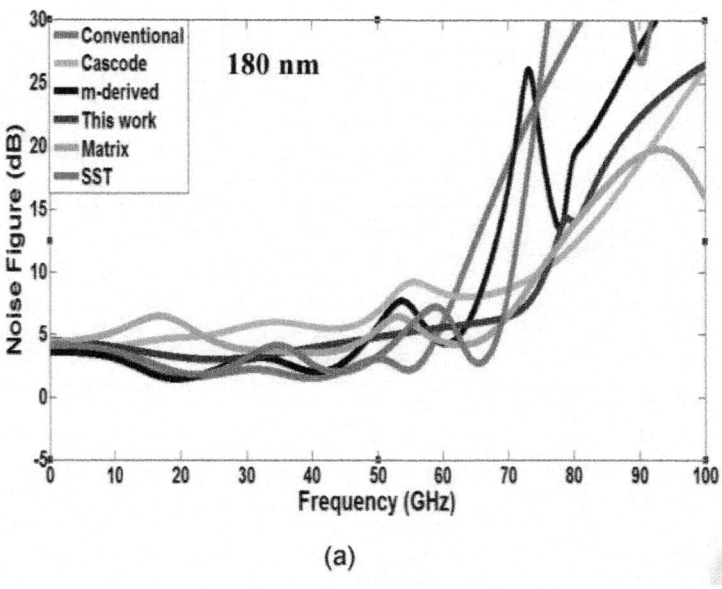

(a)

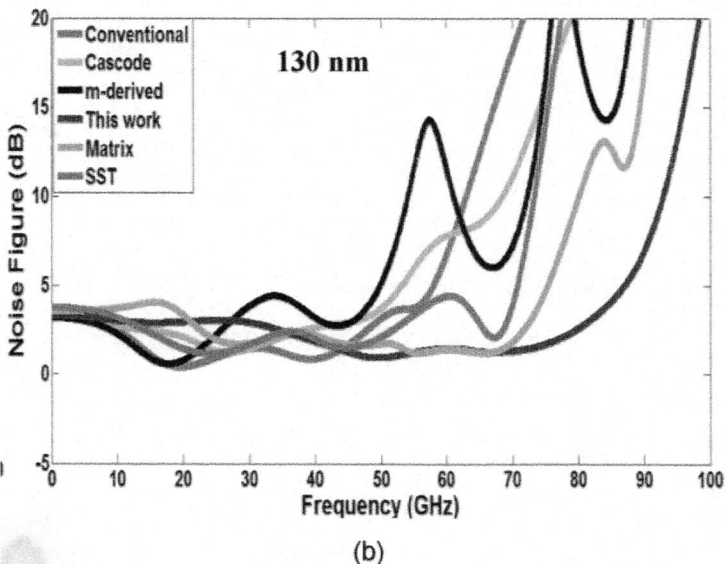

(b)

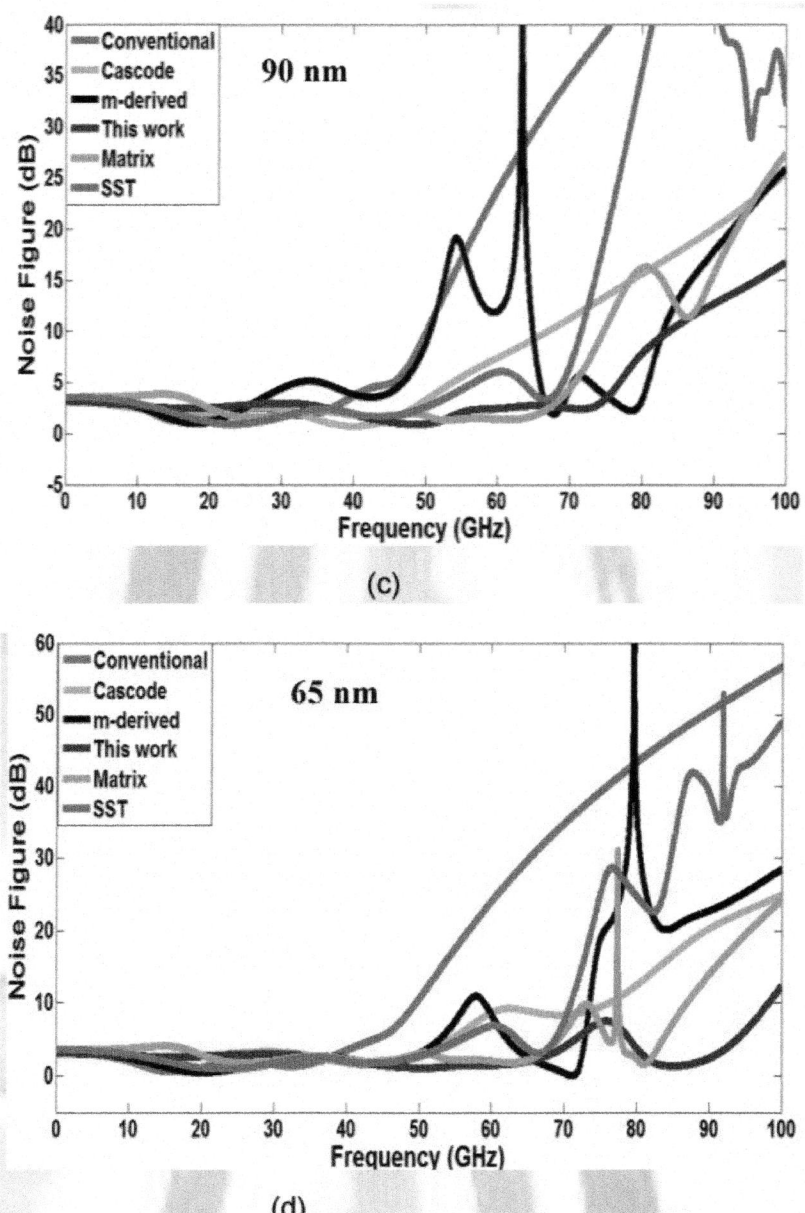

(c)

(d)

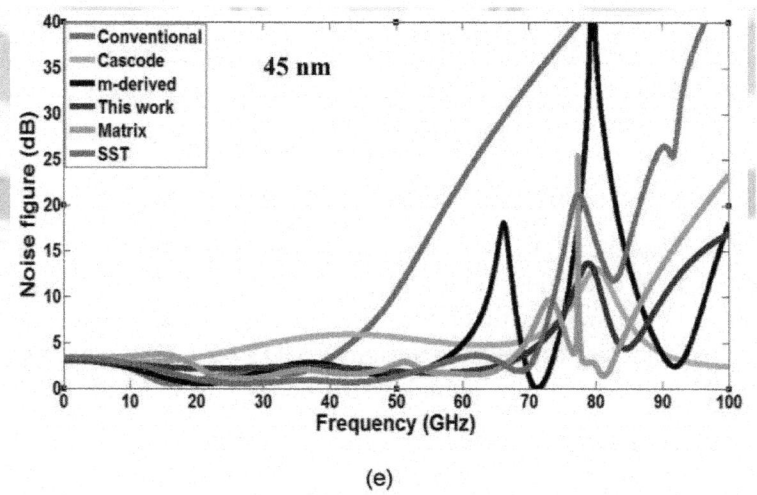

(e)

Figure14: Noise figure spectra of various DA configurations designed with 80 GHz cutoff frequency using (a) 180 nm standard, (b) 130 nm standard, (c) 90 nm standard, (d) 65 nm standard, (e) 45 nm standard.

Table 8: Noise figure of various DAs designed with 35 GHz cutoff frequency

Distributed Amplifier	DC Noise figure (dB)				
	180 nm	130 nm	90 nm	65 nm	45 nm
Conventional	3 dB	4 dB	16 dB	30 dB	3 dB
Cascode	15 dB	25 dB	14 dB	10 dB	9 dB
m-derived	5.1 dB	4.5 dB	8 dB	18 dB	1.7 dB
Matrix	2.4 dB	2.2 dB	3.7 dB	2.2 dB	4 dB
Shifted second tier	3 dB	3.6 dB	10 dB	5 dB	2 dB
This work	4 dB	1.5 dB	5 dB	1.7 dB	2 dB

Table 9: Noise figure of various DAs designed with 80 GHz cutoff frequency

Distributed Amplifier	DC Noise figure (dB)				
	180 nm	130 nm	90 nm	65 nm	45 nm
Conventional	18 dB	12 dB	26 dB	30 dB	36 dB
Cascode	7 dB	7.3 dB	10.1 dB	8 dB	5 dB
m-derived	18 dB	5 dB	0.8 dB	1 dB	0.3 dB
Matrix	2.5 dB	2.5 dB	1.5 dB	4 dB	2 dB
Shifted second tier	2 dB	3 dB	3 dB	4.5 dB	4 dB
This work	6 dB	1.7 dB	3 dB	3.5 dB	3.5 dB

From the results illustrated in Figs. 5 and 6 and Tables 8 and 9 one can find out that proposed DA gives low NF with almost flat spectrum when compared with other DAs. The value of the NF of this DA doesn't exceed 6 dB for all standards (180, 130, 90, 65 and 45 nm).

Effect of Transmission Line Cutoff Frequency

The cutoff frequency f_c of drain and gate transmission lines is usually used as one of the main entry design parameters for DAs. This section illustrates the depends of DA characteristic on the cutoff frequency. The results are reported for various DA architectures and various CMOS standards. The investigation is focused on low-frequency gain, low-frequency NF, and 3 dB BW.

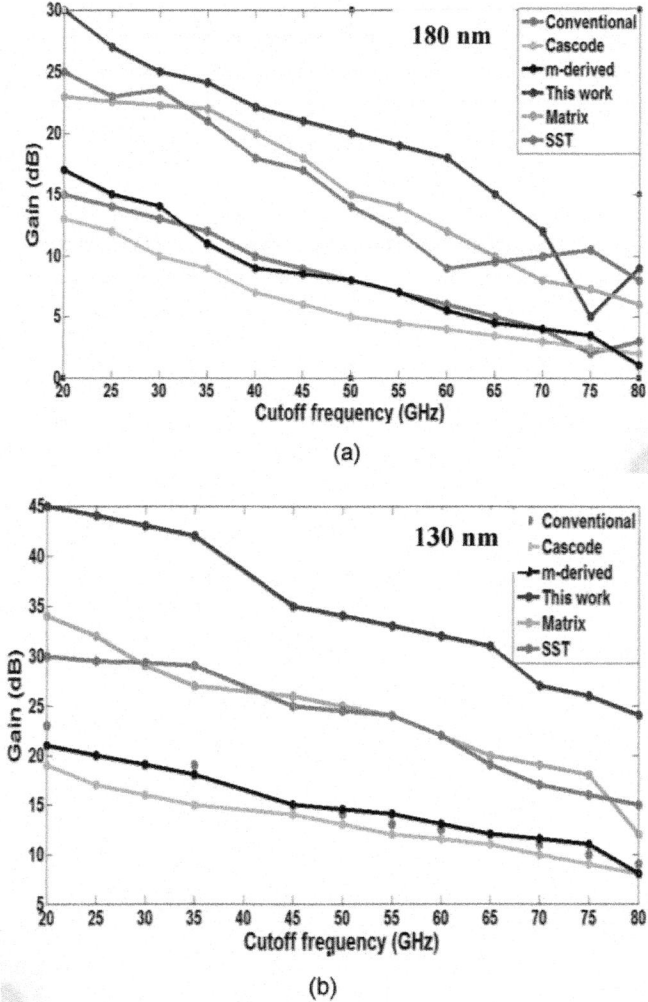

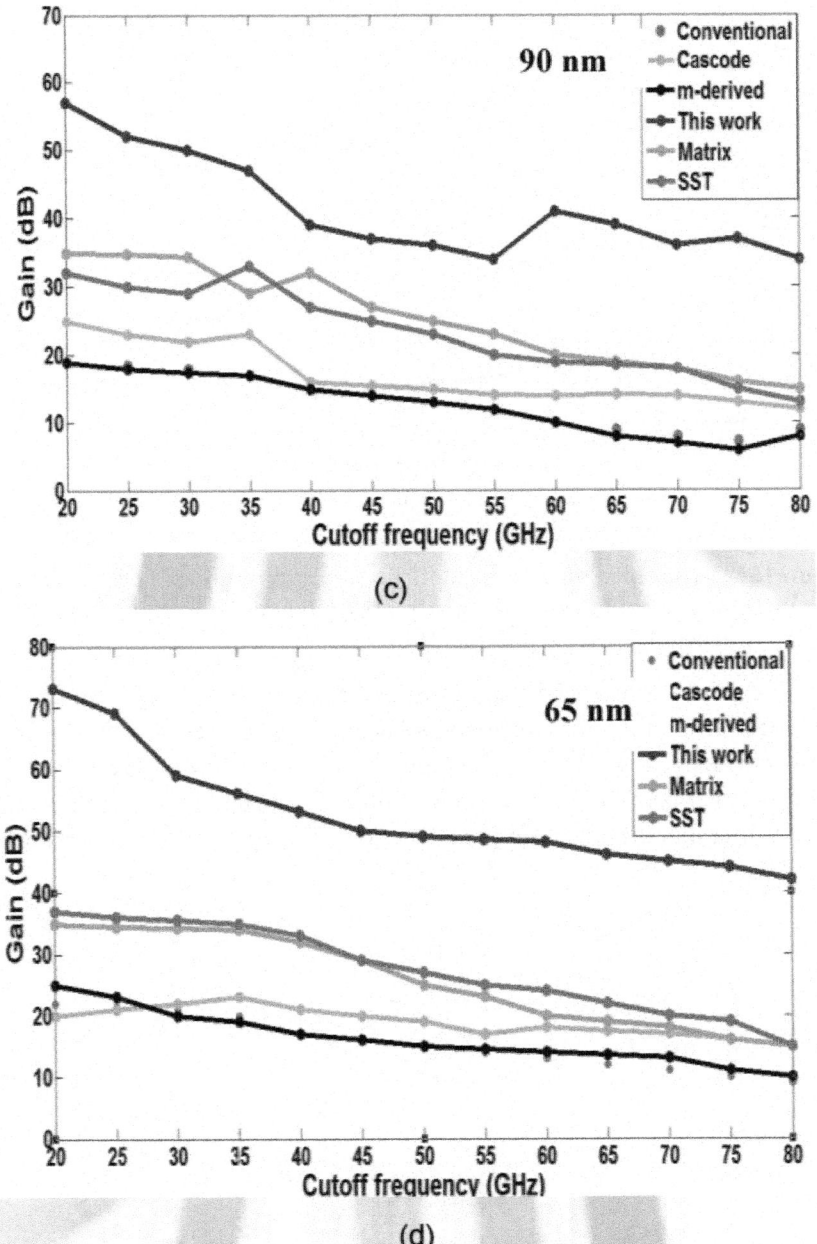

(c)

(d)

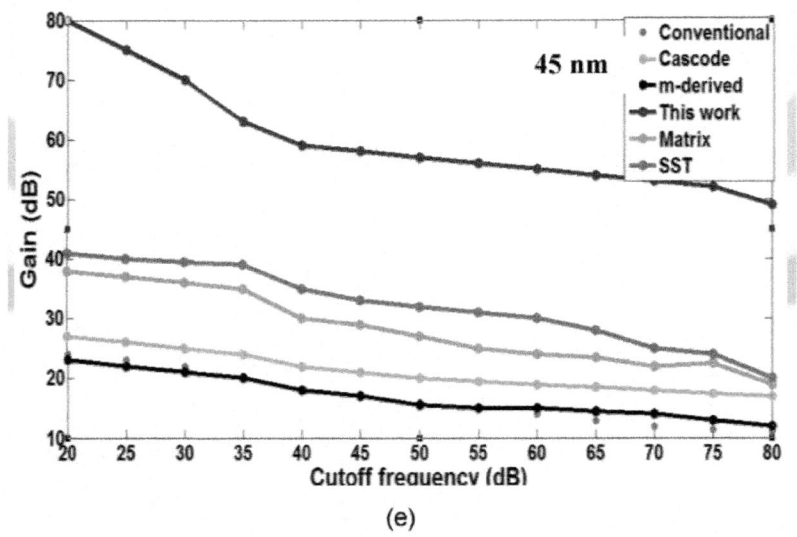

(e)

Figure15: Dependence ofgain on cutoff frequency for various DAs designed using CMOS technology of (a) 180 nm standard, (b) 130 nm standard, (c) 90 nm standard, (d) 65 nm standard and (e) 45 nm standard.

Figs. 15 a-e shows the dependence of the low-frequency gain on the line cutoff frequency for CMOS standards 180, 130, 90, 65 and 45 nm, respectively. Note that the proposed DA has the highest gain among the DAs considered here and this conclusion holds true for all cutoff frequency and CMOS standards. Note further that the amplifier gain decreases as the cutoff frequency increases.

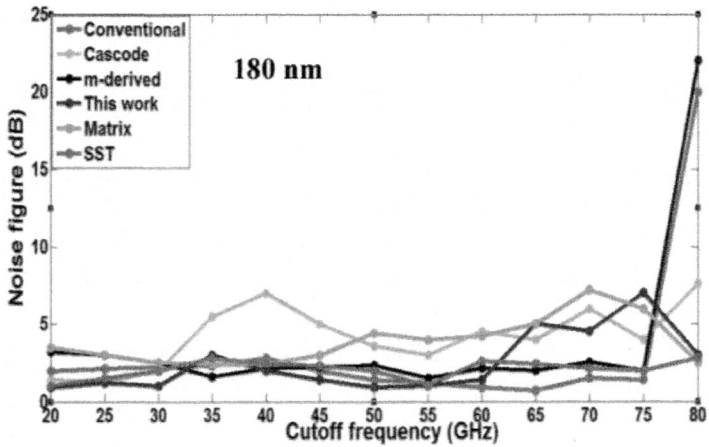

(a)

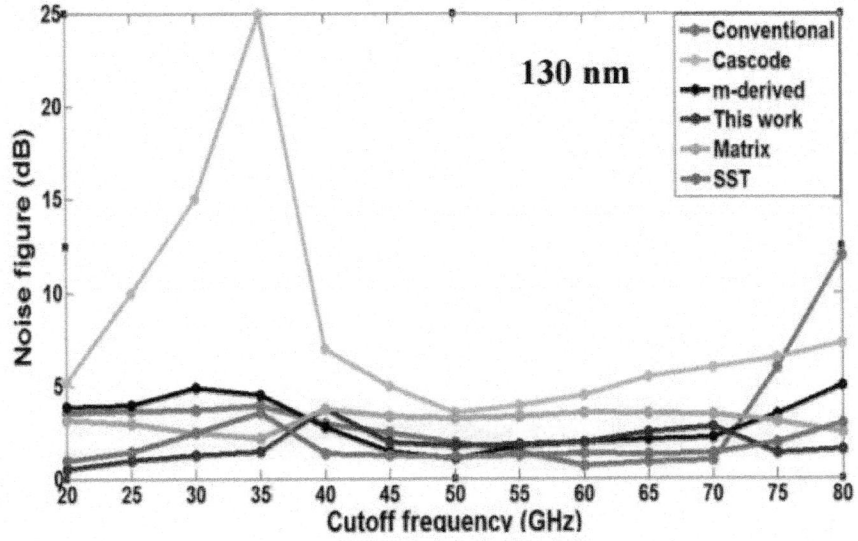

(b)

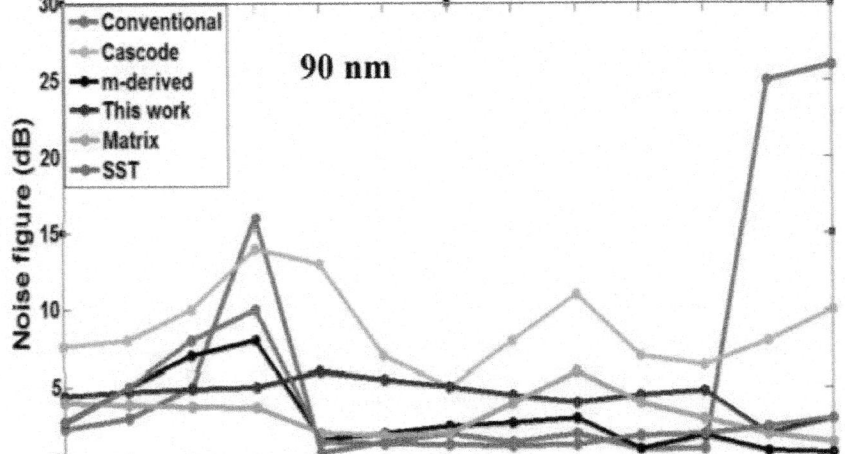

(c)

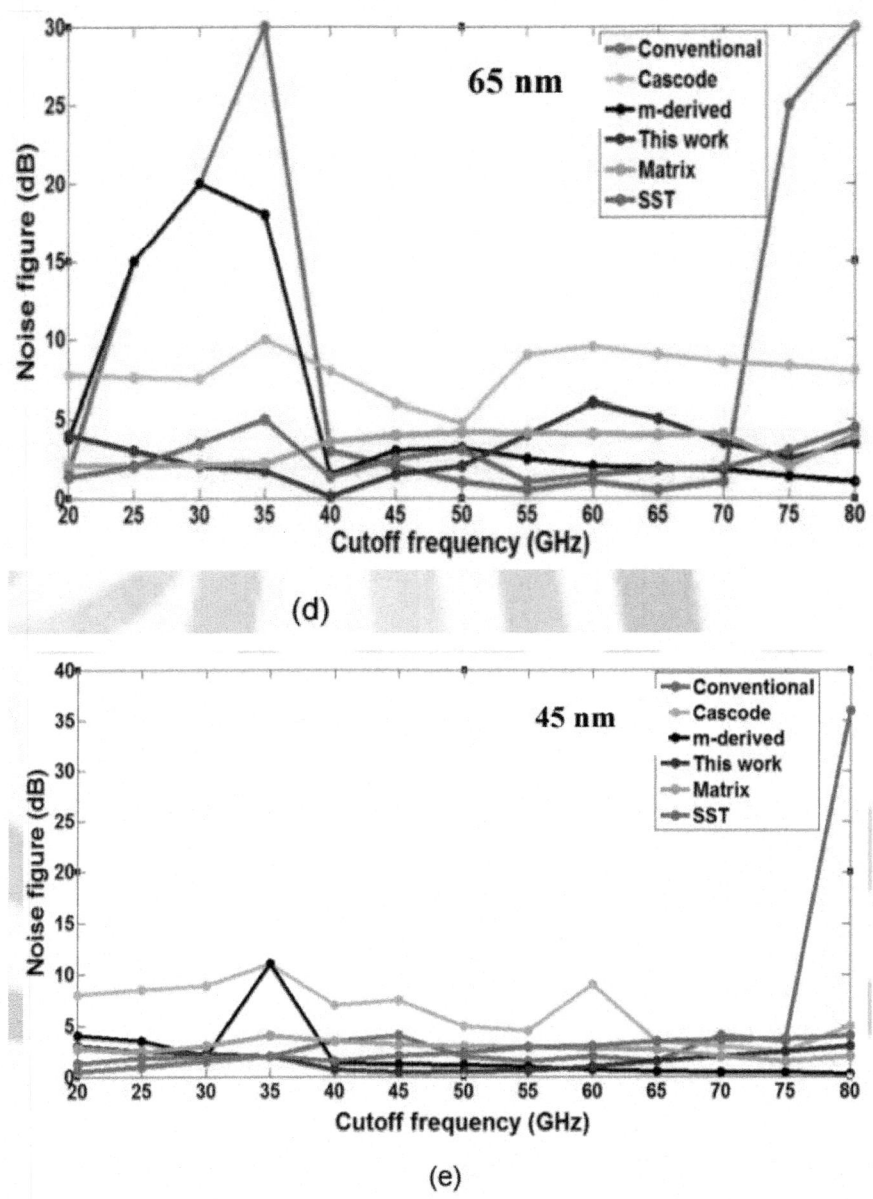

Figure 16: Dependence of noise figure on cutoff frequency for various DAs designed using CMOS technology of (a) 180 nm standard, (b) 130 nm standard, (c) 90 nm standard, (d) 65 nm standard and (e) 45 nm standard.

Variation of low-frequency NF with cutoff frequency is given in Fig. 16. The results are reported for various DA configuration and CMOS standards. Investigation the results in this figure highlights the following findings. The proposed DA is characterized by relatively low NF over all values of cutoff frequency and CMOs standards.

The calculation is carried further to estimate the 3 dB BW of the DAs and the results are displayed in Fig. 17 for various CMOS standards. The main conclusions drawn from this figure are

1. For a given cutoff frequency, the proposed DA generally offers the highest BW compared with other DAs and this effect is more pronounced for high CMOS standards. For example, at 35 GHz cutoff frequency and 180 nm standard, the BWs of the proposed, cascode, SST, m-derived, matrix and conventional DAs are 31, 30, 9.7, 7, 3.7 and 1.2 GHz, respectively. These values are to be compared with 31, 31, 10, 8, 4 and 3 GHz for 130 nm DAs designed with 35 GHz cutoff frequency.

2. In general, the BW of all DAs increases with cutoff frequency and this effect is more pronounced with the proposed DA. For example, if the DAs are designed with $f_c = 80$ GHz and 180 nm standard, then the BW are enhanced by 2.25, 1.66, 2.7, 3.85, 2.70 and 15.83 for the proposed, cascode, SST, m-derived, matrix and conventional DAs when compared with 35 GHz cutoff frequency counterparts.

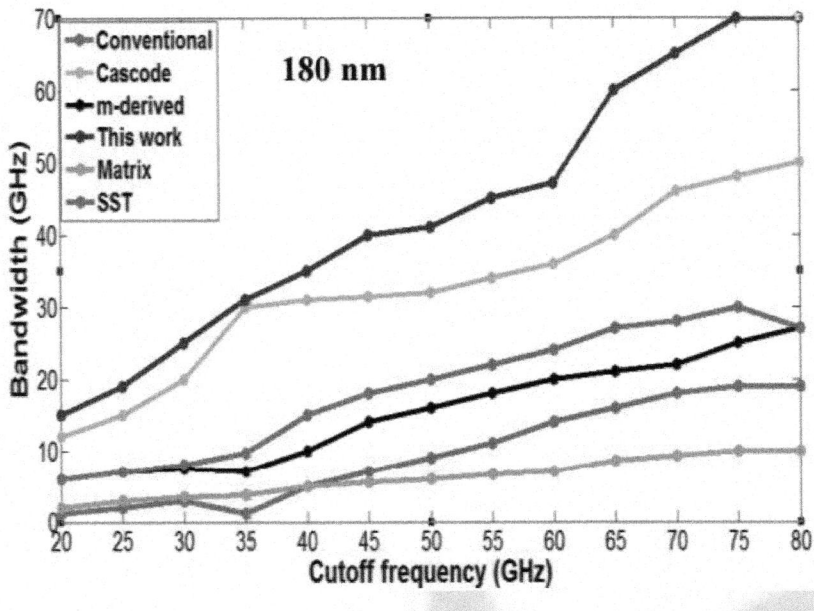

(a)

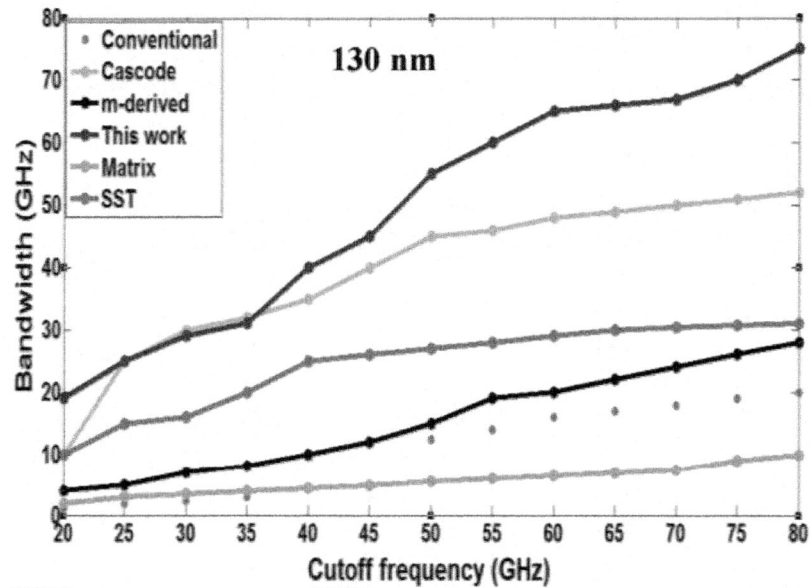

(b)

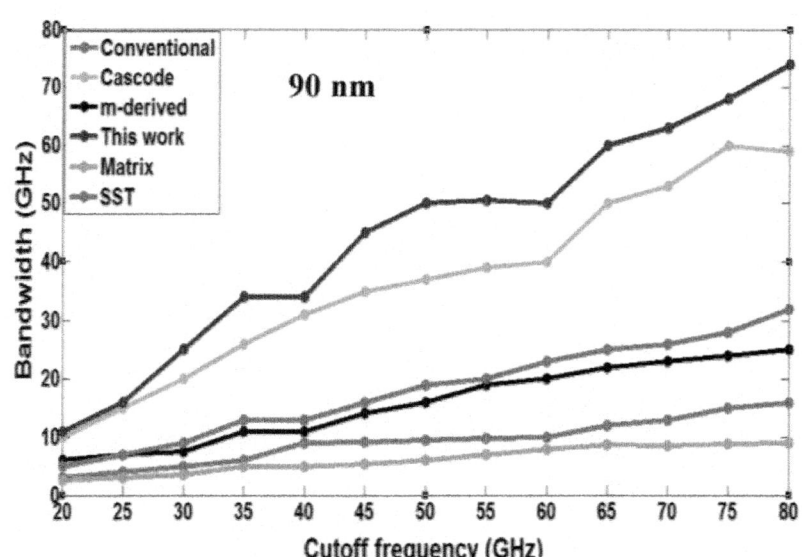

(c)

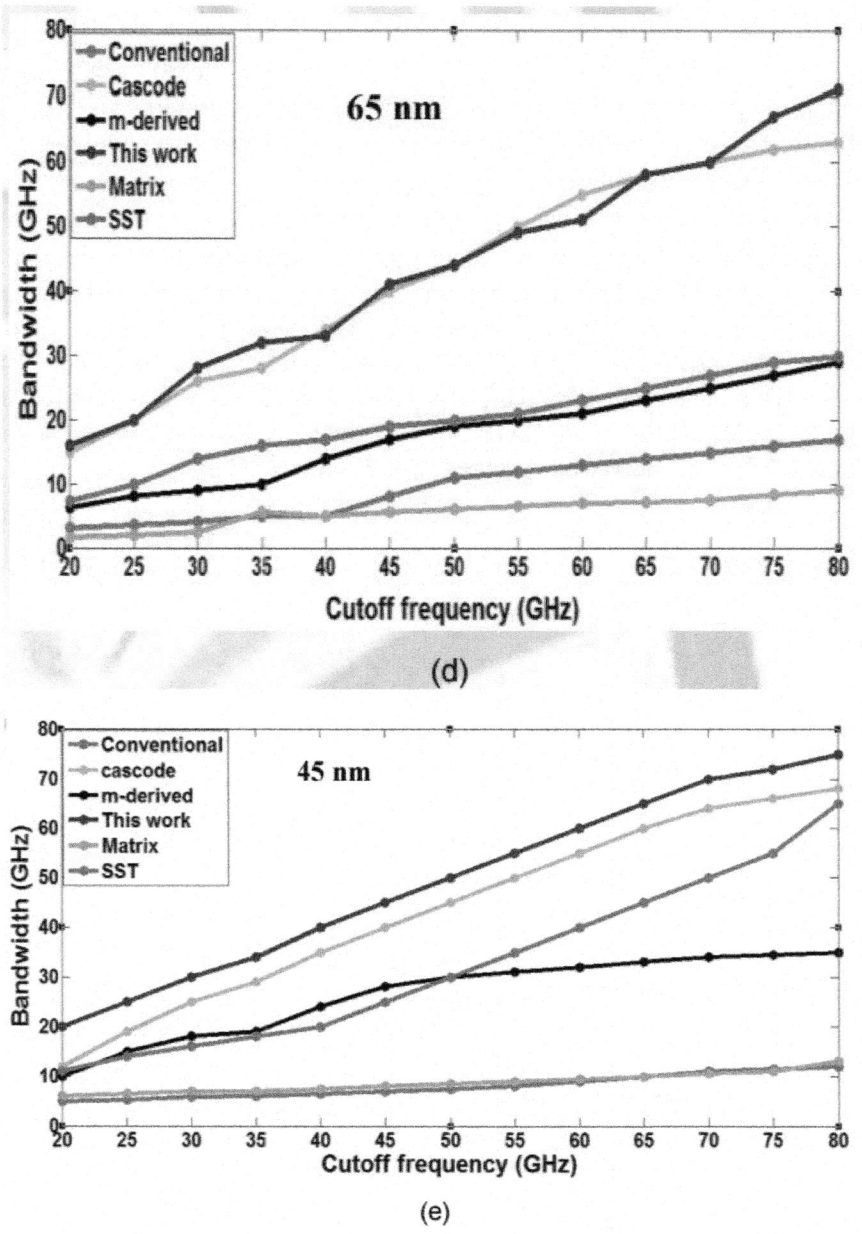

Figure17: Dependence of bandwidth on cutoff frequency for various DAs designed using CMOS technology of (a) 180 nm standard, (b) 130 nm standard, (c) 90 nm standard, (d) 65 nm standard and (e) 45 nm standard.

Designing the Proposed DAs for 40 and 100 Gb/s Operation

The results reported in the previous sections can be used as a guideline to design the proposed DA for front-end amplification in 40 Gb/s and 100 Gb/s optical receivers. The BW of the optical receiver is usually set equal to 0.7x bit rate as a hand of thumb estimate.

From Fig. 17, one can deduced the cutoff frequencies for various CMOs standards that can be used as a design parameter to achieve BWs corresponding to 40 and 100 Gb/s. This design parameter is used to deduce both geometric and characteristics parameters of the DAs. Tables 10 and 11 list the obtained results, for both 40 and 100 Gb/s DAs, respectively. Again various CMOS standards are used to estimate the DA parameters for both operating bit rates. The thickness of the oxide layer t_{ox} is calculated from the gate capacitance using a simple model based on parallel-plate capacitance,

$$t_{ox} = \frac{\varepsilon_{ox}}{C_{ox}}$$

$$\text{(11.a)}$$

$$\varepsilon_{ox} = 3.9 \; \varepsilon_o = 3.45 \times 10^{-11} \text{F/m}$$

$$\text{(11.b)}$$

Figs. 18a and 18b show the gain and NF spectra for 40 Gb/s DAs designed with different CMOS standards. The calculations are repeated in Figs. 19a and 19b for the designed 100 Gb/s DAs.

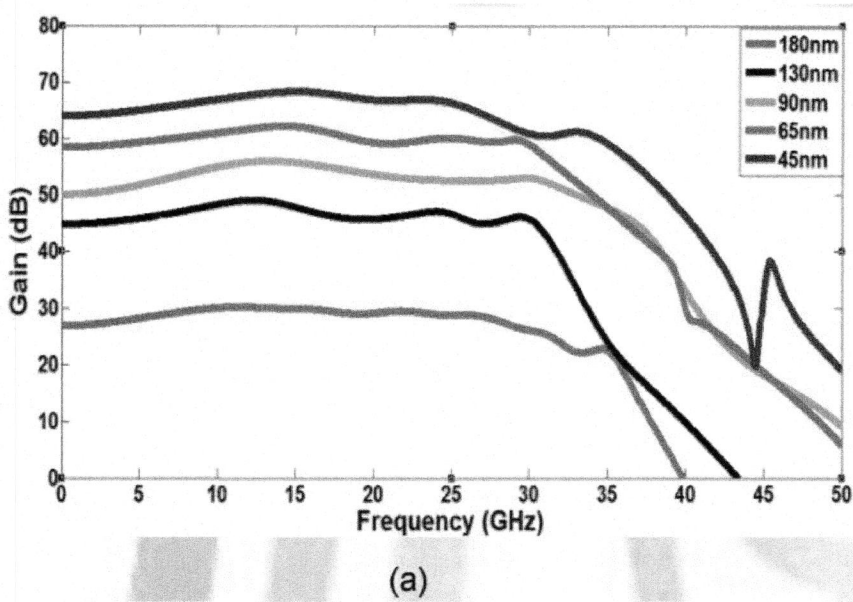

(a)

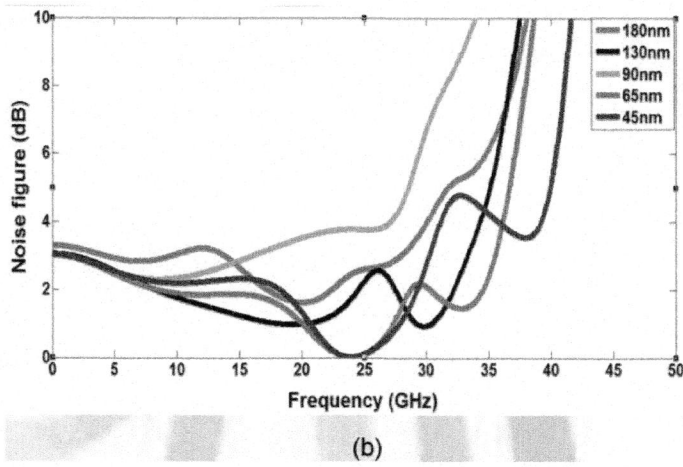

(b)

Figure 18: Gain and noise figure spectra for the proposed 40 Gb/s DA designed using different CMOS standards.

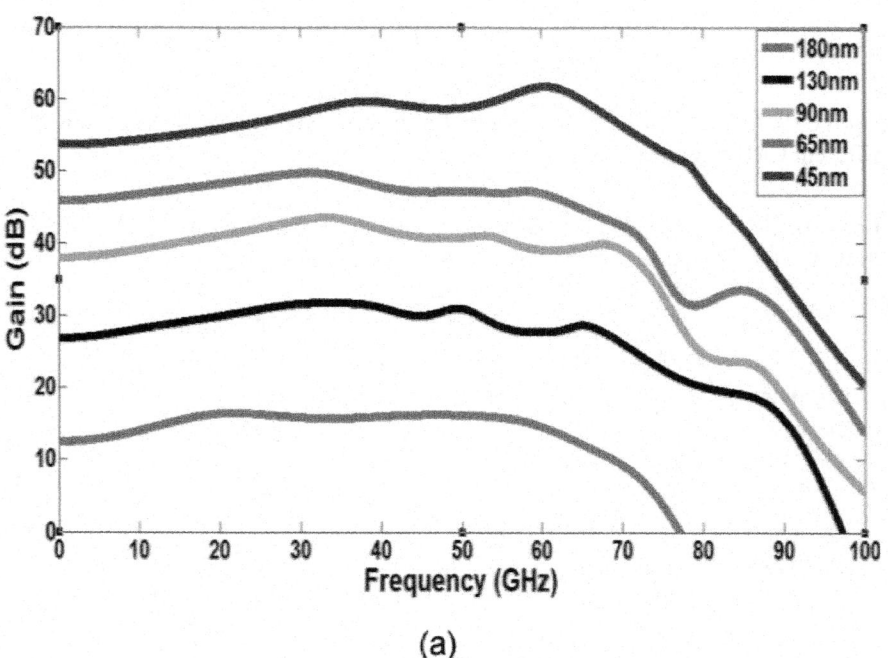

(a)

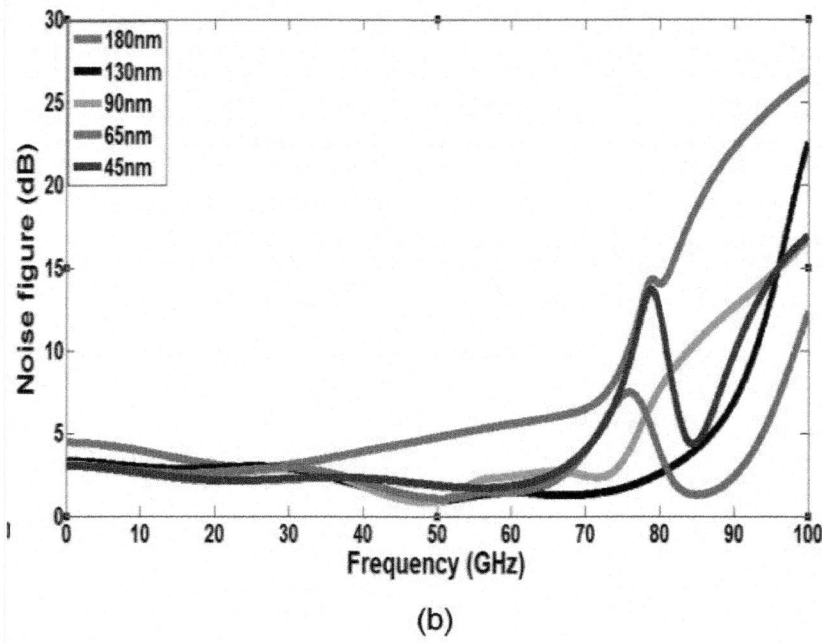

(b)

Figure 19: Gain and noise figure spectra for the proposed 100 Gb/s DA designed using different CMOS standards.

Table 10: Design and characteristics parameters of the proposed DA for 40 Gb/s operation. Cutoff frequency of 35 GHz is used in the design

Parameter	Value				
	180 nm	130 nm	90 nm	65 nm	45 nm
Gate length (nm)	180	130	90	65	45
Gate width (μm)	158	233	306	399	551
Oxide thickness (nm)	4	3.8	3.7	3.4	3.1
Transconductance (S)	0.061	0.142	0.293	0.559	1.159
Gain (dB)	25	42	47	59	62
Noise figure (dB)	3	1.5	3.5	1.7	2
Gain - BandwidthProduct (GHz)	551	3902	7611	28512	42772

From the results illustrated in graphs and tables above, one can find that the gain is increased by increasing the standard values, this can applied at both 35 and 80 GHz cutoff frequency, and the value of NF remains low. Gate width also increased by increasing the standard values, oxide thickens decrease by increasing the standard value, while transconductance increase by increasing standard value.

Table 11: Design and characteristics parameters of the proposed DA for 100 Gb/s operation. Cutoff frequency of 80 GHz is used in the design

Parameter	Value				
	180 nm	130 nm	90 nm	65 nm	45 nm
Gate length (nm)	180	130	90	65	45
Gate width (μm)	69	101	134	174	241
Oxide thickness (nm)	4	3.8	3.7	3.4	3.1
Transconductance (S)	0.026	0.062	0.128	0.244	0.507
Gain (dB)	10	23	34	42	49
Noise figure (dB)	6	1.2	3	3.5	3.5
Gain x Bandwidth Product (GHz)	221	1059	3708	8938	21137

CONCLUSIONS

A new distributed amplifier architecture has been introduced to achieve high flat gain and low noise figure over ultra wideband bandwidth. Investigation has been carried out to assess the performance of the proposed DA when it is implemented in various submicron CMOS standards process. The results have been used to design DAs for front-end amplification in 40 and 100 Gb/s optical receivers. The main conclusions drawn from this work are

1. The proposed DA offers the highest gain among various DAs investigated in this work.

2. The proposed DA offers a nearly flat gain over the wide bandwidth. The degree of gain flatness is the highest among the investigated DAs.

3. Designing the DAs with high cutoff frequency leads to gain reduction.

4. The noise figure of the proposed DA does not exceed 6 dB for various CMOS standards.

REFERENCES

1. B. Razavil, "Integrated circuit for optical communication", Second edition, Johenwiely and Sons Inc, 2012.

2. K. Kurokawa, "Optical fiber for high power optical communication", Journal of Crystals, Vol. 2, No. 2, PP. 1382-1392, 2012.

3. V.S. Bagad, "Optical fiber communication", First edition, MC. Grow. Hill, 2009.

4. R. Zhou, M. Anandarujah, R. Maher, M. Paskov, D. Lavery, B.C. Thomsen, S. J. Savory and L.P. Barry, "80 -km coherent DWDM-PON on

20-GHz grid with injected gain switched comb source", IEEE Photonics Technology Letters, Vol. 26, No. 4, PP. 364-366, 2014.

5.   G. Cossu, F. Bottoni, R. Corsini, M. Presi, and E. Ciaramella, "40 Gb/s single R-SOA transmission by optical equalization and adaptive of DM", IEEE Photonics Technology Letters, Vol. 25, No. 21, PP. 2119 - 2122, 2013.

6.   Q. Guo and V. Tran, "Demonstration of a 40 Gb/s wave length reused WDM – PON using coding and equalization", IEEE Journal of Optical Communication and Networking, Vol. 5, No. 10, PP. 119 – 124, 2013.

7.   J. Yu, and X. Zhou, "Ultra–high–capacity DWDM transmission systems for 100 G and beyond", IEEE Communication Magazine, PP. 556 – 563, 2010.

8.   J.C. Cartledge and A.S. karar, "100 Gb/s intensity modulation and direct detection", Journal of Lightwave Technology, Vol. 32, No. 16, PP. 2809-2813, 2014.

9.   N.A. Quadir, P.D. Townsend and P. Ossieur, "An inductorless linear optical receiver for Gbaud/s (40Gb/s) PAM-4 modulation using 28nm CMOS", IEEE International Symposium on Circuits and Systems, Melbourne, PP. 2473-247, 2014.

10. J.M. Gene Bernaus, "Fiber-optic communication", Departament de Teoria del Senyal i Comunicacions, UniversitatPolitecnica de Catalunya, 2010.   (http://ocw.upc.edu/sites/default/files/materials/15011956/oc_chapter_i_100224-2804.pdf)

11. G. Wolf, S. Demichel, R. Leblanc, F. Bdache, R. Lefevre, G. Dambrine and H. Happy, "A metamorphic GaAS HEMT Distribut Amplifier with 50 GHZ bandwidth and low noise for 40 Gb/s optical receivers" , InstitutElectronique de Microelectronique de Nanotechnologie (IEMN), DepartementHyperfrequences et semiconductors, PP. 93-94, 2005.

12. J. C. Chien and L. –H. Lu, "40-Gb/s high-gain distributed amplifiers with cascaded gain stages in 0.18-μm CMOS", IEEE Journal of Solid-State Circuits, Vol. 42, No. 12, PP. 2715-2724, 2007.

13. K. Moez and M. Elmusry, "A new loss compensation technique for CMOS distributed amplifiers", IEEE Transactions on Crcuits and Systems, Vol. 56, No. 3, PP. 185-188, 2009.

14. K. Entesari, A.R. Tavakoli and A. Helmy, "CMOS distributed amplifier with extended flat bandwidth and improved input matching using gate line with coupled inductors", IEEE Transactions on Microwave Theory and Techniques, Vol. 57, No. 12, PP. 2862 – 2870, 2009.

15. A. Ghadiri and K. Moez, "Gain-Enhanced distributed amplifier using negative capacitance", IEEE Transactions on Circuit and Systems, Vol. 57, No. 11, PP. 2834 – 2841, 2010.

16. C. K. Chien, H. H. Hsich, H. S. Chen and L. H. Lu, "A transimpedance amplifier with tunable bandwidth in 0.18 μm CMOS", IEEE Transactions on Microwave Theory and Techniques, Vol. 58, No. 3, PP. 498-504, 2010.

17. J. Kim and J. F. Buckwalter, "A 92 GHz bandwidth distributed amplifier in 45nm SOI CMOS Technology", IEEE Microwave and Wireless Components Letters, Vol. 21, No. 6, PP. 329 – 331, 2011.

18. A. Jahanian and P. Heydari, "A CMOS distributed amplifier with distributed active input balun using GBW and linearity enhancing techniques", IEEE Transactions on Microwave Theory and Techniques, Vol. 60, No. 5, PP. 1331 – 1340, 2012.

19. C. Feng, W. M. Lim and K. S. Yeo, "A compact 2.1 – 39 GHz self – biased low – niose amplifier in 65nm CMOS technology", IEEE Microwave and Wireless Components Letters, Vol. 23, No. 12, PP. 662 – 669, 2013.

20. J. C. Kao, P. C. Huang and H. W. fellow, "A novel distributed amplifier with high gain, Low noise, and high output power in 0.18 m CMOS technology", IEEE Transactions on Microwave Theory and Techniques, Vol. 61, No. 4, PP. 1533 – 1541, 2013.

21. M. K. Cho, J. G. Kim and D. Baek, "A switch less CMOS Bi- Directional distributed gain amplifier with multi – octave bandwidth" , IEEE Microwave and Wireless Component Letters, Vol. 23, No. 11, PP. 611 – 613, 2013.

22. K. Kim and C. N. Fellow, "A concurrent Ku / K / Ka Tri-band distributed power amplifier with negative – resistance active notch using SiGeBiCMOS process" , IEEE Transactions on Microwave Theory and Techniques, Vol. 62, No. 1, PP. 125-135, 2014.

23. J. Kim, J.F.Buckwalter, "A 92 GHz bandwidth distributed amplifier in 45 nm SOI CMOS technology" ,IEEE Microwave and Wireless Component Letters, Vol.21, No.6,June 2011.

24. Y. Cao, "Predictive technology model for robust nanoelectronic design", 1st Edition, Predictive Process Design Kids, 2011.

25. S. Serdar and C. Smith, "Microelectronic circuits", 6th Edition, Oxford University Press, 2010.

Chapter 6

VOLTAGE DIFFERENCING BUFFERED/INVERTED AMPLIFIERS AND THEIR APPLICATIONS FOR SIGNAL GENERATION

Roman SOTNER[1], Jan JERABEK[2], Norbert HERENCSAR[2]

[1]Dept. of Radio Electronics, Brno University of Technology, Technicka 3082/12, 616 00 Brno, Czech Republic

[2]Dept. of Telecommunications, Brno University of Technology, Technicka 3082/12, 616 00 Brno, Czech Republic

ABSTRACT

This paper presents some interesting new applications in the field of analog signal processing focused on signal generation. A novel modifications of recently developed and studied family of active elements, called voltage differencing buffered amplifier (VDBA) and voltage differencing inverted buffered amplifier (VDIBA) are discussed. Our attention is focused on simple application of active elements like dual output VDBAs (DO-VDBAs) and fully balanced VDBAs (FB-VDBAs), where one or two z terminals and always voltage outputs of both polarities are present. The last modification of VDBA allows additional electronic control of voltage gain in frame of active element except standard transconductance control. Discussed active elements were used to build very simple multiphase oscillators with minimal complexity as a simple non-tunable alternative to classical conceptions utilizing lossy integrators in phase-shifted loop. Linearly tunable quadrature differential mode (balanced) oscillator or balanced simple triangle and square wave generator were chosen as other useful examples. Features of proposed circuits are discussed and selected examples verified and evaluated by computer simulations with appropriate low-voltage TSMC 0.18 μm CMOS technology models.

INTRODUCTION

There are many active elements in the field of analog signal processing, however new ideas in this area help to provide further improvements in order to obtain more effective and interesting circuits. Plenty of novel active elements were introduced by Biolek et al. [1]. Nevertheless, many of them are only hypothetical elements and offer further research mainly from practical point of view. Main aim of this paper is to show simple alternatives allowing multiphase generation, simple quadrature generation and square and triangle wave differential mode generation with help of modifications of novel active elements known as voltage differencing buffered amplifier (VDBA) [1]-[4] or voltage differencing inverted amplifier (VDIBA) [5]. Presented active elements and their modifications allow interesting utilization and design of more profitable or more challenging application (differential mode operation, advanced electronic control, etc.) in comparison to classical VDBA or VDIBA elements.

This paper is divided to three parts. The first part deals with behavioral principles of used active elements. The second part discusses possible applications in analog signal generation (there are three areas: multiphase oscillators, quadrature/differential mode oscillator and simple functional generator). The third part deals with possible structures of discussed active elements that are mostly used in presented designs, their behavior and simulation results, and features of selected of proposed applications (multiphase oscillator and functional generator).

Several approaches to design of multiphase oscillators are available in the open literature. The first way uses integrators or similar selective sections in the loop [6]-[8]. Classical integrator phase shifted loops [6]-[8] were key circuits for generation signals with several phase shifts between them for many years. However, such circuits require many lossy integrators or selective sections (it depends on the number of multiplicand of basic shift - p/2, p/4 or p/6 typically). Such structures were based on classical simple active elements like current conveyors (CC) [6], [8], current amplifiers [7], etc. However, complexity and number of sections required for such type of oscillators is high. It means at least 4 sections (4 capacitors and 4 active elements) for phase shifts 45, 90, 135 and 180 degrees for example in classical loop way of synthesis. The second possibility of construction of multiphase oscillators (unfortunately number of phases is more limited) utilizes allpass sections (for example [9]-[11]). Here simple (operational amplifiers for example [9]) or more complex active elements like current differencing transconductance amplifiers (CDTAs) have been employed [10]. These already studied circuits usually allow quadrature phase shifts (ormultiple of p/2) [11]. However, they usually do not allow nonstandard phase shifts such as 45, 90, 135 and 180 degrees simultaneously in comparison to classical types employing phase shifted integrator loops. The last used approach is based on combination of active all-pass section with lossy or lossless integrators or differentiators or similar sections. This approach is also in field of our interest in design of multiphase oscillator used in this paper. Unfortunately, our study below shows that not all published oscillator structures have been analyzed carefully in the past and many other circuits with similar features probably exist in the literature. In addition, the analysis of possible outputs of oscillator circuits or analysis of the relation between them is also not provided in many papers. In the following paragraphs we summarized the most important features of approaches used in recent literature and also by us:

a) Phase Shifted Loops with Integrators or Similar Selective Sections and Typical Examples

Advantages: tunability; easy synthesis; easy CO (condition of oscillation) control by gain of amplifier cascaded in the loop; even and odd phases of p/2, p/4 or p/6.

Disadvantages: many sections (one section = the lowest available phase shift) for required number of phase shifts; higher power consumption (many active elements); in some cases simultaneously matched control of parameter of each elementary transfer in each section - complication of FO (frequency of oscillation) and CO control (some types are controllable by simultaneous changes of each time constant [7], some require special matching conditions [6], [8]).

Abuelmaatti et al. [6] proposed multiphase currentmode oscillator employing controllable current conveyors in lossy integrators. Loop structure utilizes two capacitors per section and FO, CO control require matching condition between capacitor values. Voltage-mode multiphase oscillator based on classical operational amplifiers was presented by Gift [9] where all-pass sections with adjustable time constants (by R and C values only) were used for construction. However, number of passive elements seems to be very high for large number of phase shifts. Souliotis et al. utilized current amplifiers in cascade of lossy integrators creating current-mode multiphase oscillator [7]. Adjusting of FO is quite simple by bias current of active elements. Kumngern et al. [8] also built their current-mode oscillator with lossy integrators utilizing current conveyors, where intrinsic resistance of the x-terminal and gain between z and x terminal simultaneously is possible. However, matching between capacitors is also required for realization of higher number of phases and therefore accuracy of such matching condition influences also accuracy of the phase shift.

b) All-Pass Section in the Loop Based Oscillators and Typical Examples

Advantages: simpler circuits; lower number of passive and active elements (in many cases two sections are sufficient [10]); lower power consumption.

Disadvantages: complicated tunability or CO control and matching in parameter values is required in many cases; accuracy of relation (transfers) between available outputs for accurate phase shifts required; phase shifts are available as multiple of p/2 almost in all cases.

Keskin et al. [10] published the perfect example of oscillator based on two all-pass sections employing two CDTAs and 6 passive elements that serve for CO and FO control. The oscillator produces output signals also in form of currents. The discussed circuit provides quadrature phase shifts only. Similarly, Songsuwankit et al. [11] also deal with phase shifter-based (all-pass section consists of three OTAs and one floating capacitor) oscillator design. Here, arbitrary setting of phase shift is possible, but the proposed oscillator provides only two outputs.

c) All-Pass Sections in Combination with Integrators or Integrators/Differentiators in Simple Loops and Typical Examples

Advantages: similar to the previous group; produced phase shifts are available similarly as in phase shifted loop oscillators (it depends on construction of particular circuit); CO control independent on FO in some cases.

Disadvantages: similar to the previous discussion; if FO is controllable by specific parameter it influences phase or at least amplitude relations.

Keawon et al. [12] proposed an oscillator employing single current controlled current differencing transconductance amplifier (CCCDTA), where two capacitors are required. Control of FO and CO is established by intrinsic resistance and tranconductance control. Output signals are in form of currents. Jaikla et al. [13] designed a circuit utilizing single CDTA and three passive elements. The independent CO and FO control is not possible and their adjustment is possible only by passive elements (also floating capacitor). Quadrature oscillator produces signals in form of currents. Also Pandey et al. [14] presented a circuit, which produces currents in quadrature phase shift with two CDTAs and two passive elements, where control is given by transconductances. Herencsar et al. [5] also discussed an oscillator, where one simple all-pass section and lossy integrator employing two VDIBA elements and three passive elements with control of CO were used. Keskin et al. [15] also presented an oscillator producing voltage signals, which is based on two current differencing buffered amplifiers (CDBAs) in which 8 passive elements are required and independent control of CO is difficult. Songkla et al. [16] presented an oscillator, where three current controlled current conveyors (controlled by intrinsic resistance) of second generation and two passive elements generate current output signals. Minaei et al. [17] utilized differential voltage current conveyor (DVCC) and the circuit requires three DVCCs, four passive elements, and produces voltage output responses. All discussedexamples provide quadrature outputs. None of them show production of phase shifts such as 45, 90, 135, 180 degrees, which is the main contribution of our circuits. General conclusions from above discussions are that tunability of cascaded phase shifted loop is better, but complexity is several times higher. Therefore, simpler structures that utilize another design approaches are more interesting in some cases. Therefore, our intention in this paper is to design simpler solutions that have not been developed so far. The next part of our contribution deals with quadrature oscillator design. Important quadrature oscillators with independent CO and FO adjusting, grounded capacitors, and voltage output responses are compared in the following text.

Many from recently investigated structures use passive elements for control of CO and FO, therefore their replacement by electronically adjustable equivalents is necessary. Soliman [18] utilizes current conveyors (two or three) and 5-6 passive elements in solution of an oscillator, where differential output signals are not easily available. Oscillator presented by Herencsar et al. based on generalized current follower transconductance amplifier (GCFTA) and voltage buffer employs two active and four passive elements. However, it does not provide linear control of FO (produced amplitude is dependent on tuning process) [19]. Tuning is realized by changes of passive elements. Similar type of FO and CO control is used in works [20]- [22] as well. Gupta et al. [20] developed an oscillator based on current and voltage followers as active elements (2-4 in the proposed circuits) and 5 passive elements. Amplitude of generated signal was influenced by FO adjusting without possibility of linear control. Oscillator with differential output signals and quadrature phase shift is proposed by Biolkova et al. in [21].

The circuit is based on two dual output current inverter buffered amplifiers (DO-CIBAs) and 5-6 passive elements. The FO control without influence on produced amplitudes is linear, but simultaneous change of floating resistor values is necessary. Lahiri [22] proposed an oscillator employing three current feedback amplifiers (CFAs) and 6 passive elements, where adjusting of resistor values is the only way how to control CO and FO. Lahiri et al. [23] also proposed another oscillator based on single current conveyor transconductance amplifier (CCTA) and four passive elements, where electronic control is realized by g_m. Nevertheless, generated amplitude is influenced by tuning of FO and dependence of control is not linear. The same features were achieved in solution presented in [24], where two CDTAs and three passive elements were used. Rodriguez-Vazquez et al. [25] presented an oscillator based on 3-4 transconductors (OTAs) and two capacitors. Linear control of FO without influence on generated amplitudes is possible electronically by transconductances. The oscillator in [26] utilizes special configuration of three specially modified CFAs and five passive elements, which allows linear control of FO without impact on generated amplitudes. Digital control of such application has been also discussed in the past, for example by Alzaher et al. [27], where 3 or 5 active elements (current amplifiers and voltage buffers) and 6 passive elements were used and linear control of FO without impact on generated amplitudes is allowed. Interesting digitally controllable solution was presented by Biolek et al. [28], where two active elements - so called z-copy controlled gain current differencing buffered amplifiers (ZC-CG-CDBAs), five passive elements are used and allow linear control of FO.

Several solutions of quadrature oscillators, where control procedure was focused on current and voltage gain adjusting in frame of active elements called controlled gain current follower differential output buffered amplifiers (CG-CFDOBA), controlled gain current inverter buffered amplifiers (CG-CIBAs) and controlled gain current amplified voltage amplifier (CG-CVA), are discussed in [29] (five passive and two active elements are used in the proposed oscillators). Some of them allow differential output responses and linear control of FO. Galan et al. [30] utilizes four dual output OTAs (g_m control) and four capacitors to achieve fully differential output oscillator. The first note about utilization of FB-VDBA in differential (balanced) quadrature was discussed by Bajer et al. [31].

The oscillator consists of two FB-VDBAs, two resistors, four grounded capacitors and allows linear electronic control of FO and independent control of CO (unfortunately by a floating resistor). Oscillator structure presented in this contribution seems to be very economical in comparison to above discussed circuits, because only single DO-VDBA and dual output controlled gain voltage differencing buffered voltage amplifier (DO-CG-VDBVA) and three passive elements (only one of them is floating) are required. The proposed oscillator provides differential outputs, linear control of FO by simultaneous adjusting of both g_m, and CO control by adjustable voltage gain. In some above discussed solutions utilizing other type of active elements the potential possibility to obtain voltage or even voltage differential (balanced) outputs also exists, but additional current to voltage conversion or voltage buffering/inversion (out of active element) is required. However, if it is possible, than these discussed oscillators have other drawbacks such as control by passive elements [21] or high number of active elements [30].

Principle of triangle and square generator (called functional) is known very well. There were interesting attempts to build very simple generator from active elements like basic operational amplifiers, current conveyors, etc. [1]. However, these simple circuits have some drawbacks (many passive elements) and lack of electronic controllability. The following discussion deals with typical examples.

Biolek et al. [32] introduced an interesting circuit with single CDTA, 3 resistors and one capacitor which provides FO in range of MHz tunable by resistor value. De Marcellis et al. [33] proposed generator employing 2 CCIIs, 6 resistors and one capacitor. Control of FO is alsopossible by adjusting of resistor value. Chien et al. [34] presented solution based on two differential voltage current conveyors (DVCCs) [1], three resistors and control of duty cycle is also allowed. Almashary et al. [35] presented a generator, where 2 CCIIs, 3 resistors, and 2 capacitors are required. Tunability of FO is possible by value of resistor. Pal et al. [36] employed two CCIIs, three resistors and floating capacitor in his approach. Two current feedback operational amplifiers (CFOAs) [1] with two capacitors and two resistor utilized generator presented by Saque et al. [37]. Minaei et al. [38] introduced similar approach based on CFOAs and DVCCs. Two operational transresistance amplifiers (OTRAs), three resistors and one floating capacitor based generator is shown by Lo et al. [39].

Electronically controllable active elements allow better performance in these types of generators. Works [40]- [42] brought key information for design of generators with transconductance (OTA) sections (controlled by DC bias currents) and special comparators with hysteresis (so called Schmitt trigger [40]), for example). Our presented topology of triangular and square wave generator is based on similar principles. Kim et al. [41] and Chung et al. [42] utilized 3 OTA sections, one capacitor and two resistors, similarly Siripruchynanun et al. [43].

Several generators were designed with current-mode outputs. Kumbun et al. [44] employed two multiple outputs through transconductance amplifiers (MO-CTTAs) only to realize adjustable generator. Two multiple output current controlled current differencing amplifiers (MO-CCCDTAs) were used for design of generator by Silapan et al. [45] and Sristakul et al. [46]. Silapan et al. [45] published a very excellent work (fundamentally very similar to our solution), where two controllable multi-output CDTA (MOCCCDTA) elements, and one capacitor is sufficient. In fact, there were used two independently adjustable OTA sections in each MO-CCCDTA and it allows to "integrate" resistor inside of the active element. The generator is designed with current output responses.

In our topology only two OTA sections, one capacitor and one resistor are sufficient (reasons for the second are explained in a specific chapter - Schmitt comparator in our contribution is adapted for differential output purposes). Differential output means two times higher output signals and immunity to common mode disturbances, which is really important in modern low-voltage CMOS technologies. The resulting conclusion from hitherto published works is clear - to the best of the authors' knowledge no generator (with electronic control of repeating frequency and duty cycle) for differential (balanced output) signal generation that is simpler exists in the literature and many presented single-ended solutions seem to be quite complicated and use also floating passive elements ([33], [36] for example).

CONTROLLABLE VOLTAGE DIFFERENCING BUFFERED AMPLIFIERS

Conceptions of VDBA/VDIBA [1]-[5] and their behavioral model are discussed in this chapter. Advantageous differences from classical VDBA [1]-[4] or VDIBA approach based on OTA and inverter section [5] are discussed and explained. Classical VDBA/VDIBA employs high-impedance voltage differencing input terminals (in this paper labeled as p and n), auxiliary high-impedance z terminal, and low-impedance output of voltage buffer/inverter noted as +w/-w. Modified conception uses transconductance section with one output polarity (one z terminal) and cascade of two voltage inverters. Therefore, both inverting (-w) and direct voltage outputs (+w) are available. However, some applications require two auxiliary terminals z for special purposes of circuit synthesis. Internal structure of FB-VDBA [2] is more complicated, because different transconductance section with current mirrors is required. Advanced VDBA/VDIBA modification allows several types of electronic control as will be shown in more details.

DO-VDBA

The first possible modification of VDBA/VDIBA [1]- [6] is created very simply by additional voltage inverter. Therefore, we have now also direct voltage buffered output, which is very useful for differential mode signal operations. Symbol and behavioral model is shown in Fig. 1.

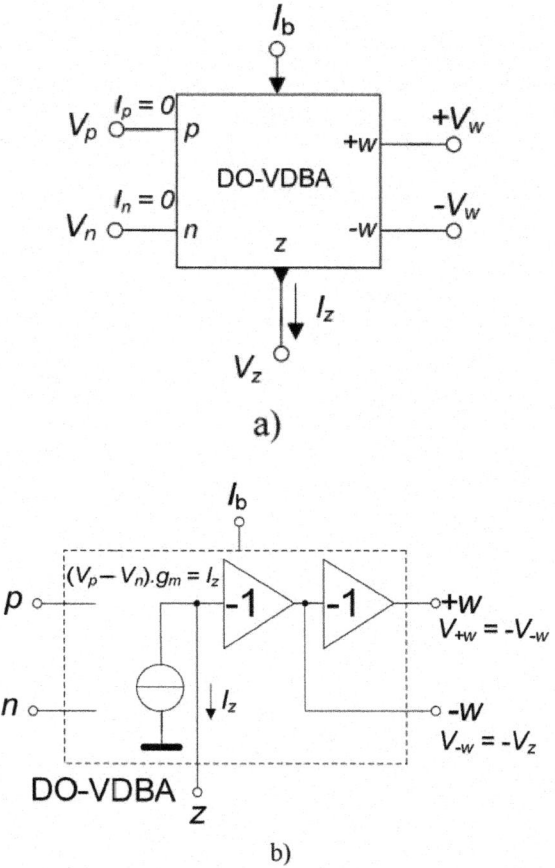

Figure 1: Dual output voltage differencing buffered amplifier (DO-VDBA): a) symbol, b) behavioral model.

We called this element dual output voltage differencing buffered amplifier (DO-VDBA) in accordance to [1]. The main principle is obvious from behavioral model (see Fig. 1b). Control of transconductance is possible by external biasing (I_b). Relation between terminals can be written in hybrid matrix as follows:

$$\begin{bmatrix} I_p \\ I_n \\ I_z \\ V_{+w} \\ V_{-w} \end{bmatrix} = \begin{bmatrix} 0 & 0 & 0 & 0 & 0 \\ 0 & 0 & 0 & 0 & 0 \\ g_m & -g_m & 0 & 0 & 0 \\ 0 & 0 & 1 & 0 & 0 \\ 0 & 0 & -1 & 0 & 0 \end{bmatrix} \begin{bmatrix} V_p \\ V_n \\ V_z \\ I_{+w} \\ I_{-w} \end{bmatrix}.$$

(1)

FB-VDBA

In some cases of circuit synthesis two auxiliary z terminals (both polarities for example) are required. A very useful and easily obtainable version [2], [3] is shown in Fig. 2.

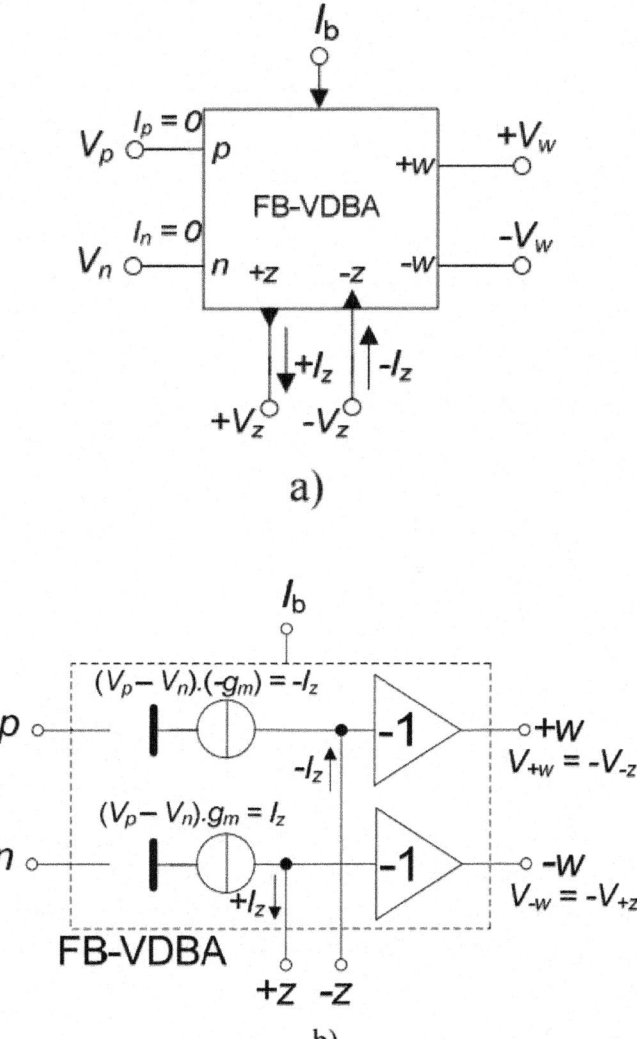

Figure 2: Fully balanced voltage differencing buffered amplifier (FB-VDBA): a) symbol, b) behavioral model.

$$
\begin{bmatrix} I_p \\ I_n \\ I_{+z} \\ I_{-z} \\ V_{+w} \\ V_{-w} \end{bmatrix} = \begin{bmatrix} 0 & 0 & 0 & 0 & 0 & 0 \\ 0 & 0 & 0 & 0 & 0 & 0 \\ g_m & -g_m & 0 & 0 & 0 & 0 \\ -g_m & g_m & 0 & 0 & 0 & 0 \\ 0 & 0 & 0 & -1 & 0 & 0 \\ 0 & 0 & -1 & 0 & 0 & 0 \end{bmatrix} \cdot \begin{bmatrix} V_p \\ V_n \\ V_{+z} \\ V_{-z} \\ I_{+w} \\ I_{-w} \end{bmatrix}.
$$

(2)

Matrix description is very similar to (1), but main difference is in additional equation for the second output current from the second z terminal, see (2).

DO-CG-VDBVA

Previous types of VDBAs (DO-VDBA and FBVDBA) [1]-[6] allow only one possibility of electronic control. However, in many applications various possibilities of controls are required. As an example, sinusoidal oscillators can be mentioned, where one controllable parameter serves for control of oscillation condition and the second one adjusts oscillation frequency. Therefore, introduced modification contains two adjustable parameters. These parameters can be obtained by additional voltage amplifier together with buffer/inverter in behavioral structure or by replacement of inverter/buffer by this controllable voltage amplifier. This active element received a typical name consisting of main features typical for classical VDBA or VDIBA [1]-[6] but considering also controllability of voltage gain (A). Symbol and behavioral model of the dual output controlled gain voltage differencing buffered voltage amplifier (DO-CG-VDBVA) is depicted in Fig. 3.

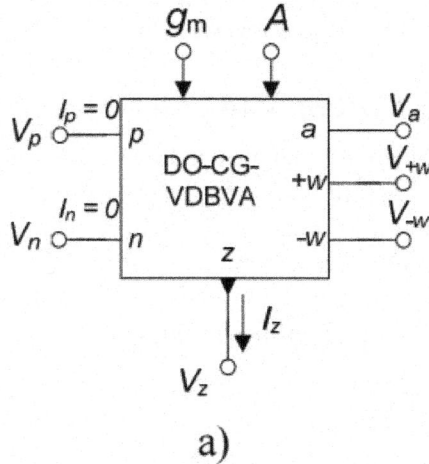

a)

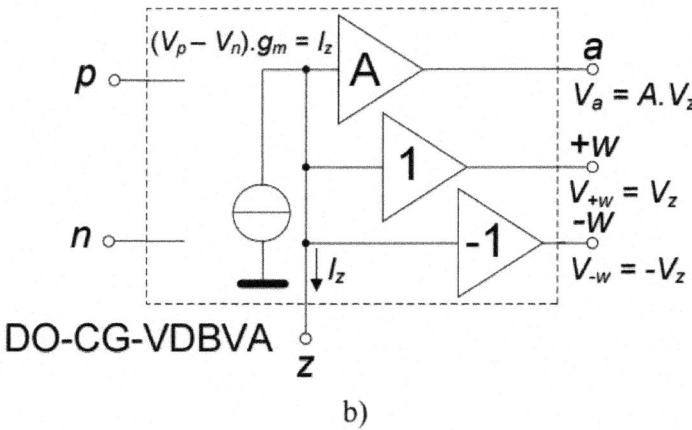

b)

Figure 3: Dual output controlled gain voltage differencing buffered voltage amplifier: a) symbol, b) behavioral model.

Matrix equations also include a new parameter - adjustable voltage gain as follows:

$$
\begin{bmatrix} I_p \\ I_n \\ I_z \\ V_a \\ V_{+w} \\ V_{-w} \end{bmatrix} = \begin{bmatrix} 0 & 0 & 0 & 0 & 0 & 0 \\ 0 & 0 & 0 & 0 & 0 & 0 \\ g_m & -g_m & 0 & 0 & 0 & 0 \\ 0 & 0 & A & 0 & 0 & 0 \\ 0 & 0 & 1 & 0 & 0 & 0 \\ 0 & 0 & -1 & 0 & 0 & 0 \end{bmatrix} \begin{bmatrix} V_p \\ V_n \\ V_z \\ I_a \\ I_{+w} \\ I_{-w} \end{bmatrix},
$$

(3)

from which it is obvious that direct voltage buffering can be achieved by a simple way shown in Fig. 1 (two inverters in cascade).

APPLICATIONS FOR SIGNAL GENERATION

The discussed active elements can be used in interesting applications in analog signal processing and mainly in signal generation. Several types of sinusoidal oscillators and functional (triangle and square wave) generators based on the discussed active elements are presented in this section.

Multiphase Harmonic Oscillator SRCO

So-called single resistance controllable oscillators (SRCO) [20] were very popular for many years due to their simplicity (one-two active elements maximally) and simultaneously independent control of CO and FO by resistor values. The basic conception employs lossy integrator (R_1, C_1), one DO-VDBA and special impedance constructed from $-R_2$ and C_2) in an interesting circuit configuration (Fig. 4).

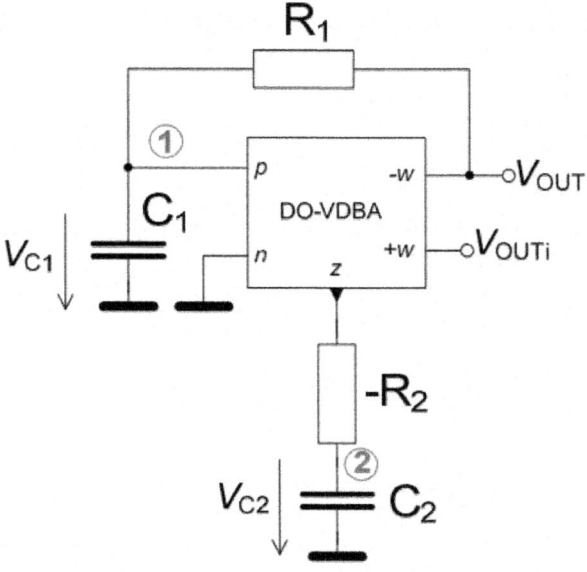

Figure 4: Simple multiphase oscillator using single DO-VDBA.

Characteristic equation of the discussed oscillator has form:

$$s^2 + \frac{\left(1-g_m R_2\right)}{R_1 C_1} s + \frac{g_m}{R_1 C_1 C_2} = 0,$$

(4)

where CO is controllable by $-R_2$ value and FO by value of R_1. Therefore, specifications for SRCO type of oscillator are fulfilled. Practical utilization of floating resistance $-R_2$ is different, because Fig. 4 serves only for theoretical and simple explanation of the principle. Controllability of $-R_2$ value in order to ensure soft CO control is not very comfortable. Therefore, we implemented the second active element to replace inconvenient floating negative passive element from another FB-VDBA. Transconductance control now realizes CO adjusting by a simple electronic way. Replacement of the impedance connected to z terminal of DO-VDBA is shown in Fig. 5. The principle of the design of the basic (Fig. 4) and modified circuit is clear from the diagram in Fig. 5. Two blocks with specific transfers (lossy integrator and special synthetic function) between current and voltage (transadmitance Y_T and transimpedance Z_T) were used for synthesis of the discussed circuit. Impedance of serial $R_2 C_2$ combination in Fig. 5 has form:

$$Z_{INP\_RC} = \frac{1 - sR_2C_2}{sC_2},$$

(5)

and from equivalent circuit resulting impedance:

$$Z_{INP\_EQ} = \frac{1 - s\dfrac{C_2}{g_{m2}}}{sC_2}.$$

(6)

The improved characteristic equation is now:

$$s^2 + \frac{g_{m2} - g_{m1}}{R_1C_1g_{m2}}s + \frac{g_{m1}}{R_1C_1C_2} = 0$$

(7)

where CO and FO can be easily obtained as:

$$g_{m2} \le g_{m1}, \quad \omega_0 = \sqrt{\frac{g_{m1}}{R_1C_1C_2}}.$$

(8),(9)

We can determine transfers between outputs of the oscillator and get relations between produced amplitudes and phase shifts respectively as:

$$\frac{V_{OUT\_0}}{V_{C1}} = \frac{g_{m1}\left(-1 + s\dfrac{C_2}{g_{m2}}\right)}{sC_2},$$

(10)

$$\frac{V_{OUT\_0}}{V_{C2}} = -\frac{1 + sC_1R_1}{1 - \dfrac{g_{m1}}{g_{m2}} + sC_1R_1},$$

(11)

$$\frac{V_{C2}}{V_{C1}} = -\frac{g_{m1}}{sC_2}.$$

(12)

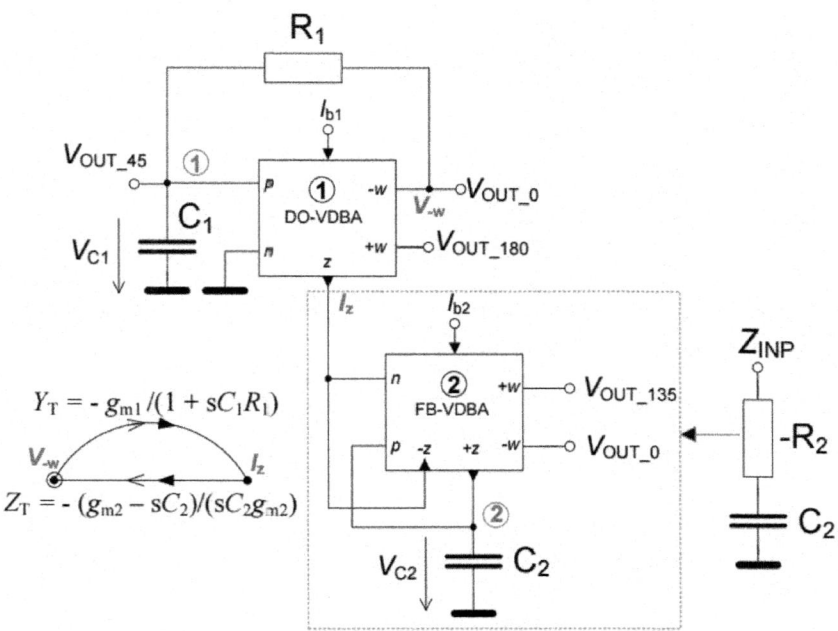

Figure 5: Modification of the oscillator in Fig. 4 with electronically controllable CO.

We can simplify these equations considering fulfilled CO (8) and equality of $C_1 = C_2 = C$ and determine relations as:

$$\frac{V_{OUT\_0}}{V_{C1}} = \frac{g_{m1}(-1+sC)}{sC},$$

(13)

$$V_{OUT\_0} = \sqrt{1+g_{m1}R_1}\, e^{\sqrt{g_{m1}R_1}\, j}\, V_{C1},$$

(14)

$$\varphi_{VOUT\_0\_C1} = \operatorname{arctg}\!\left(\sqrt{g_{m1}R_1}\right),$$

(15)

which leads to 45° phase shift in case of equality $g_{m1} = 1/R_1$. The next relation has form:

$$\frac{V_{OUT\_0}}{V_{C2}} = -\frac{1+sCR_1}{sCR_1},$$

(16)

$$V_{OUT\_0} = -\sqrt{1 + \frac{1}{g_{m1}R_1}} e^{-\frac{1}{\sqrt{g_{m1}R_1}}j} V_{C2},$$

(17)

$$\varphi_{VOUT\_0\_C2} = 180 + \text{arctg}\left(-\frac{1}{\sqrt{g_{m1}R_1}}\right),$$

(18)

which leads to 135° for fulfilled $g_{m1} = 1/R_1$. The last relation that we can find between voltages across capacitors is:

$$\frac{V_{C2}}{V_{C1}} = -\frac{g_{m1}}{sC},$$

(19)

$$V_{C2} = \sqrt{g_m R_1} e^{-\frac{\pi}{2}j} V_{C1},$$

(20)

$$\varphi_{C2\_C1} = \text{arctg}\left(\frac{\sqrt{g_m R_1}}{0}\right) = \frac{\pi}{2}$$

(21)

where only amplitude relation depends on R_1 value (FO control). We can find out that the proposed oscillator produces signals with phase shifts 45, 90, 135 and 180 degrees for fulfilled CO and $g_{m1} = 1/R_1$ as conclusion of this analysis.

This type of oscillators is not very suitable for FO adjusting, if multiphase outputs are also required. Any change of FO leads to inequality of $g_{m1} = 1/R_1$ through R1 and causes disturbance of phase and amplitude proportions in the circuit. Only phase relation between V_{C1} and V_{C2} keeps preserved (but not amplitude). Additional voltage buffering of V_{OUT_45} is also a complication for practical utilization. However, obtaining of four-phase outputs is possible by quite a simple way (similarly as in [5], where one disadvantage remains – floating capacitor) without necessity of many active elements in comparison to [6]-[8], for example. Therefore, the presented solution offers some benefits and improvements.

Multiphase Oscillator - Special Requirements

The previous type of the oscillator requires synthetic replacement of floating negative resistance. Therefore, in Fig. 6 another solution with similar phase shifts like the above discussed circuit is introduced, but usage of floating negative resistance is not required. This oscillator uses conversion (V ® I, I ®

V) transfer sections in the loop based on only lossy integrators and additional positive feedback in comparison to the previous type (diagram in Fig. 6). The second DO-VDBA$_2$ with R$_3$ serves as a subtracter, which allows to obtain required outputs with appropriate phase shifts. Characteristic equation, CO and elementary FO have forms:

$$s^2 + \frac{R_2C_2 + R_1C_1\left(1 - R_2g_{m1}\right)}{R_1R_2C_1C_2}s + \frac{1}{R_1R_2C_1C_2} = 0,$$

(22)

$$g_{m1} \geq \frac{R_1C_1 + R_2C_2}{R_1R_2C_1}, \quad \omega_0 = \sqrt{\frac{1}{R_1R_2C_1C_2}}.$$

(23), (24)

Relations between generated signals are:

$$\frac{V_{C1}}{V_{C2}} = \frac{1}{1 + sC_1R_1},$$

(25)

$$\frac{V_{OUT\_135}}{V_{C1}} = R_3g_{m2}\left(\frac{1 + sC_2R_2}{-R_2g_{m1} + 1 + sC_2R_2}\right),$$

(26)

$$\frac{V_{OUT\_135}}{V_{C2}} = -R_3g_{m2}\left(\frac{sC_1R_1}{1 + sC_1R_1}\right),$$

(27)

which leads to (considering C$_1$ = C$_2$ = C, fulfilled CO: R$_2$g$_{m1}$ = 2 and R$_3$g$_{m2}$ = 1):

$$V_{C1} = \frac{\sqrt{2}}{2}e^{-\frac{\pi}{4}j}V_{C2},$$

(28)

$$V_{OUT\_135} = e^{-\frac{\pi}{2}j}V_{C1},$$

(29)

$$V_{OUT\_135} = -\frac{\sqrt{2}}{2}e^{-\frac{\pi}{4}j}V_{C2} = \frac{\sqrt{2}}{2}e^{-\frac{3\pi}{4}j}V_{C2}.$$

(30)

Simple solution of both DO-VDBAs (only one positive z terminal) is sufficient in this modification of the oscillator in comparison to the previous type.

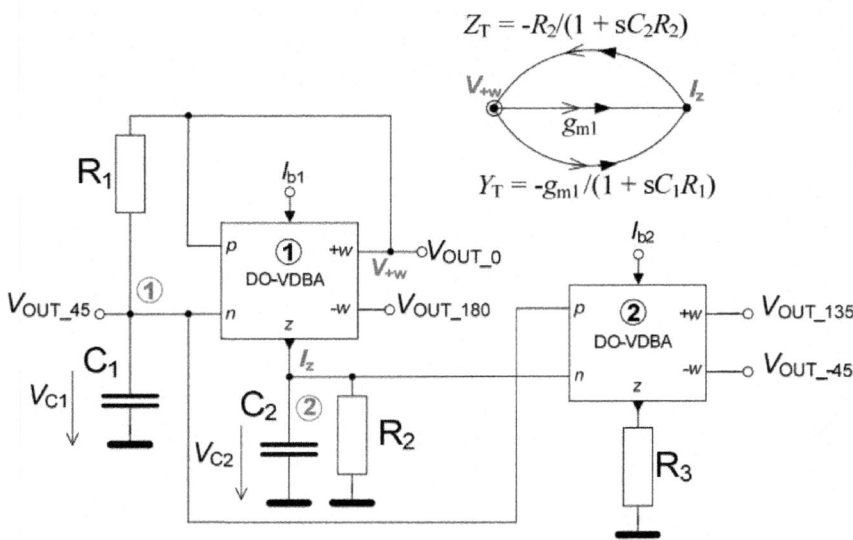

Figure 6: Modification of the proposed oscillator without necessity of employing floating negative resistance.

Adjustable Harmonic Quadrature Oscillator

The solution presented in this section offers more benefits in comparison to the previous multiphase type, where FO adjusting was limited or not possible due to request of four multiphase outputs (45, 90, 135, 180 degrees). The main requirement of the intended synthesis is the design of fully electronically controllable (CO and linear control of FO) oscillator with multiphase purposes (90, 180, 270 degrees) or quadrature differential (balanced) mode oscillator. The presented circuit consists of lossy and loss-less integrators in one loop complemented by negative resistance controllable by voltage gain. This application (Fig. 7) is the perfect example of utilization of the DO-CGVDBVA. Loss-less controllable (by g_{m2}) voltage integrator forms the first part (C_2 and DO-VDBA) of the oscillator. The second important part is the lossy integrator (controllable by gm1) together with negative resistance supplementary simulating circuit (R_1, C_1 and DO-CG-VDBVA), where negative resistance is adjustable by voltage gain A_1. Characteristic equation, CO and FO have the following forms:

$$s^2 + \frac{(1 - A_1)}{R_1 C_1} s + \frac{g_{m1} g_{m2}}{C_1 C_2} = 0,$$

(31)

$$A_1 \geq 1, \quad \omega_0 = \sqrt{\frac{g_{m1}g_{m2}}{C_1 C_2}}.$$

$$(32), (33)$$

Relative sensitivities of oscillation frequency on parameters in (33) achieve typical values (± 0.5). Linear control of FO is ensured by simultaneous adjusting of g_{m1} and g_{m2} ($g_{m1} = g_{m2}$) and independent control of CO by A_1. The oscillator provides low impedance at each of the outputs, therefore, easy connection to low-resistance loads is allowed. The relation between produced signals across the capacitors has form:

$$\frac{V_{C2}}{V_{C1}} = -\frac{g_{m2}}{sC_2} = j\sqrt{\frac{g_{m2}C_1}{g_{m1}C_2}},$$

$$(34)$$

which means quadrature phase shift and equal amplitudes during the tuning process (FO) at all outputs in case of simultaneous control of both g_m and when capacitors have equal values. Single-ended operation mode allows to obtain the oscillator with unchangeable amplitudes providing four phase shifts (multiples of p/2). Differential operation mode has benefit of two-times higher output amplitudes, but in quadrature form only.

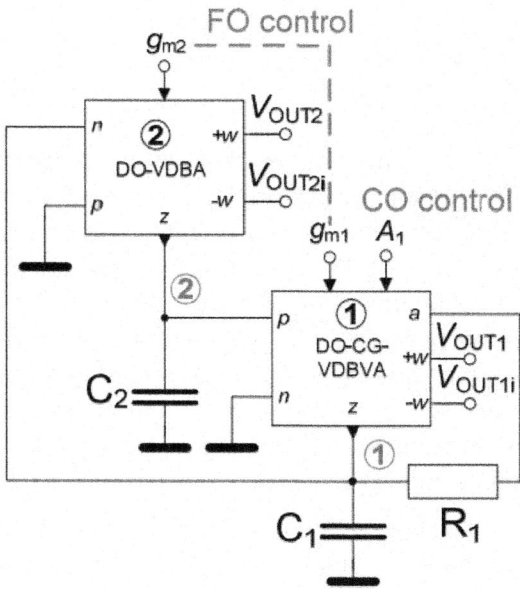

Figure 7: Fully controllable multiphase/quadrature differential mode (balanced) oscillator.

Triangle and Square Wave Generator

The last presented useful and quite simple application of the discussed active elements is a differential output (balanced) triangle and square wave generator. Explanations of processes in the function of the generator are provided by charging of a capacitor and switching (turn-over) at reference levels. The lossless integrator and comparator with hysteresis (Schmitt trigger) in the feedback loop are main building blocks of these types of non-harmonic signal sources [32]-[46]. We show possibilities how to build lossless integrator with VDBA elements (partial block of the oscillator in Fig. 7).

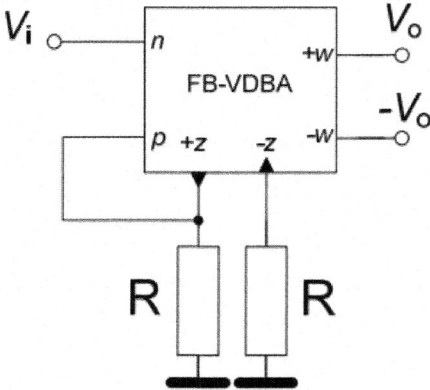

Figure 8: Schmitt comparator with hysteresis employing FBVDBA.

For construction of a dual-output Schmitt comparator (Fig. 8) one FB-VDBA with two z terminals was employed. Theoretically, the DO-VDBA (one z terminal) is also possible for construction of a comparator (we can save one resistor). However, the FB-VDBA (with current multiplying performed by current mirrors in internal topology - we will present it in experimental/ simulation part of this paper) allows higher gain of the whole voltage feedback system, which is the key factor (in low-power and low-supply voltage technology) for precise flip-over process of the comparator. Quality of the comparator has a direct impact on accuracy of oscillation/repeating frequency (also noted by abbreviation FO) and accuracy of reference levels.

The following equation is valid between the input and output voltage of FB-VDBA in the comparator (Fig. 8):

$$\left(\mp V_{o\_sat}\right) = \frac{g_m R}{g_m R - 1}\left(\pm V_i\right),$$

(35)

which leads to

$$\mp V_{o\_sat} = \pm V_i \,,$$

(36)

validity of $g_m R \gg 1$ is ensured. We found referencing value (V_{ref}) for very high gains (given by transconductance g_m and resistor R) which is necessary for turnover of the output of the comparator from a high output voltage level to a low output voltage level respectively in (36). The output voltage (in ideal case also reference threshold voltage) is given by maximum output current $\pm I_{+z\_max}$, which means maximum of positive or negative output saturation: $\pm V_{+z} = \pm R.I_{+z\_max}$. Dynamical characteristic of the comparator has two comparative reference voltages determined as $\pm R.I_{+z\_max}$, thanks to positive feedback from voltage across R and very high gain ($g_m R$).

Fig. 9 shows the complete generator, where the lossless integrator employs the first type of DO-VDBA and the comparator presents the same circuit as we discussed in Fig. 8.

Linear charging of the capacitor C starts at the negative reference voltage level ($-V_{C\_max}$) and is given by $+I_{C\_max} = +I_{z1\_max} + I_d = k_1.I_{b1} + n.k_1.I_{b1}$, where constant k represents current gain in the internal topology of VDBA (multiplying by current mirrors), Id is auxiliary controlled DC current and n is the ratio between Id and Ib1 ($I_d = n.I_{b1}$). The input voltage linear range of DO-VDBA is very small for higher g_m and the slope of the DC transfer characteristic very sharp.

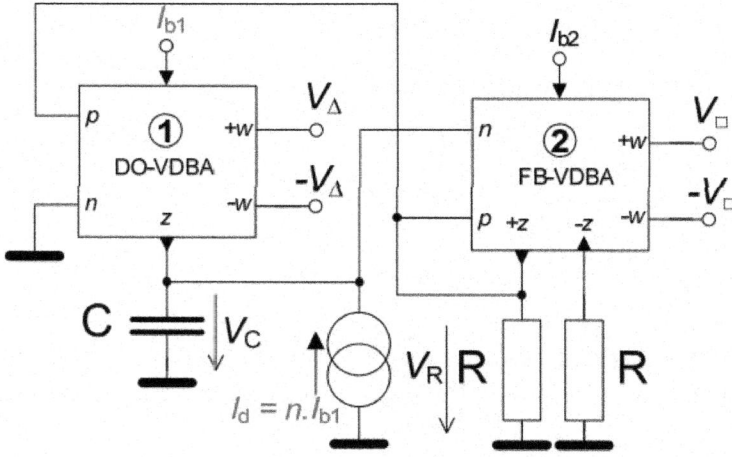

Figure 9: Differential mode triangle and square wave generator employing DO-VD-BA and FB-VDBA.

Therefore, the output currents are given by one from both saturation corners limited by bias current I_b as $\pm I_{z1\_max} = \pm k_1.I_{b1}$ (DO-VDBA) and $\pm I_{z2\_max} = \pm k_2.I_{b2}$ (FBVDBA), see Fig. 11a and Fig. 13a. Discharging that starts at $+V_{C\_max}$ is given similarly, because $-I_{C\_max} = -I_{z1\_max} + I_d$. Both time intervals per one signal period are obtained from:

$$\Delta V_C = \frac{I_{z1\_max} + I_d}{C} T_1,$$

(37)

$$\Delta V_C = \frac{-\left(-I_{z1\_max} + I_d\right)}{C} T_2.$$

(38)

The distance between the negative and positive threshold values ($-V_{C\_max} = +RI_{z2\_max}$ and $+V_{C\_max} = -R.I_{z2\_max}$) can be expressed as:

$$\Delta V_C = \Delta V_R = +V_{C\_max} - \left(-V_{C\_,max}\right) = 2V_{C\_max},$$

(39)

$$2V_{R\_max} = +RI_{z2\_max} - \left(-RI_{z2\_max}\right).$$

(40)

Both time intervals are stated as:

$$T_1 = \frac{\Delta V_C C}{I_{z1\_max} + I_d} = \frac{2RCk_2 I_{b2}}{k_1 I_{b1} + n.k_1 I_{b1}},$$

(41)

$$T_2 = \frac{\Delta V_C C}{I_{z1\_max} - I_d} = \frac{2RCk_2 I_{b2}}{k_1 I_{b1} - n.k_1 I_{b1}}.$$

(42)

The repeating period and frequency have forms:

$$T = T_1 + T_2 = \frac{4RCk_1 I_{b1} k_2 I_{b2}}{\left(k_1 I_{b1} + n.k_1 I_{b1}\right)\left(k_1 I_{b1} - n.k_1 I_{b1}\right)},$$

(43)

$$f_0 = \frac{k_1 I_{b1}(1+n)(1-n)}{4RCk_2 I_{b2}} = \frac{k_1 I_{b1}(1-D)D}{RCk_2 I_{b2}}.$$

(44)

It is obvious that current I_d is DC component, which shifts linear trace (triangular signal), i.e. offset. It influences the duty cycle of the produced wave as:

$$D = \frac{T_1}{T} = \frac{1}{2}\left(1 - \frac{I_d}{I_{b1}}\right) = \frac{1}{2}(1-n).$$

(45)

The maximal theoretical limits of I_d are given by $\pm I_{b1}$ (n = 1 for D = 0% and n = -1 for D = 100%). Therefore, change of the polarity of Id is required. The repeating frequency can be controlled independently with respect to the duty cycle if ratio I_d/I_{b1} = n is kept strictly constant while frequency f_0 is tuned by I_{b1}.

SIMULATION RESULTS

Possible CMOS Implementations of Selected DO-VDBA Solutions

We designed internal topologies of DO-VDBA and FB-VDBA to demonstrate functionality and practical features of the designed applications. Models of TSMC 0.18 μm CMOS technology parameters [47] were used for our simulations. The first topology uses classical transcondutance section with active load (PMOS mirror) and two simple voltage inverters, see Fig. 10. Lower gain of one transconductance section (higher gain requires cascading) and only one z terminal are the main disadvantages of this solution. Nevertheless, it is a useful solution in a specific situation and power consumption is lower in comparison to the second solution (it will be discussed later). PSpice analyses provided the following results (I_b = 50 mA, V_{CC} = ± 1.2 V): g_m = 500 μS, R_z » 170 kW, C_z » 0.35 pF, $R_{\pm w}$ » 53 W, voltage gain between z and +w has value 0.926 and gain between z and -w is 0.962. Differential input resistance of OTA section is high - at 1 MHz has values higher than 1.1 MW (parasitic capacitance » 0.27 pF). Adjusting of I_b influences R_z value, for I_b = 200 mA R_z value is approx. 46 kW (value decreases with higher I_b).

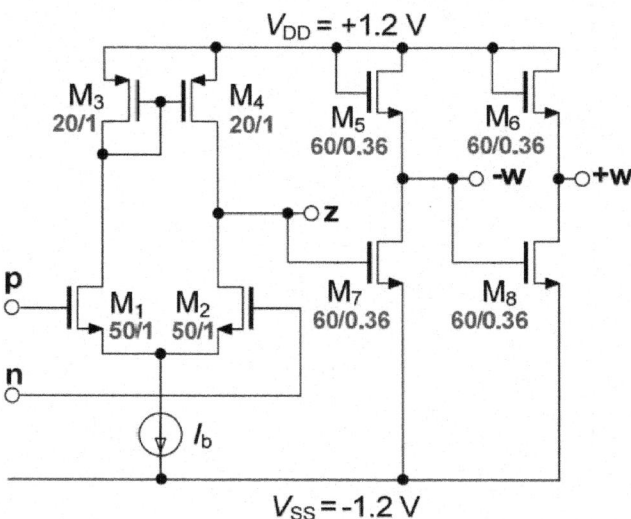

Figure 10: DO-VDBA with one z terminal.

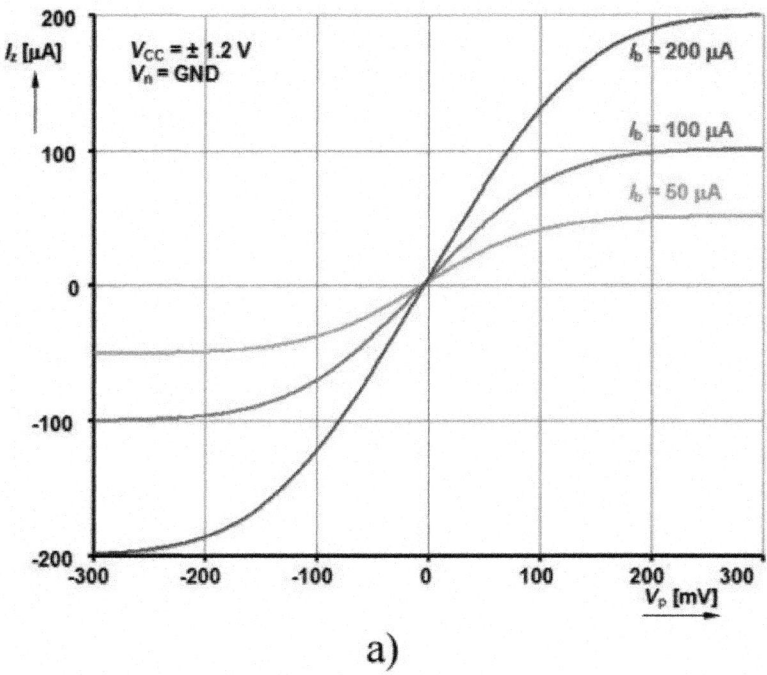

a)

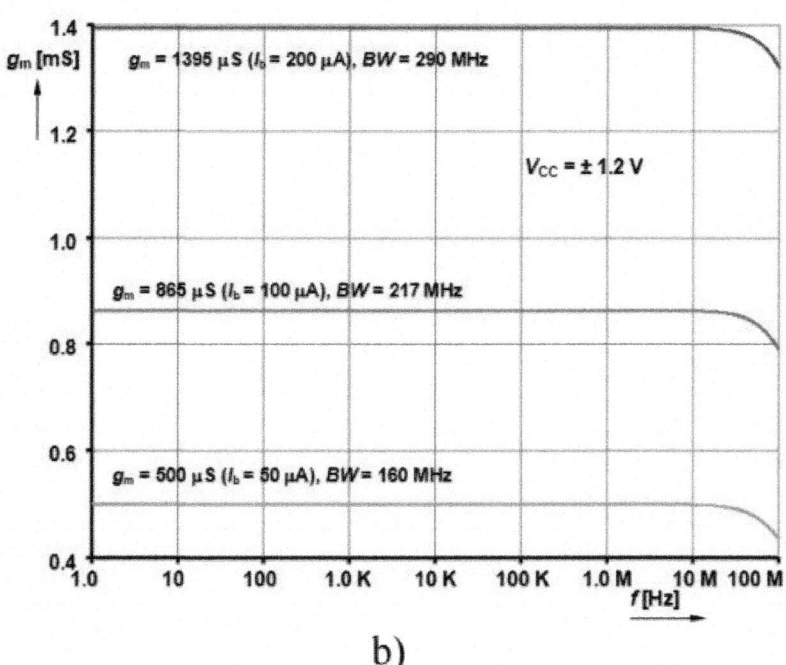

b)

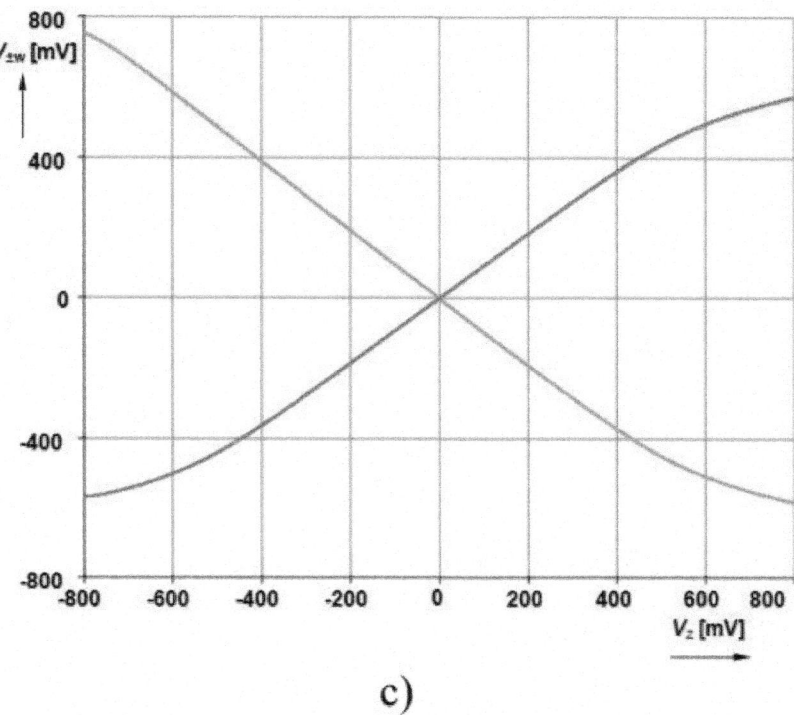

c)

Figure 11: Selected features of the proposed DO-VDBA: a) DC transfer characteristic of OTA section, b) dependence of g_m on frequency, c) DC transfer characteristics of inverter/buffer.

Adjusting of I_b in the range of 5 - 200 mA causes changes of g_m from 60 µS to 1.4 mS. Some of the simulation results are documented graphically in Fig. 11.

The FB-VDBA has a different construction (Fig. 12). Function of the second type (FB-VDBA) is practically similar, only internal topology is slightly complicated due to necessity of z terminals of both polarities and k-times higher gain of transconductance section and mirrors of FBVDBA. Many parameters of the second type of DO-VDBA are practically identical to the previously discussed type (parameters of voltage inverters). Simulations provided the following results ($I_b = 50$ mA, $V_{CC} = \pm 1.2$V): $g_m = 1030$ µS, $R_{\pm w} » 130$ kW, $C_{\pm z}$ » 0.38 pF. The differential input resistance of OTA section has higher values than 0.64 MW at 1 MHz (parasitic capacitance » 0.25 pF). Adjusting of gm from 132 µS to 2.76 mS was verified (I_b between 5 and 200 mA). The highest tested $I_b = 200$ mA decreases $R_{\pm z}$ value to 53 kW approximately. Selected results are documented in Fig. 13.

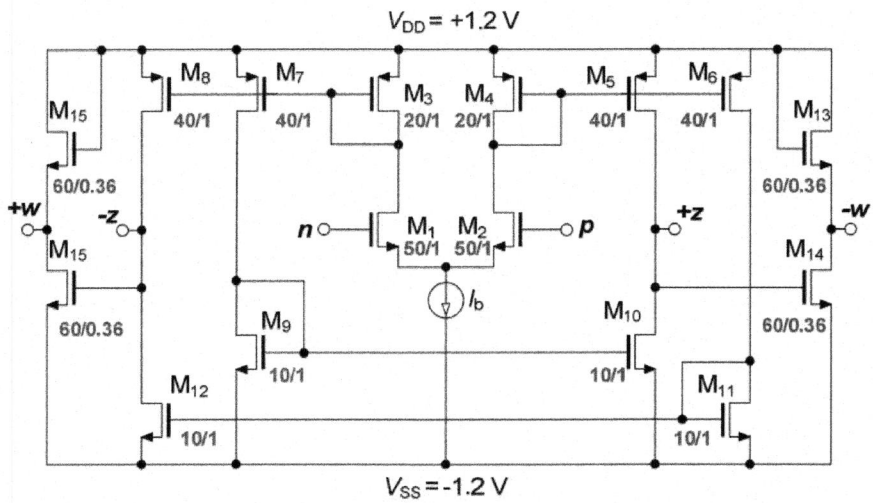

Figure 12: FB-VDBA with z terminals of both polarities.

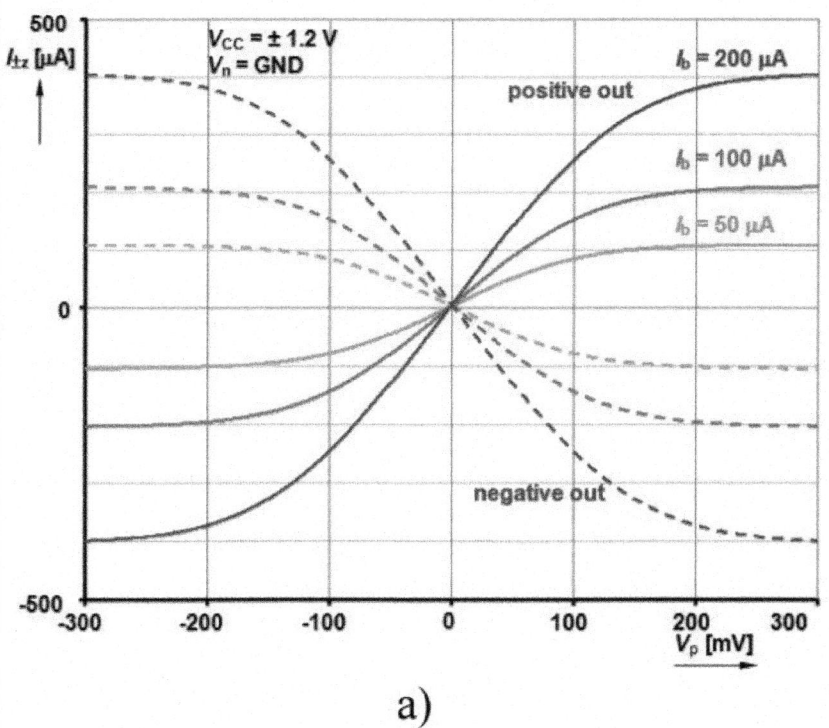

a)

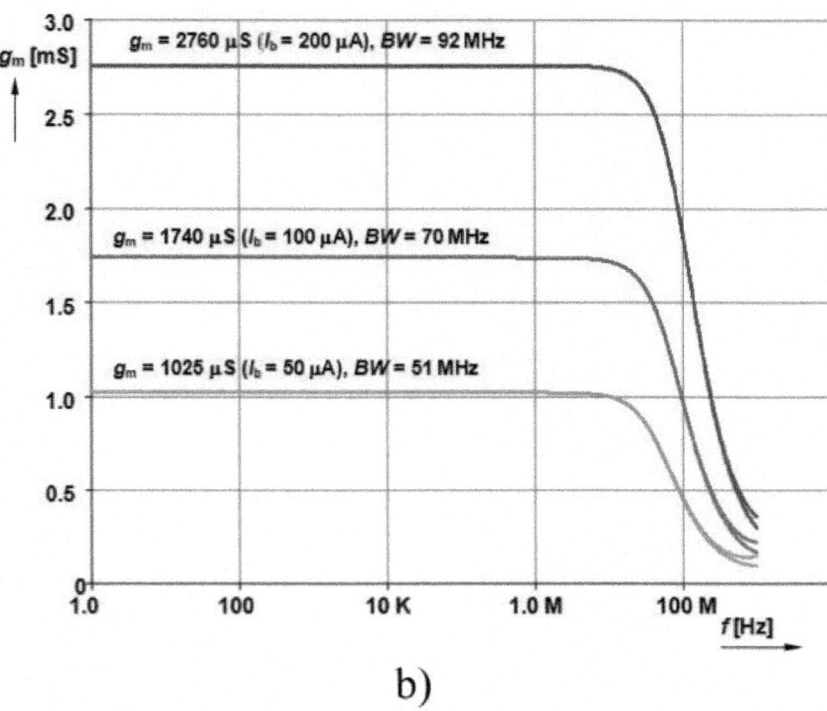

b)

Figure 13: Selected features of FB-VDBA: a) DC transfer characteristic of OTA section, b) dependence of g_m on frequency.

Detailed PSpice Analysis of Some Proposed Applications

Multiphase Oscillator - Type without Necessity of Floating Negative Resistance

Oscillator from Fig. 6 we designed for operation at $f_0 = 1.539$ MHz with parameters: $R_1 = R_2 = R = 2.2$ kW, $C_1 = C_2 = C = 47$ pF, $R_3 = 1$ kW, and $g_{m2} = 1$ mS. Obtained results of simulation are shown in Fig. 14. Voltage gain of subtracting point created by DO-VDBA$_2$ and R$_3$ is approximately equal to 1. Real value of gm1 necessary for start of oscillation was increased to 1.133 mS ($I_{b1} ==146$ mA). Oscillation frequency 1.506 MHz was obtained, which is very close to the ideal value (error about 2%). Achieved total harmonic distortion (THD) was 0.40%, 0.39%, 0.26%, 0.81% and 0.81% for all outputs namely: V_{OUT0}, V_{OUT180}, V_{OUT45}, V_{OUT135}, V_{OUT-45}.

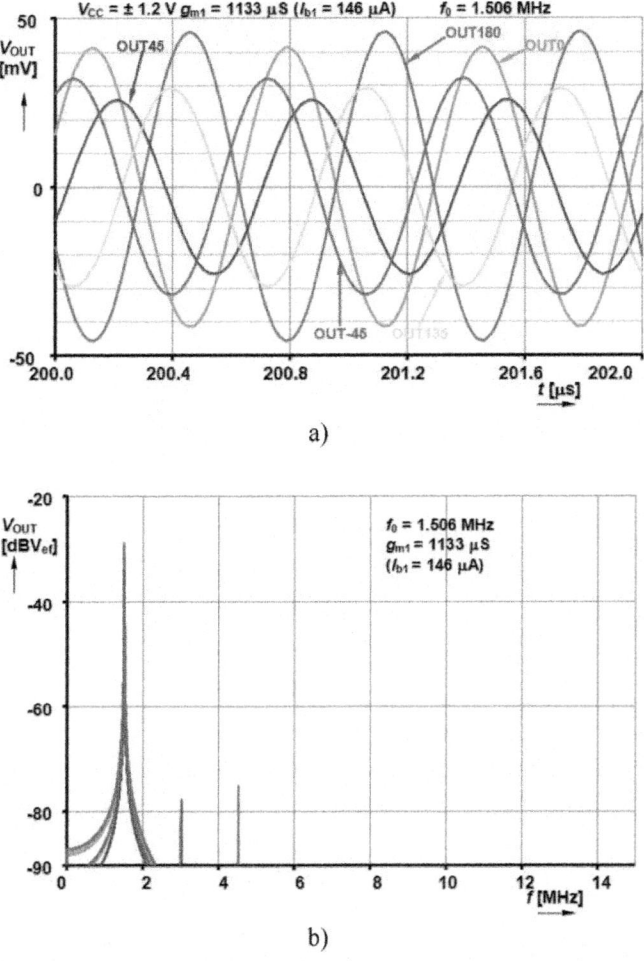

Figure 14: Results of simulation of the oscillator from Fig. 6: a) transient responses, b) frequency spectrum.

Functional Generator

We require operation in hundreds of kHz, therefore parameters of the design (considering solution shown in Fig. 9) are the following: $C = 22$ pF, $R = 10$ kΩ, $I_{b2} = 20$ mA, $k_2 = 2$ and suitable range of maximal output current (given by $k_1.I_{b1}$, where $k_1 = 1$) adjusting between 5 mA and 100 mA. Duty cycle was set to 50% $(n = 0)$. Load resistances $R_L = 1$ kΩ were connected at the output ports (w) in simulations. We used FB-VDBA based comparator. This solution has one disadvantage, which is the necessity of the second resistor. However, such OTA section with mirrors in FB-VDBA structure has higher available voltage gain (bias mirroring with gain $k_2 = 2$, see Fig. 12) than simpler OTA section in the DO-VDBA. Of course, there is possibility to use simpler DO-VDBA, but validity of (36) is highly influenced by insufficient gain of the comparator in low-power solution. In fact, we are discussing a specific design for our requirements in the following text. Quality of the comparator directly influences oscillation frequency (threshold referencing voltage for changing of output polarity). The transconductance of OTA section of FBVDBA has value 470 µS for $I_{b2} = 20$ mA. This value is quite low for sufficient voltage amplification in the comparator (it is better than for the same solution with DOVDBA). We expected equality between input threshold voltage and output voltage of the comparator. Unfortunately, gain is not sufficient and therefore we achieved only $V_\Delta \gg 1.3 V_\Delta$. In addition this gain is not valid in the whole DC transfer characteristic due to its nonlinearity in frame of the OTA section as part of the FB-VDBA basedcomparator. AC analysis allows finding small-signal g_m for ideal calculation (35) of the comparator behavior. Nevertheless, g_m has a lower value than in the origin of the characteristic (around 0) in case of comparator in overturn corners (reference voltages) of DC transfer characteristic. The new equation for repeating frequency considering the above discussed inequality of input (reference) and output voltage of the comparator and D = 50% has the following form:

$$
f_{0\_exp} = \cfrac{I_{b1}}{4C \cfrac{V_R}{\left(\cfrac{g_{m2}R}{g_{m2}R-1}\right)}} = \cfrac{I_{b1}}{4RCk_2 I_{b2}\left(\cfrac{g_{m2}R}{g_{m2}R-1}\right)} \cong \cfrac{I_{b1}}{8RCI_{b2}}.
$$

$$\text{(46)}$$

Maximal output voltage at terminal $+z_2$ (V_R) of FBVDBA has amplitude value $V_R = V_\Delta = I_{b2}.k_2.R \gg 400$ mV. This value is also expected for output amplitude of square wave signal (in case of single-ended solution). Considering the practical inequality of (36) given by (35) means that input reference voltage (causing overturn of the comparator) is 1.3 times lower (approximately 300 mV) than V_Δ. This value is expected for amplitude of triangle (symmetrical ramp) wave signal. All these presumptions expect unity-gain followers/inverters and symmetrical referencing voltage (it is given by quality of the comparator), in real case it can be slightly different (non-unity gains of followers/inverters causes changes of expected amplitudes). Repeating frequency f_0 is expected for selected parameters and $I_{b1} = 30$ mA and with help of (46) as $f_0 = 1.108$ MHz. Simulated value of generator with models of VDBAs discussed above was obtained as 1.031 MHz. Results are in Fig. 15.

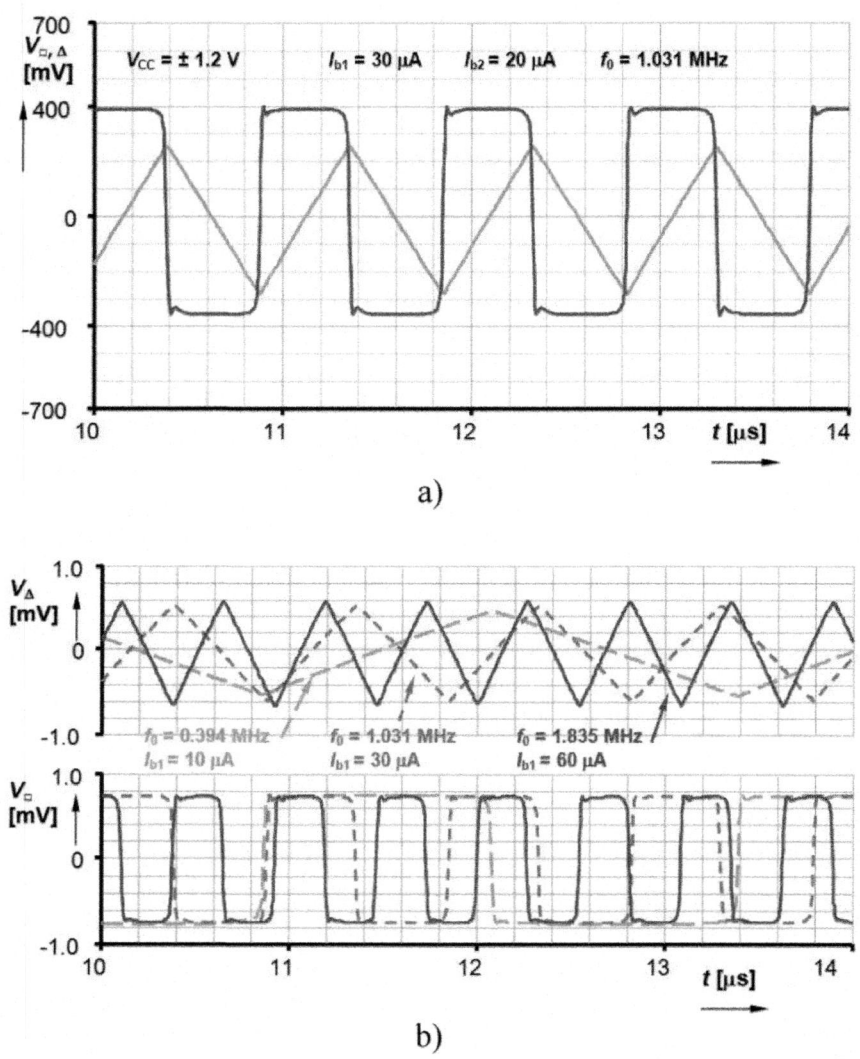

Figure 15: Simulated transient responses of the proposed generator: a) single-ended mode, b) example of electronic control of f_0 (differential-balanced output mode).

Differential mode of operation has advantage of two-times higher produced amplitude. Electronic adjusting of f_0 was verified between 211 kHz and 2.83 MHz (I_{b1} changed from 5 mA to 100 mA), see Fig. 16 where comparison of calculated f_0 (46) and simulation results are depicted.

An example of operation of the generator with D = 24.5% at f_0 = 1 MHz was provided. Expected f_0 can be expressed by the following equation considering a finite gain similarly as (46):

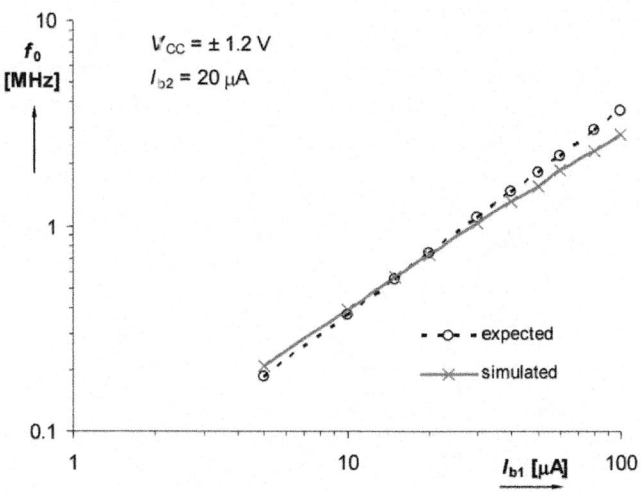

Figure 16: Calculated and simulated dependence of f_0 on I_b.

$$f_{0\_exp} = \frac{I_{b1}(1-D)D}{2RCI_{b2}}\left(\frac{g_{m2}R-1}{g_{m2}R}\right).$$

(47)

Analysis used the same values of the rest of parameters as in the previous case. The bias current $I_{b1} = 39$ mA was set in accordance to (47) for the above discussed assignment. Results in differential output mode are shown in Fig. 17 where possibility of adjusting of f_0 between two values is documented while D keeps unchangeable.

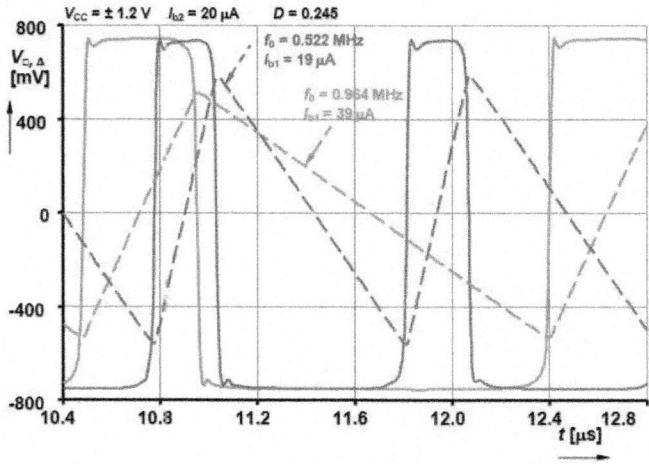

Figure 17: Examples of variability f0 in simulated transient responses with D = 24.5 %.

EXPERIMENTAL RESULTS

The comparator in Fig. 8 was tested experimentally with behavioral model (Fig. 18) based on commercially available devices. All parameters and values are noted in Fig. 18. The circuit was tested for ramp-pulse excitation (±0.7 V, 1 kHz) and the results are in Fig. 19.

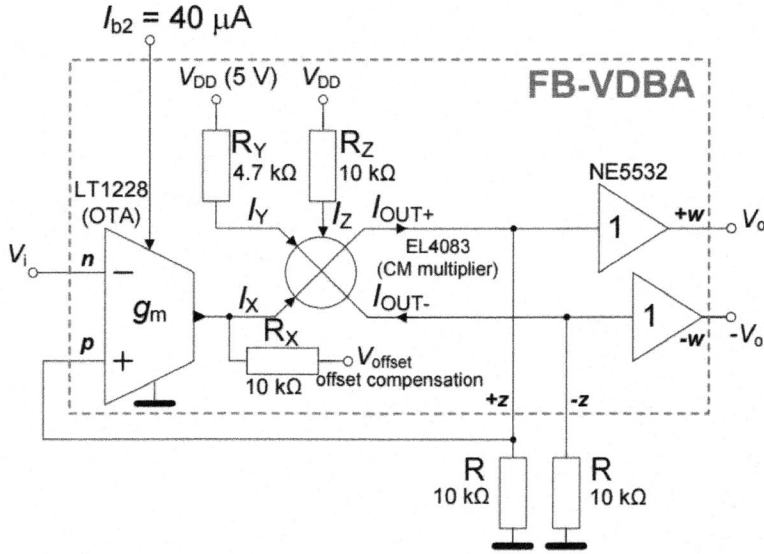

Figure 18: Schmitt comparator employing behavioral model of FB-VDBA (based on commercially available devices).

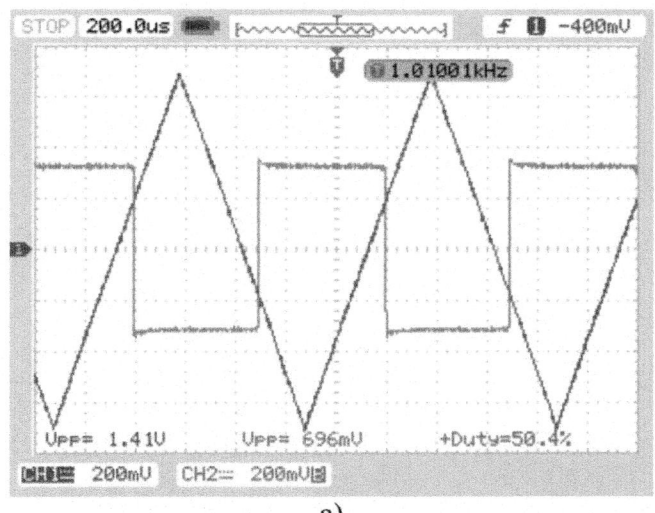

a)

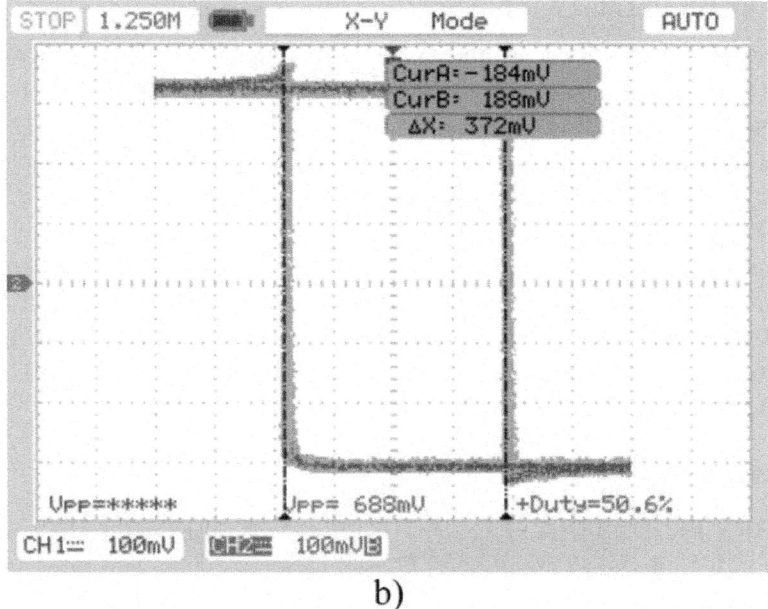

b)

Figure 19: Measurement results of the comparator: a) transient responses Vi (triangular), Vo, b) hysteresis characteristic of the comparator.

CONCLUSION

The designed active elements allow beneficial features in specific applications. They were intended for utilization in multiphase oscillators, quadrature oscillators (single-ended or differential mode) or triangle and square wave generators for example, where provided modifications of VDBA and VDIBA elements [1]-[5] with standard transconductance control or also additional electronic control of voltage gain in frame of active element were employed. Some of the presented circuits allow benefits of simplicity (multiphase oscillators for example) in comparison to classical design methods based on cascading lossy integrators or similar sections in phase shifted loop [6]-[8], where higher number of active and passive elements is necessary. Higher demands on precise accuracy of relations between generated outputs and necessity of several matching conditions is cost of these benefits. Nevertheless, matching conditions are also drawback of some more sophisticated phase-shifted solutions. Both introduced quadrature oscillators and triangle and square wave generator offer advantages of simplicity and differential output responses. The proposed circuits were verified by simulations at frequencies of several MHz and complemented by detailed discussions.

ACKNOWLEDGEMENTS

The research described in the paper was supported by Czech Science Foundation projects under No. 102/09/1681 and No. 102/11/P489 and by project (Brno University of Technology) of specific research FEKT-S-11-15. Dr. Herencsar was supported by the project of the Brno University of Technology CZ.1.07/2.3.00/ 30.0039. The support of the project CZ.1.07/2.3.00/20.0007 WICOMT, financed from the operational program Education for Competitiveness is gratefully acknowledged. The described research was performed in laboratories supported by the SIX project; registration number CZ.1.05/2.1.00/03.0072, the operational program Research and Development for Innovation. This research work is funded also by projects EU ECOP EE.2.3.20.0094, CZ.1.07/2.2.00/28.0062.

REFERENCES

1.   BIOLEK, D., SENANI, R., BIOLKOVA, V., KOLKA, Z. Active elements for analog signal processing: Classification, review, and new proposal. Radioengineering, 2008, vol. 17, no. 4, p. 15-32.

2.   BIOLKOVA, V., KOLKA, Z., BIOLEK, D. Fully balanced voltage differencing buffered amplifier and its applications. In Proc. of 52nd MWSCAS. Cancún (Mexico), 2009, p. 45-48.

3.   BIOLEK, D., BIOLKOVA, V., KOLKA, Z. All-pass filter employing fully balanced voltage differencing buffered amplifier. In Proc. of the IEEE Latin American Symposium on Circuits and Systems (LASCAS 2010). Iguacu (Brazil), 2010. p. 232-235.

4.   KACAR, F., YESIL, A., NOORI, A. New CMOS realization of voltage differencing buffered amplifier and its biquad filter applications. Radioengineering, 2012, vol. 21, no. 1, p. 333-339.

5.   HERENCSAR, N., MINAEI, S., KOTON, J., YUCE, E., VRBA, K. New resistorless and electronically tunable realization of dualoutput VM all-pass filter using VDIBA. Analog Integrated Circuits and Signal Processing, 2013, vol. 74, no. 1, p. 141-154.

6.   ABUELMAATTI, M. T., AL-QAHTANI, M. A. A new currentcontrolled multiphase sinusoidal oscillator using translinear current conveyors. IEEE Transactions on Circuits and Systems II: Analog and Digital Signal Processing, 1998, vol. 45, no. 7, p. 881-885.

7.   SOULIOTIS, G., PSYCHALINOS, C. Electronically controlled multiphase sinusoidal oscillators using current amplifiers. International Journal of Circuit Theory and Applications, 2009, vol. 37, no. 1, p. 43-52.

8. KUMNGERN, M., CHANWUTIUM, J., DEJHAN, K. Electronically tunable multiphase sinusoidal oscillator using translinear current conveyors. Analog Integrated Circuits and Signal Processing, 2010, vol. 65, no. 2, p. 327–334.

9. GIFT, S. J. G. The application of all-pass filters in the design of multiphase sinusoidal systems. Microelectronics Journal, 2000, vol. 31, no. 1, p. 9-13.

10. KESKIN, A. U., BIOLEK, D. Current mode quadrature oscillator using current differencing transconductance amplifiers (CDTA). IEE Proc. Circuits Devices and Systems, 2006, vol. 153, no. 3, p. 214-218.

11. SONGSUWANKIT, K., PETCHMANEELUMKA, W., RIEWRUJA, V. Electronically adjustable phase shifter using OTAs. In Proc. of the International Conference on Control, Automation and Systems (ICCAS2010). Gyeongii-do (Korea), 2010, p. 1622-1625.

12. KEAWON, R., JAIKLA, W. A resistor-less current-mode quadrature sinusoidal oscillator employing single CCCDTA and grounded capacitors. Przeglad Elektrotechniczny, 2011, vol. 87, no. 8, p. 138-141.

13. JAIKLA, W., SIRIPRUCHYANUN, M., BAJER, J., BIOLEK, D. A simple current-mode quadrature oscillator using single CDTA. Radioengineering, 2008, vol. 17, no. 4, p. 33-40.

14. PANDEY, N., PAUL, S. K. Single CDTA-based current mode allpass filter and its applications. Journal of Electrical and Computer Engineering, 2011, vol. 2011, p. 1-5.

15. KESKIN, A. U., AYDIN, C., HANCIOGLU, E., ACAR, C. Quadrature oscillator using current differencing buffered amplifiers (CDBA). Frequenz, 2006, vol. 60, no. 3, p. 21-23.

16. SONGKLA, S. N., JAIKLA, W. Realization of electronically tunable current-mode first-order allpass filter and its application. International Journal of Electronics and Electrical Engineering, 2012, vol. 2012, no. 6, p. 40-43.

17. MINAEI, S., YUCE, E. Novel voltage-mode all-pass filter based on using DVCCs. Circuits, Systems and Signal Processing, 2010, vol. 29, no. 3, p. 391-402.

18. SOLIMAN, A. M. Synthesis of grounded capacitor and grounded resistor oscillators. Journal of the Franklin Institute, 1999, vol. 336, no. 4, p. 735-746.

19. HERENCSAR, N., VRBA, K., KOTON, J., LAHIRI, A. Realizations of single-resistance-controlled quadrature oscillators using a generalized

current follower transconductance amplifier and a unity gain voltage-follower. Int. Journal of Electronics, 2010, vol. 97, no. 8, p. 879-906.

20. GUPTA, S. S., SENANI, R. New single-resistance-controlled oscillator configurations using unity-gain cells. Analog Integrated Circuits and Signal Processing, 2006, vol. 46, no. 2, p. 111-119.

21. BIOLKOVA, V., BAJER, J., BIOLEK, D. Four-phase oscillators employing two active elements. Radioengineering, 2011, vol. 20, no. 1, p. 334-339.

22. LAHIRI, A., GUPTA, M. Realizations of grounded negative capacitance using CFOAs. Circuits, Systems and Signal Processing, 2011, vol. 30, no. 1, p. 134-155.

23. LAHIRI, A. Explicit-current-output quadrature oscillator using second-generation current conveyor transconductance amplifier. Radioengineering, 2009, vol. 18, no. 4, p. 522-526.

24. LAHIRI, A. Novel voltage/current-mode quadrature oscillator using current differencing transconductance amplifier. Analog Integrated Circuits and Signal Processing, 2009, vol. 61, no. 2, p. 199-203.

25. RODRIGUEZ-VAZQUEZ, A., LINAREZ-BARRANCO, B., HUERTAS, L., SANCHEZ-SINENCIO, E. On the design of volt-age-controlled sinusoidal oscillators using OTA′s. IEEE Transaction on Circuits and Systems, 1990, vol. 37, no. 2, p. 198-211.

26. SOTNER, R., HERENCSAR, N., JERABEK, J., KOTON, J., DOSTAL, T., VRBA, K. Quadrature oscillator based on modified double current controlled current feedback amplifier. In Proc. of 22nd Int. Conf. Radioelektronika. Brno (Czech Republic), 2012, p. 275-278.

27. ALZAHER, H. CMOS digitally programmable quadrature oscillators. International Journal of Circuit Theory and Applications, 2008, vol. 36, no. 8, p. 953-966.

28. BIOLEK, D., LAHIRI, A., JAIKLA, W., SIRIPRUCHYANUN, M., BAJER, J. Realization of electronically tunable voltagemode/current-mode quadrature sinusoidal oscillator using ZC-CGCDBA. Microelectronics Journal, 2011, vol. 42, no. 10, p. 1116- 1123.

29. SOTNER, R., JERABEK, J., HERENCSAR, N., HRUBOS, Z., DOSTAL, T., VRBA, K. Study of adjustable gains for control of oscillation frequency and oscillation condition in 3R-2C oscillator. Radioengineering, 2012, vol. 21, no. 1, p. 392-402.

30. GALAN, J., CARVALAJ, R. G., TORRALBA, A., MUNOZ, F., RAMIREZ-ANGULO, J. A low-power low-voltage OTA-C sinusoidal

oscillator with large tuning range. IEEE Transaction on Circuits and Systems I: Fundamental Theory and Applications, 2005, vol. 52, no. 2, p. 283-291.

31. BAJER, J., BIOLEK, D., BIOLKOVA, V., KOLKA, Z. Voltagemode balanced-outputs quadrature oscillator using FB-VDBAs. In Proc. of Int. Conf. on Microelectronics (ICM), 2010, p. 491-494.

32. BIOLEK, D., BIOLKOVA, V. Current-mode CDTA-based comparators. In Proc. of the 13th International Conference on Electronic Devices and Systems EDS IMAPS. Brno (Czech Republic), 2006, p. 6-10.

33. DE-MARCELLIS, A., DI-CARLO, C., FERRI, G., STORNELLI, V. A CCII-based wide frequency range square waveform generator. International Journal of Circuit Theory and Applications, 2013, vol. 41, no. 1, p. 1-13.

34. CHIEN, H-CH. Voltage-controlled dual slope operation square/triangular wave generator and its application as a dual mode operation pulse width modulator employing differential voltage current conveyors. Microelectronics Journal, 2012, vol. 43, no. 12, p. 962-974.

35. ALMASHARY, B., ALHOKAIL, H. Current-mode triangular wave generator using CCIIs. Microelectronics Journal, 2000, vol. 31, no. 4, p. 239-243.

36. PAL, D., SRINIVASULU, B. B., PAL, A., DEMOSTHENOUS, B., DAS, N. Current conveyor-based square/triangular waveform generators with improved linearity. IEEE Transaction on Instrumentation and Measurement, 2009, vol. 58, no. 7, p. 2174-2180.

37. SAQUE, A. S., HOSSAIN, M. M., DAVIS, W. A., RUSSELL, H. T., CARTER, R. L. Design of sinusoidal, triangular, and square wave generator using current feedback amplifier (CFOA). In Proc. of IEEE Region 5 Conference. Kansas City (USA), 2008, p. 1–5.

38. MINAEI, S., YUCE, E. A simple Schmitt trigger circuit with grounded passive elements and its application to square/triangular wave generator. Circuits, Systems, and Signal Processing, 2012, vol. 31, no. 3, p. 877-888.

39. LO, Y. K., CHIEN, H. C. Switch-controllable OTRA-based square/triangular waveform generator. IEEE Transaction on Circuits, Systems and Signal Processing II, 2007, vol. 54, no. 12, p. 1110-1114.

40. KIM, K., CHA, H. W., CHUNG, W. S. OTA-R Schmitt trigger with independently controllable threshold and output voltage levels. Electronics Letters, 1997, vol. 33, no. 13, p. 1103-1105.

41. CHUNG, W. S., KIM, H., CHA, H. W., KIM, H. J. Triangular/ square-wave generator with independently controllable frequency and amplitude. IEEE Transactions on Instrumentation and Measurement, 2005, vol. 54, no. 1, p. 105-109.

42. CHUNG, W. S., CHA, H. W., KIM, H. J. Current-controllable monostable multivibrator using OTAs. IEEE Transactions on Circuits and Systems I: Fundamental Theory and Applications, 2002, vol. 49, no. 5, p. 703-705.

43. SIRIPRUCHYANUN, M., WARDKEIN, P. A full independently adjustable, integrable simple current controlled oscillator and derivative PWM signal generator. IEICE Trans. Fundam. Electron. Commun. Comput. Sci., 2003, vol. E86-A, no. 12, p. 3119-3126.

44. KUMBUN, J., SIRIPRUCHYANUN, M. MO-CTTA-based electronically controlled current-mode square/triangular wave generator. In Proc. of the 1st International Conf. on Technical Education (ICTE2009), 2010, p. 158–162.

45. SILAPAN, P., SIRIPRUCHYANUN, M. Fully and electronically controllable current-mode Schmitt triggers employing only single MO-CCCDTA and their applications. Analog Integrated Circuits and Signal Processing, 2011, vol. 68, no. 11, p. 111-128.

46. SRISAKUL, T., SILAPAN, P., SIRIPRUCHYANUN, M. An electronically controlled current-mode triangular/square wave generator employing MO-CCCCTAs. In Proc. of the 8th Int. Conf. on Electrical Engineering/ Electronics, Computer, Telecommunications, and Information Technology. 2011, p. 82-85.

47. MOSIS parametric test results of TSMC LO EPI SCN018 technology.

48. Online. Cited 2012-05-24. Available at: ftp: //ftp.isi.edu/pub/mosis/vendors/tsmc-018/t44e lo epi-params.txt.

49. BAKER, J. CMOS Circuit Design, Layout and Simulation. IEEE Press Series on Microelectronic Systems, 2005.

Chapter 7

DIFFERENTIAL DIFFERENCE CURRENT CONVEYOR TRANSCONDUCTANCE AMPLIFIER: A NEW ANALOG BUILDING BLOCK FOR SIGNAL PROCESSING

Neeta Pandey[1] and Sajal K. Paul[2]

[1]Department of Electronics and Communication Engineering, Delhi Technological University, Delhi 110042, India

[2]Department of Electronics Engineering, Indian School of Mines, Jharkhand 826004, India

ABSTRACT

A new active building block for analog signal processing, namely, differential difference current conveyor transconductance amplifier (DDCCTA), is presented, and performance is checked through PSPICE simulations which show the usability of the proposed element is up to 201 MHz. The proposed block is implemented using 0.25 μm TSMC CMOS technology. Some of the applications are presented using the proposed DDCCTA, namely, a voltage mode multifunction filter, a current mode universal filter, an oscillator, current and voltage amplifiers, and grounded inductor simulator. The feasibility of DDCCTA and its applications is confirmed via PSPICE simulations.

INTRODUCTION

The analog integrated circuit design in current mode is receiving increased attention due to some potential performance features like wide bandwidth, less circuit complexity, wide dynamic range, low power consumption, and high operating speed [1]. The current mode approach has emerged as an alternate method besides the traditional voltage mode circuits. The current mode active elements are appropriate to operate with signals in current or voltage or mixed mode and are gaining acceptance as building blocks in high-performance circuit designs. A number of current mode active elements such as operational transconductance amplifier (OTA) [2], current conveyors (CCs)

[3–5], differential voltage current conveyor (DVCC) [6], differential difference current conveyor (DDCC) [7], current feedback operational amplifier (CFOA) [8] are available in the literature.

Recently some new analog building blocks, such as current conveyor transconductance amplifier (CCTA) [9, 10], current controlled current conveyor transconductance amplifier (CCCCTA) [11], current difference transconductance amplifier (CDTA) [12], current controlled current difference transconductance amplifier (CCCDTA) [13], differen tial voltage current conveyor transconductance ampli- fier (DVCCTA) [14], and differential voltage current controlled conveyor transconductance amplifier (DVCCCTA) [15], are reported in the literature. These may be constructed by cascading of current mode building blocks with transconductance amplifier (TA) analog building blocks in monolithic chip for compact implementation of signal processing circuits and systems. It is well known that DDCC has some advantages [7, 16, 17] specially for applications demanding differential and floating inputs, over CCII or CCCII owing to three high input impedance terminals for DDCC compared to one high input impedance terminal for CCII or CCCII. However, DDCC does not have a powerful inbuilt tuning property in contrast to CCCII. The DDCC is more versatile than DVCC as it has an extra high input impedance terminal.

The main intention of this paper is to propose a new active building block, namely, differential difference current conveyor transconductance amplifier (DDCCTA), which has DDCC [7] as input block and is followed by a TA. The DDCCTA has all the good properties of CCTA, CCCCTA, and DVCCTA including the possibility of inbuilt tuningof the parameters of the signal processing circuits to be implemented and also all the versatile and special properties of DDCC such as easy implementation of differential and floating input circuits.

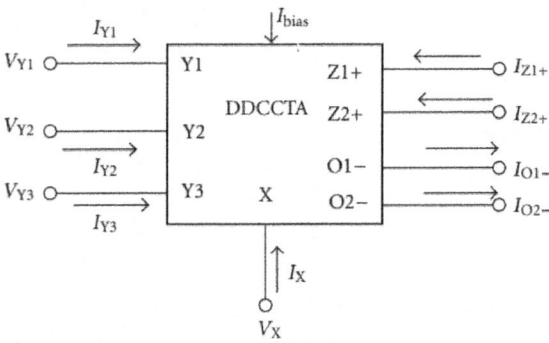

Figure 1: Circuit symbol of the proposed DDCCTA.

However, the same may be implemented using separate DDCC and OTA analog building blocks, but it will be more convenient and useful if DDCCTA is implemented in monolithic chip which will result in compact implementation of signal processing circuits and systems. Section 2 deals with the proposed DDCCTA circuit and some of its properties. Section 3 is devoted for some of its applications in developing signal processing circuits such as voltage mode (VM) filter, current mode (CM) filter, oscillator, current and voltage amplifier, and grounded inductor simulator. The functionality of all the proposed circuits has been verified using SPICE simulations. The conclusion is given in Section 4.

PROPOSED DDCCTA

The DDCCTA is based on DDCC [7] and consists of differential amplifier, current mirrors, and TA. The port relationships of the DDCCTA as shown in Figure 1 can be characterized by the following matrix:

$$
\begin{bmatrix} I_{Y1} \\ I_{Y2} \\ I_{Y3} \\ V_X \\ I_{Z1+} \\ I_{Z2+} \\ I_{01} \\ I_{02-} \end{bmatrix} = \begin{bmatrix} 0 & 0 & 0 & 0 & 0 & 0 & 0 & 0 \\ 0 & 0 & 0 & 0 & 0 & 0 & 0 & 0 \\ 0 & 0 & 0 & 0 & 0 & 0 & 0 & 0 \\ 1 & -1 & 1 & 0 & 0 & 0 & 0 & 0 \\ 0 & 0 & 0 & 1 & 0 & 0 & 0 & 0 \\ 0 & 0 & 0 & 1 & 0 & 0 & 0 & 0 \\ 0 & 0 & 0 & 0 & -g_m & 0 & 0 & 0 \\ 0 & 0 & 0 & 0 & -g_m & 0 & 0 & 0 \end{bmatrix} \times \begin{bmatrix} V_{Y1} \\ V_{Y2} \\ V_{Y3} \\ I_X \\ V_{Z1+} \\ V_{Z2+} \\ V_{O1-} \\ V_{O2-} \end{bmatrix},
$$

(1)

where g_m is transcon/ductance of the DDCCTA.

The CMOS-based internal circuit of DDCCTA in CMOS is depicted in Figure 2. It consists of the circuit of DDCC [7] (transistors M1 to M14) followed by a transconductance amplifier (transistors M15 to M24). The derivation of port relationships is given in Sections 2.1 to 2.3 [13].

Relationship between Voltages of X Port and Y1, Y2, and Y3 Ports

The voltage at X port may be found by analyzing the differential difference part (comprising of transistors M1 to M10) of the circuit of Figure 2 as follows:

$$V_X = \beta_1 V_{Y1} - \beta_2 V_{Y2} + \beta V_{Y3} + \varepsilon_V,$$

(2)

where

$$\beta_1 = \frac{1}{P_1}\left(g_{m3}g_{m6} + \frac{g_{m3}\left(g_{m4}g_{m5} - g_{m3}g_{m6}\right)}{g_{m3} + g_{m4}}\right),$$

$$\beta_2 = \frac{1}{P_1}\left(g_{m1}g_{m5} + \frac{g_{m1}\left(g_{m1}g_{m5} - g_{m2}g_{m6}\right)}{g_{m1} + g_{m2}}\right),$$

$$\beta_3 = \frac{1}{P_1}\left(g_{m2}g_{m6} + \frac{g_{m2}\left(g_{m1}g_{m5} - g_{m2}g_{m6}\right)}{g_{m1} + g_{m2}}\right),$$

$$\varepsilon_V = -\frac{I_B}{P_1}\left(\frac{g_{m1}g_{m5} - g_{m2}g_{m6}}{g_{m1} + g_{m2}} + \frac{g_{m4}g_{m5} - g_{m3}g_{m6}}{g_{m3} + g_{m4}}\right),$$

$$P_1 = g_{m4}g_{m5} - \frac{g_{m4}\left(g_{m4}g_{m5} - g_{m3}g_{m6}\right)}{g_{m3} + g_{m4}},$$

$$(3)$$

and I_B represents current through transistor Mi (i = 7, 8, 10, 12, 14). With matched transconductances $g_{m1} = g_{m2} = g_{m3} = g_{m4}$ and $g_{m5} = g_{m6}$, V_X is obtained as

$$V_X = V_{Y1} - V_{Y2} + V_{Y3}.$$

$$(4)$$

Relationship between Currents at Z1+, Z2+, and X Ports

The analysis of the portion of the circuit comprising of transistors M9 to M14 of the circuit of Figure 2 gives

$$I_{Z1+} = \alpha_1 I_X + \varepsilon_{I1},$$

$$I_{Z2+} = \alpha_2 I_X + \varepsilon_{I2},$$

$$(5)$$

where

$$\alpha_1 = \frac{g_{m11}}{g_{m9}}, \qquad \varepsilon_{I1} = \left(1 - \frac{g_{m11}}{g_{m9}}\right)I_B,$$

$$\alpha_2 = \frac{g_{m13}}{g_{m9}}, \qquad \varepsilon_{I2} = \left(1 - \frac{g_{m13}}{g_{m9}}\right)I_B.$$

$$(6)$$

For matched transconductances $g_{m9} = g_{m11} = g_{m13}$, the port currents are simplified to

$$I_{Z1+} = I_{Z2+} = I_X. \tag{7}$$

Relation for Currents at O1– and O2– Ports

The proposed DDCCTA contains a transconductor cell comprising of transistors M15 to M24. Assuming gate voltages of transistors M17 and M18 as V_{T1} and V_{T2}, the output currents $I_{O1}–$ and $I_{O2}–$ may be found, respectively, as

$$I_{O1-} = -(\gamma_1 V_{T1} - \gamma_2 V_{T2} + \varepsilon_{T1}),$$

$$I_{O2-} = -(\gamma_3 V_{T1} - \gamma_4 V_{T2} + \varepsilon_{T2}), \tag{8}$$

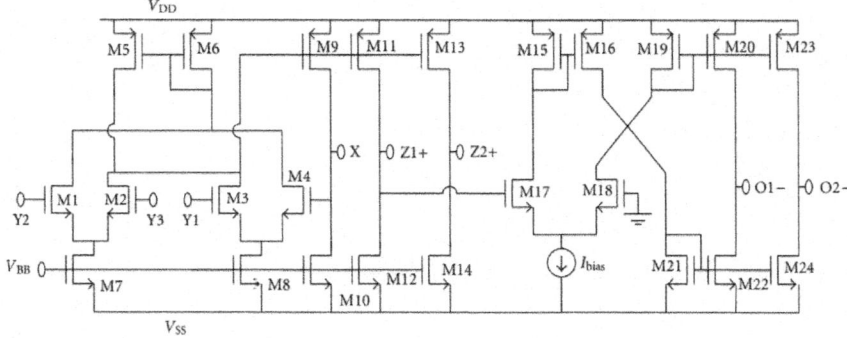

Figure 2: DDCCTA implementation.

where

$$\gamma_1 = g_{m17}\left(\frac{g_{m16}g_{m22}}{g_{m15}g_{m21}} - \frac{1}{g_{m15}g_{m19}g_{m21}}\right.$$

$$\times \left.\left(\frac{g_{m16}g_{m17}g_{m19}g_{m22} - g_{m15}g_{m18}g_{m20}g_{m21}}{g_{m17} + g_{m18}}\right)\right),$$

$$\gamma_2 = g_{m18}\left(\frac{g_{m20}}{g_{m19}} + \frac{1}{g_{m15}g_{m19}g_{m21}}\right.$$

$$\times \left.\left(\frac{g_{m16}g_{m17}g_{m19}g_{m22} - g_{m15}g_{m18}g_{m20}g_{m21}}{g_{m17} + g_{m18}}\right)\right),$$

$$\gamma_3 = g_{m17}\left(\frac{g_{m16}g_{m24}}{g_{m15}g_{m21}} - \frac{1}{g_{m15}g_{m19}g_{m21}}\right.$$

$$\left.\times\left(\frac{g_{m16}g_{m17}g_{m19}g_{m24} - g_{m15}g_{m18}g_{m23}g_{m21}}{g_{m17} + g_{m18}}\right)\right),$$

$$\gamma_4 = g_{m18}\left(\frac{g_{m23}}{g_{m19}} + \frac{1}{g_{m15}g_{m19}g_{m21}}\right.$$

$$\left.\times\left(\frac{g_{m16}g_{m17}g_{m19}g_{m24} - g_{m15}g_{m18}g_{m23}g_{m21}}{g_{m17} + g_{m18}}\right)\right),$$

$$\varepsilon_{T1} = -\frac{I_{Bias}}{g_{m15}g_{m19}g_{m21}}$$

$$\times\left(\frac{g_{m16}g_{m17}g_{m19}g_{m22} - g_{m15}g_{m18}g_{m20}g_{m21}}{g_{m17} + g_{m18}}\right),$$

$$\varepsilon_{T2} = -\frac{I_{Bias}}{g_{m15}g_{m19}g_{m21}}$$

$$\times\left(\frac{g_{m16}g_{m17}g_{m19}g_{m24} - g_{m15}g_{m18}g_{m23}g_{m21}}{g_{m17} + g_{m18}}\right).$$

(9)

With $g_{m17} = g_{m18}$, $g_{m21} = g_{m22} = g_{m24}$, $g_{m15} = g_{m16} = g_{m19} = g_{m20} = g_{m23}$, the output currents $I_{O1}-$ and $I_{O2}-$ reduce to

$$I_{O1-} = -(g_{m17}V_{T1} - g_{m18}V_{T2}),$$

(10)

$$I_{O2-} = -(g_{m17}V_{T1} - g_{m18}V_{T2}).$$

(11)

In the circuit, $V_{T1} = V_{Z1}+$ and $V_{T2} = 0$; hence the output currents are simplified to

$$I_{O1-} = -g_{m17}V_{Z1+}, \qquad I_{O2-} = -g_{m17}V_{Z1+}.$$

$$(12)$$

The value of g_{m17} is obtained as $\sqrt{2\mu C_{ox}(W/L)_{17}I_{Bias}}$ if transistors are biased in strong inversion region and $2I_{Bias}/V_T$ ($V_T = KT/q$) if transistors are biased in subthreshold region which can be adjusted by bias current I_{Bias}.

Simulation

To validate the behaviour of the proposed element, PSPICE simulations have been carried out using TSMC 0.25 μm CMOS process model parameters. The supply voltages of $V_{DD} = -V_{SS} = 1.25$ V and $V_{BB} = -0.8$ V are used. The aspect ratio of various transistors for DDCCTA is given in Table 1. The DC transfer characteristics of the proposed DDCCTA from Y1, Y2, and Y3 terminals to X terminal are shown in the Figure 3. It is clear that the voltage at X terminal follows the Y terminal voltages in the range of −200 mV to +200 mV. The variation of current at Z_{i+} and Z_{2+} terminals with X terminal current from −100 μA to 100 μA is shown in Figure 4. It may be noted that there is deviation for current below −80 μA. The variation of the transconductance value by changing I_{Bias} from 0 to 500 μA is depicted in Figure 5. The decreases in transconductance for larger bias currents than 450 μA or so is due to transistors (M_{17}, M_{18}) entering in linear region of operation from saturation region. The maximum transconductance is about 1.6 mS. The other circuit performance parameters of the DDCCTA are summarised in Table 2.

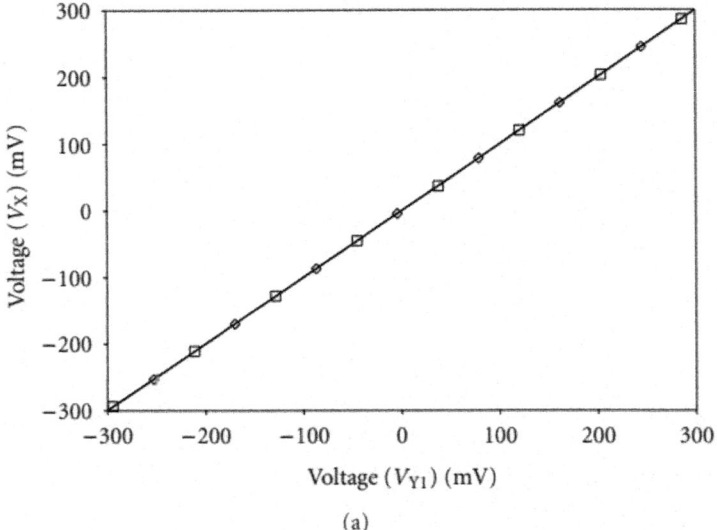

(a)

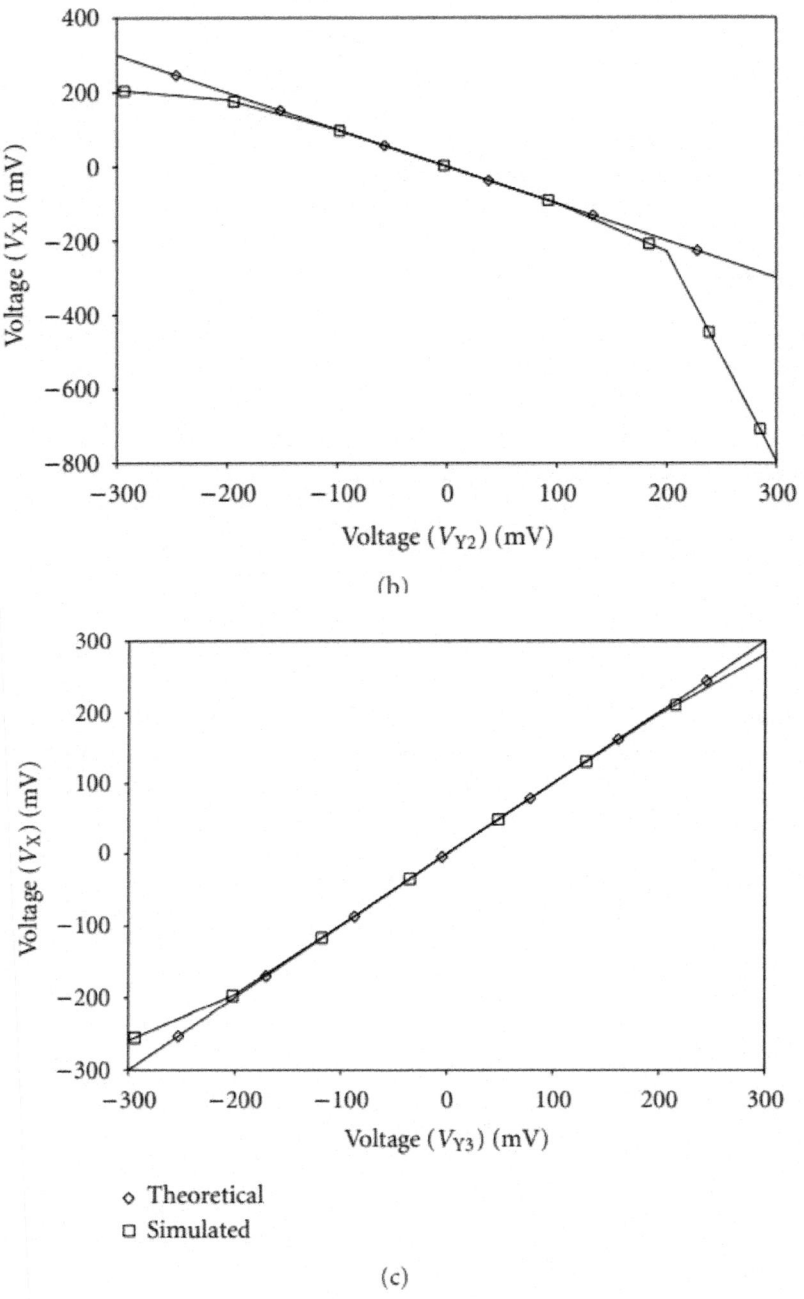

(b)

(c)

○ Theoretical
□ Simulated

Figure 3: DC transfer characteristic for voltage transfer from (a) Y1 port to X port, (b) Y2 port to X port, and (c) Y3 port to X port.

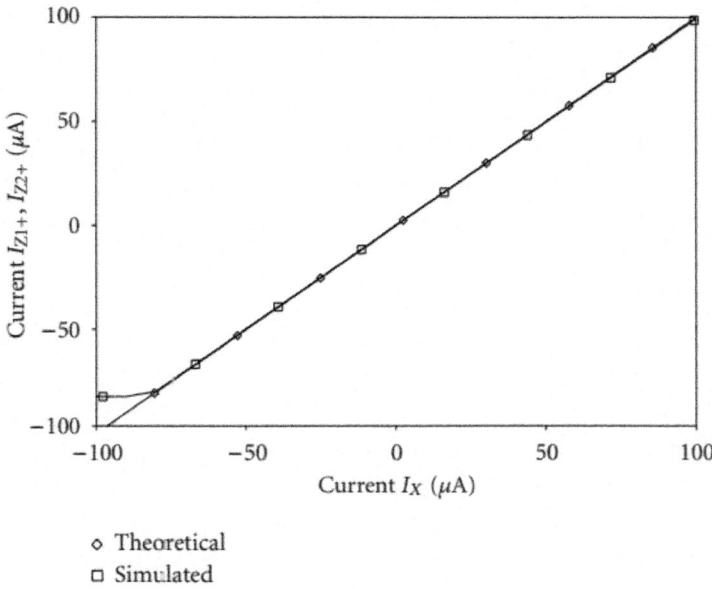

Figure 4: DC transfer characteristic for current transfer from X port to Z_{1+} and Z_{2+} ports.

Table 1: Aspect ratio of various transistors

Transistor	Aspect ratio (W(μm)/L(μm))
M1–M4	10/0.5
M5, M6	5/0.5
M7, M8	27.25/0.5
M9, M11, M13	8.5/0.5
M10, M12, M14	44/0.5
M15, M16, M19–M24	5/0.5
M17, M18	27/0.5

APPLICATIONS

Multifunction Voltage Mode Filter

In this section a multifunction voltage mode (VM) filter is proposed. It uses a single DDCCTA, two grounded capacitors, and a grounded resistor. The proposed multifunction VM filter is shown.

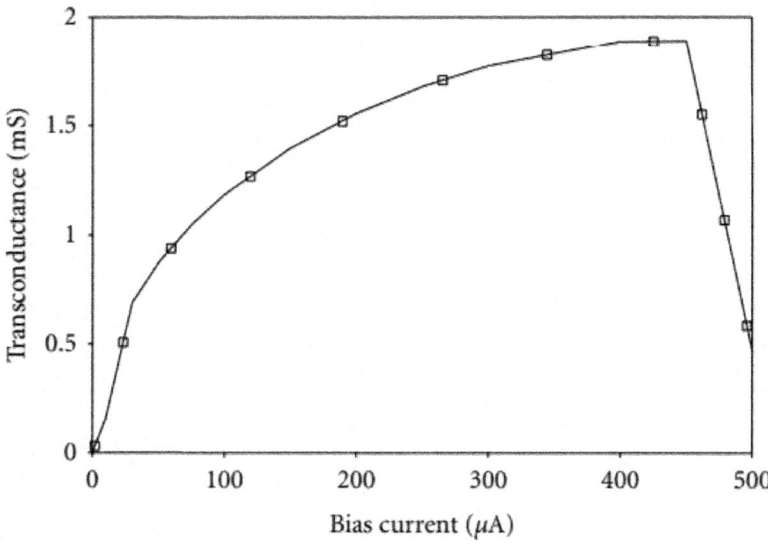

Figure 5: Variation of transconductance with bias current.

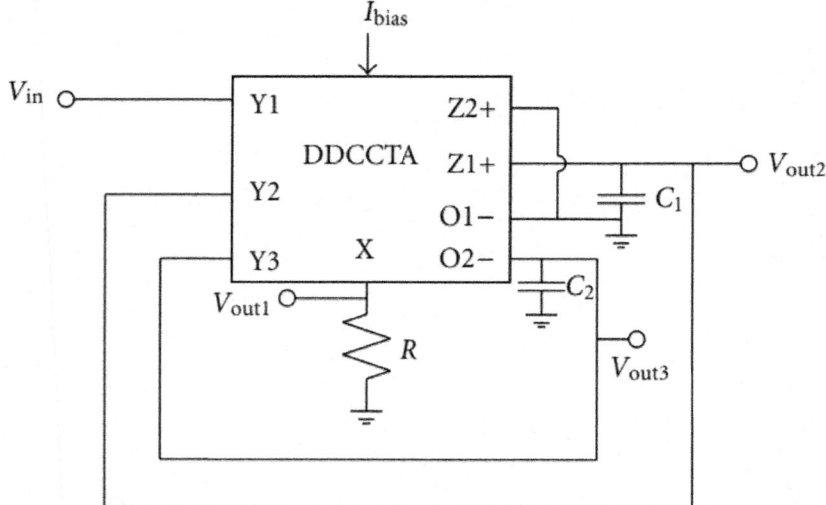

Figure 6: Proposed voltage mode filter.

Table 2: Circuit performance parameters of the DDCCTA

Power consumption	1.8 mW
Input voltage linear range (voltage inputs)	−200 mV to +200 mV
Input current linear range (current input)	−80 μA to 100 μA
Parasitic at Y ports (R_Y, C_Y)	very high, 20 fF
Parasitic at Z ports (R_Z, C_Z)	218 kΩ, 35 fF
Parasitic at $O−$ ports (R_{O-}, C_{O-})	324 kΩ, 20 fF
−3 dB bandwidth (at $I_{Bias} = 100\,\mu A$)	236 MHz for V_X/V_Y 223 MHz for I_Z/I_x 201 MHz for I_{O-}/V_z
Input bias range for controlling transconductance amplifier	10 nA to 450 μA

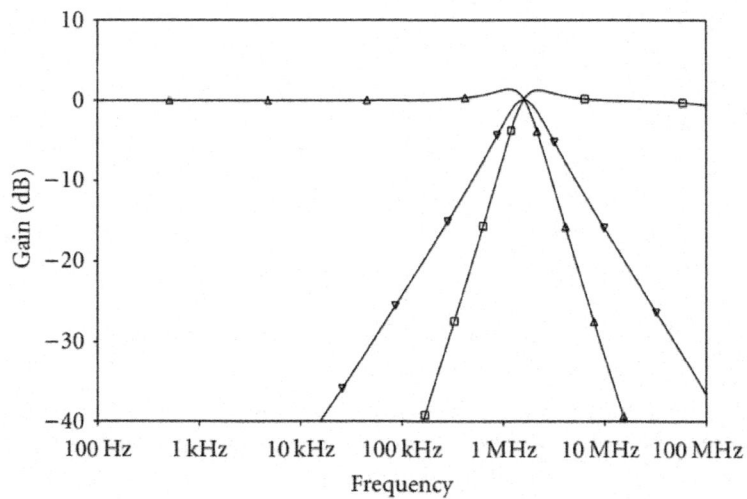

Figure 7: Simulated responses of the proposed voltage mode filter.

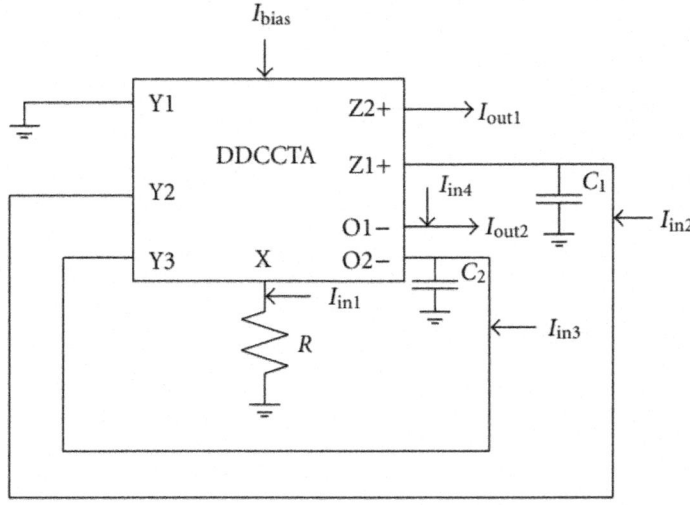

Figure 8: Current mode universal filter.

Figure 6. The analysis of circuit yields the output voltages at various nodes as

$$\frac{V_{\text{out1}}}{V_{\text{in}}} = \frac{s^2 C_1 C_2 R}{D(s)},$$

$$\frac{V_{\text{out2}}}{V_{\text{in}}} = \frac{s C_2}{D(s)},$$

$$\frac{V_{\text{out3}}}{V_{\text{in}}} = -\frac{g_m}{D(s)},$$

(13)

where

$$D(s) = s^2 C_1 C_2 R + s C_2 + g_m.$$

(14)

It may be observed from (13) that high-pass, band-pass, and low-pass responses are available simultaneously at V_{out1}, V_{out2}, and V_{out3}, respectively. Thus, the proposed structure is a single-input-and-three-output voltage mode filter. It may be noted that no component matching constraint is required. The responses are characterized by pole frequency (ω_0), bandwidth (ω_0/Q_0), and quality factor (Q_0) as

$$\omega_0 = \left(\frac{g_m}{R C_1 C_2}\right)^{1/2}, \qquad \frac{\omega_0}{Q_0} = \frac{1}{R C_1}, \qquad Q_0 = \left(\frac{g_m R C_1}{C_2}\right)^{1/2}.$$

(15)

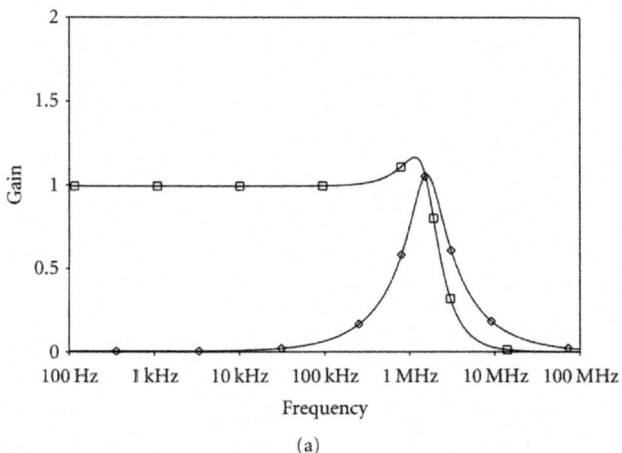

(a)

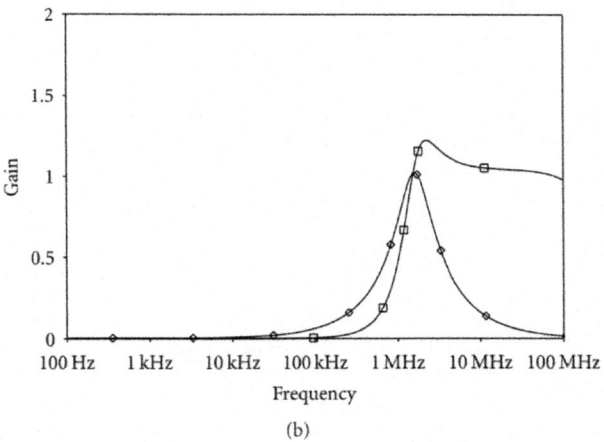

(b)

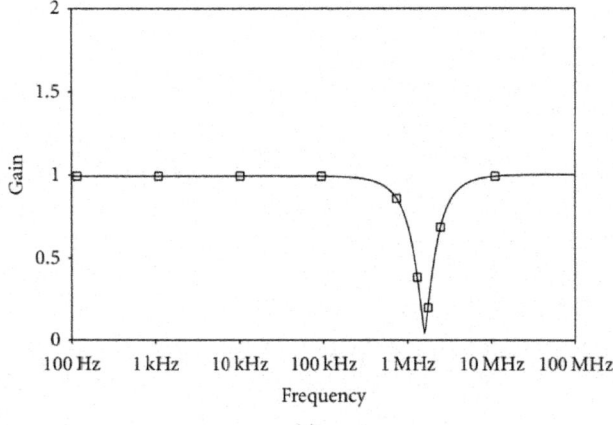

(c)

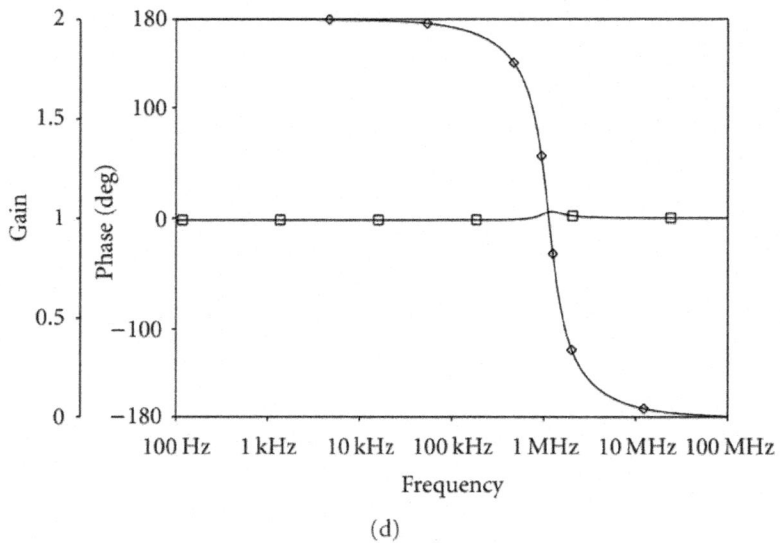

(d)

Figure 9: Simulated responses of the proposed current mode universal filter: (a) low pass and band pass; (b) high pass and band pass; (c) notch; (d) all pass.

Table 3: Comparative study of the available similar type of single-active-element-based VM filters

Reference	Active element used and number of passive components	Grounded passive components	Availability of LP, HP, BP, notch, and AP responses	Number of available simultaneous responses	No inversion of input voltage	No requirement of gain of input signal such as $2V_{in}$ and $3V_{in}$	Tunability	No matching condition
[11]	CCCCTA, 2	No	All	1	No	No	Yes	No
[15]	DVCCCTA, 2	Yes	2	2	Yes	Yes	Yes	Yes
[18]	DBTA, 5	No	LP, HP, BP, Notch	1	Yes	Yes	Yes	No
[19]	DBTA, 4	No	All	2	Yes	Yes	Yes	No
Proposed	DDCCTA, 3	Yes	LP, HP, BP	3	Yes	Yes	Yes	Yes

LP: low pass; HP: high pass; BP: band pass; Notch: notch response; AP: all pass.

Equation (15) reveals that for high-pass and band-pass responses the pole frequency (ω_0) and quality factor (Q_0) can be adjusted by gm, that is, by bias current of DDCCTA, without disturbing $\omega 0/Q_0$. The ω_0 and Q0 are orthogonally adjustable with simultaneous adjustment of gm and R such that the product $g_m R$ remains constant and the quotient gm/R varies and vice versa. The resistance R being a grounded one may easily be implemented as a variable resistance using only two MOS [17]. Equation (15) also indicates that high values of Q-factor will be obtained from moderate values of ratios of passive components, that is, from low component spread [22]. These ratios can be chosen as $g_m R = (C_1/C_2) = Q_0$. Hence, the spread of the component values becomes of the order of $\sqrt{Q_0}$. This feature of the filter related to the component spread allows the realization of high Q_0 values more accurately compared to the topologies where the spreadof passive components becomes Q_0 or Q_0^2.

Table 4: The I_{in1}, I_{in2}, I_{in3}, and I_{in4} values selection for each filter function response

| Filter responses | Inputs | | | | Output |
	I_{in1}	I_{in2}	I_{in3}	I_{in4}	
Low pass	0	0	1	0	I_{out2}
	0	0	1	0	I_{out1}
Band pass	1	0	0	0	I_{out2}
	0	1	0	0	I_{out2}
High pass	1	0	0	0	I_{out1}
Notch	0	1	0	1	$I_{out2}, R = 1/g_m$
All pass	0	1	0	1	$I_{out2}, R = 1/g_m$

Table 5: Comparative study of the available similar type of single-active-element-based CM filters

Reference	Active element used and number of passive components	Grounded passive components	Availability of LB, HP, BP, notch, and AP responses	No requirement of current inversion	No requirement of gain of input signal such as $2I_{in}$ and $3I_{in}$	Output current at high impedance	Tunability	Simple/no matching condition
[11]	CCCCTA, 2	Yes	Yes	Yes	No	Yes	Yes	Yes
[13]	CCCDTA, 2	Yes	Yes	No	Yes	No	Yes	Yes
[20]	CCCDTA, 2	Yes	Yes	Yes	No	Yes	Yes	Yes
[21]	CCCCTA, 2	Yes	No	Yes	Yes	No	Yes	Yes
Proposed	DVCCTA, 3	Yes	Yes	Yes	Yes	Yes	Yes	Yes

LP: low pass; HP: high pass; BP: band pass; Notch: notch response; AP: all pass.

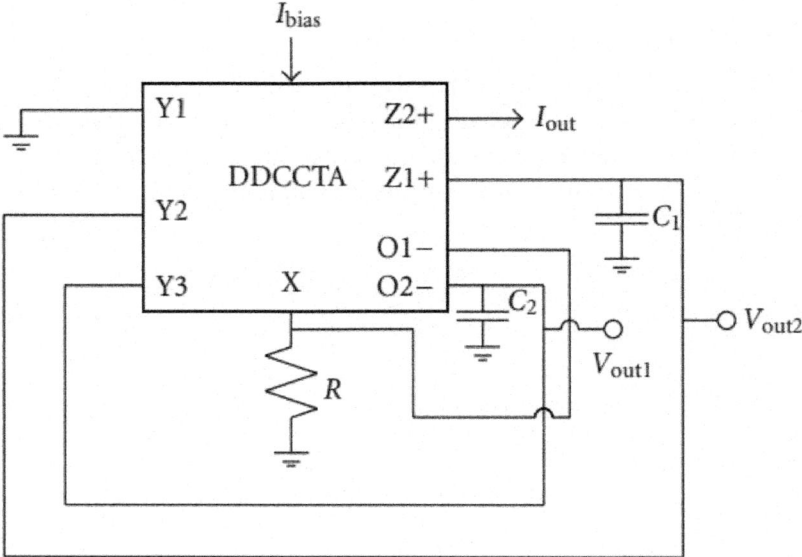

Figure 10: Proposed Oscillator.

It can also be easily evaluated to show that the sensitivities of pole ω_0 and pole Q_0 are within unity in magnitude. Thus, the proposed structures, can be classified as insensitive.

A detailed study of the available similar type of singleactive-element-based (such as CCCTA, DBTA, and DVCCCTA) voltage mode filters and the proposed one is given in Table 3. It reveals that the topology [18, 19] uses excessive number of passive components whereas the proposed topology uses one extra passive component, namely, resistor (R) than [11, 15]. The proposed topology also provides the availability of a maximum number of simultaneous responses. Structures [11, 18, 19] use floating passive components andalso use matching condition. Topology [11] needs input signal V_{in}, $-V_{in}$, and $-2V_{in}$; hence there is requirement of additional circuits. Thus, it reveals that although the proposed topology realizes only LP, HP, and BP responses, it has two or more advantages over the other available topologies [11, 15, 18, 19].

To verify the functionality of the proposed single DDCCTA-based voltage mode filter, SPICE simulations have been carried out using TSMC 0.25 μm CMOS process model parameters and supply voltages of $V_{DD} = -V_{SS} = 1.25$ V and $V_{BB} = -0.8$ V. The filter is designed for a pole frequency of $f_0 = 1.59$ MHz, $Q = 1$, the component values are found to be $C_1 = C_2 = 100$ pF, $R = 1$ kΩ, and bias current of DVCCTA equals 100 μA. Figure 7 shows the simulation results for high-pass (V_{out1}), band-pass (V_{out2}), and low-pass (V_{out3}) filter responses which are available simultaneously.

MISO Current Mode Universal Filter

A multiple-input single-output (MISO) universal current mode (CM) filter is proposed in this section which is obtained by grounding voltage input in Figure 6 and exciting it with current inputs as shown in Figure 8. It employs a single DDCCTA, two grounded capacitors, and a grounded resistor. Analysis of this circuit gives the output current as

$$I_{out1} = \frac{-s^2 C_1 C_2 R I_{in1} - (sC_2 + g_m) I_{in2} + sC_1 I_{in3}}{D(s)},$$

$$I_{out2} = \frac{(I_{in1} - I_{in2})sC_2 R g_m + g_m I_{in3} + D(s) I_{in4}}{D(s)}, \tag{16}$$

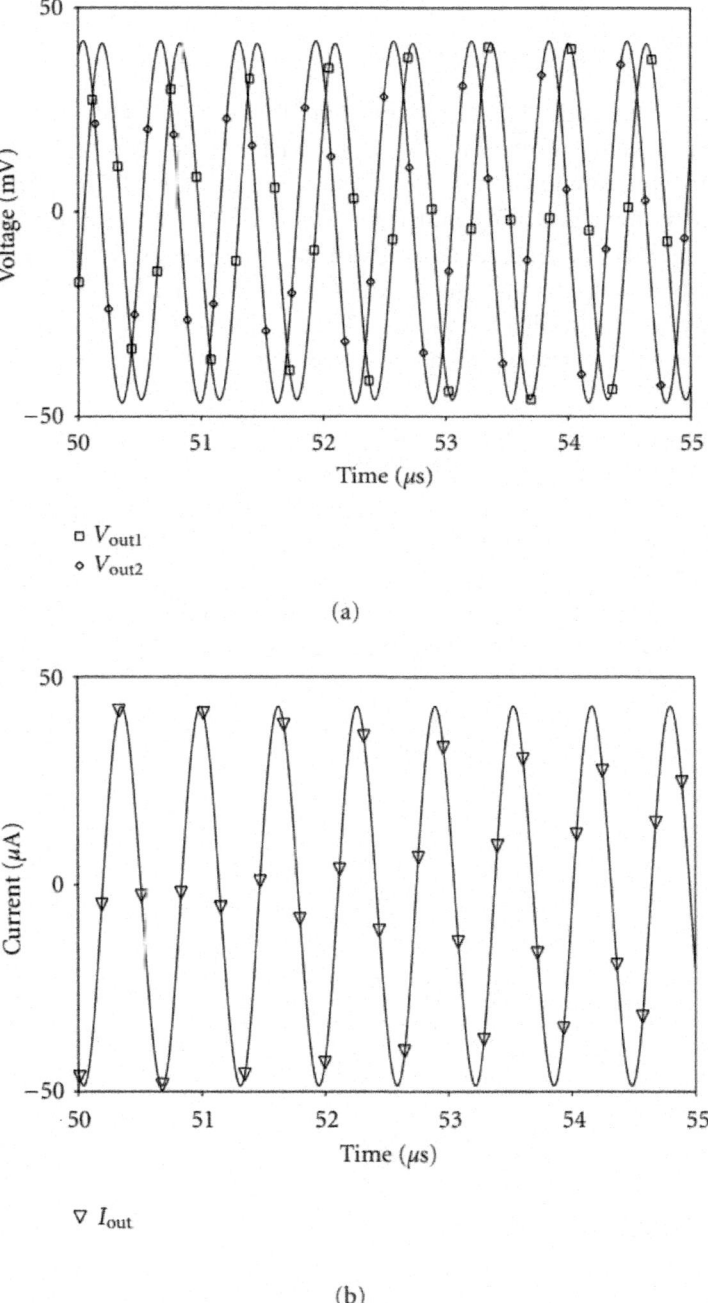

Figure 11: Outputs of the proposed oscillator: (a) quadrature output voltages at V_{out1} and V_{out2}; (b) current oscillation at I_{out}.

where

$$D(s) = s^2 C_1 C_2 R + s C_2 + g_m.$$

(17)

Table 4 shows the availability of each filter response and the corresponding selection of input currents I_{in1}, I_{in2}, I_{in3}, and I_{in4}. Thus, the proposed structure is a four-input-single output current mode filter. It may be noted that there is no component matching constraint for obtaining any filter response. The filter parameters are the same as given in (15). The grounded resistance (R) may easily be implemented as variable one using only two MOS [17] for full electronic control of filter parameters. The ω_0, Q_0, and ω_0/Q_0 can be orthogonally adjusted for low-pass, high-pass, and band-pass responses the way discussed in Section 3.1.

A detailed study of the available similar type of activeelement-based (such as CCCCTA, CCCDTA, and CCCCTA) CM filters and the proposed one is given in Table 5. It reveals that although the proposed structure needs one extra resistor, the reported structures [11, 13, 20, 21] suffer from one or more features. In addition some active elements are required to sense current in [13, 21]. Thus, structures in [11, 13, 20, 21] will require some extra circuits to compensate the shortcomings in their features in comparison to the proposed one.

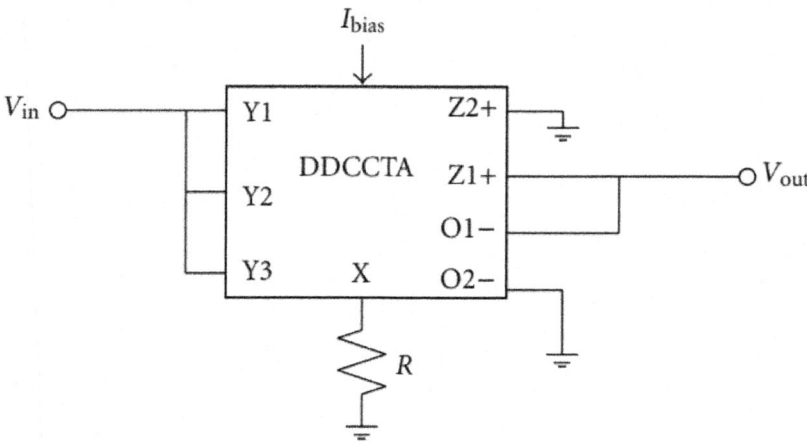

Figure 12: Voltage amplifier.

The proposed universal MISO current mode filter is validated through SPICE simulations. The circuit of Figure 8 for a pole frequency of $f_0 = 1.59$ MHz, Q = 1 has been designed with the component values of $C_1 = C_2 = 100$ pF, R = 1 kΩ, and bias current of DDCCTA equal to 100 μA. Figure 9(a) shows the simulation results for band pass (I_{out1}) and low pass (I_{out2}) filter responses which are available simultaneously for Iin = Iin3, Iin1 = Iin2 = Iin4 = 0. Figure 9(b) shows the simulation results for band pass (I_{out2}) and high-pass (I_{out1}) filter responses which are available simultaneously for $I_{in} = I_{in1}$, $I_{in2} = I_{in3} = I_{in4} = 0$. Notch and all pass responses are shown in Figures 9(c) and 9(d) with $I_{in} = I_{in2}$ = I_{in4}, $I_{in1} = I_{in3} = 0$ and R = 1 kΩ and 2 kΩ, respectively.

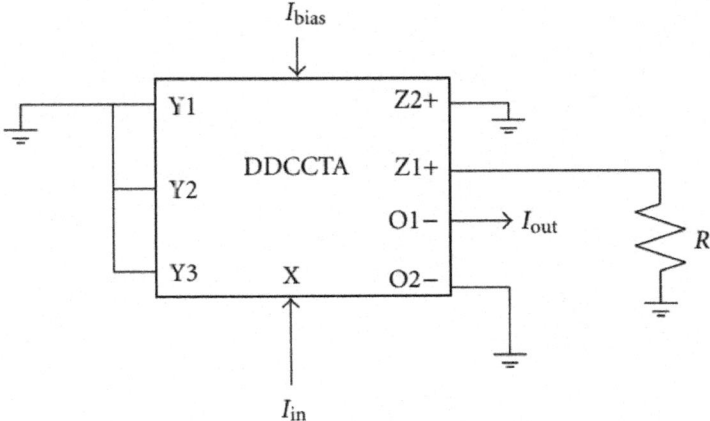

Figure 13: Current amplifier.

Oscillator

The current mode filter of Figure 8 may be used as oscillator when output I_{out2} is connected to I_{in1} as shown in Figure 10.

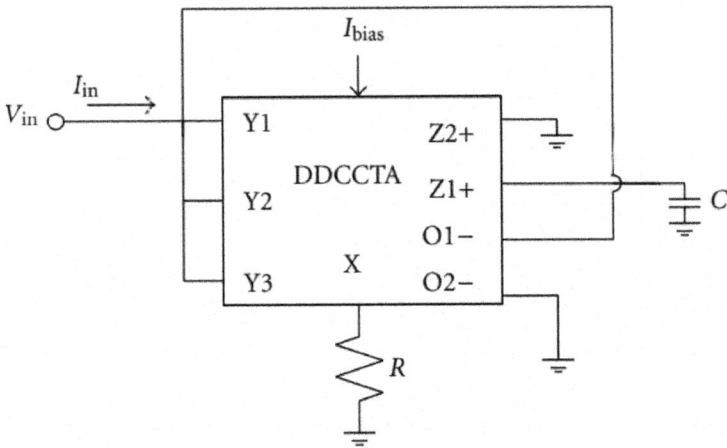

Figure 14: Grounded inductor simulator.

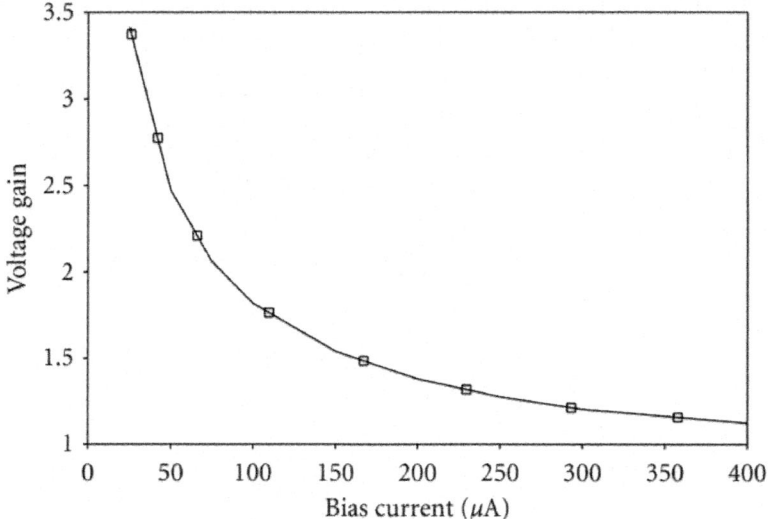

Figure 15: Voltage gain.

$$s^2 C_1 C_2 R + s C_2 (R g_m - 1) + g_m = 0.$$

$$(18)$$

The condition and frequency of oscillation may be computed as

$$\text{CO: } g_m = \frac{1}{R},$$

$$\text{FO: } \omega_0 = \sqrt{\frac{g_m}{R C_1 C_2}}.$$

$$(19)$$

The oscillations are available at outputs V_{out1} and V_{out2}, and they are related as

$$V_{out1} = -\frac{g_m}{s C_2} V_{out2}.$$

$$(20)$$

Thus, these voltages exhibit quadrature relationship. The current oscillations are also available at high output impedance at I_{out}.

To verify the proposed circuit, an oscillator was designed for 1.59 MHz with $C_1 = C_2 = 100$ pF, $R = 1$ kΩ, and bias current of 100 μA. The simulated current and quadrature voltage waveforms are shown in Figure 11 for which the total harmonic distortion is 1.28%.

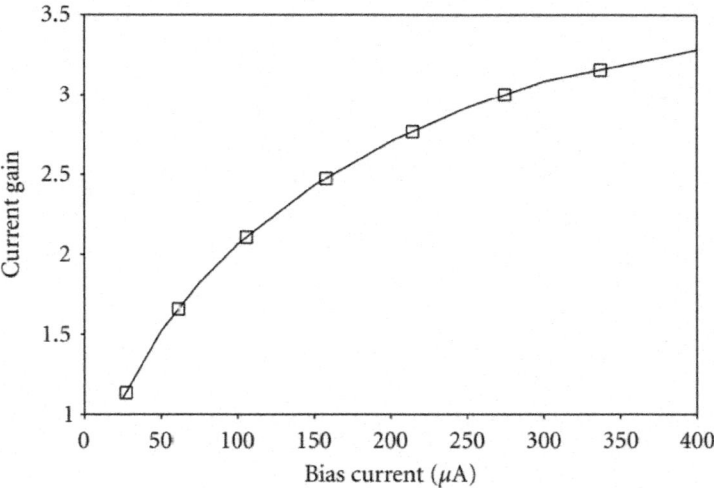

Figure 16: Current gain.

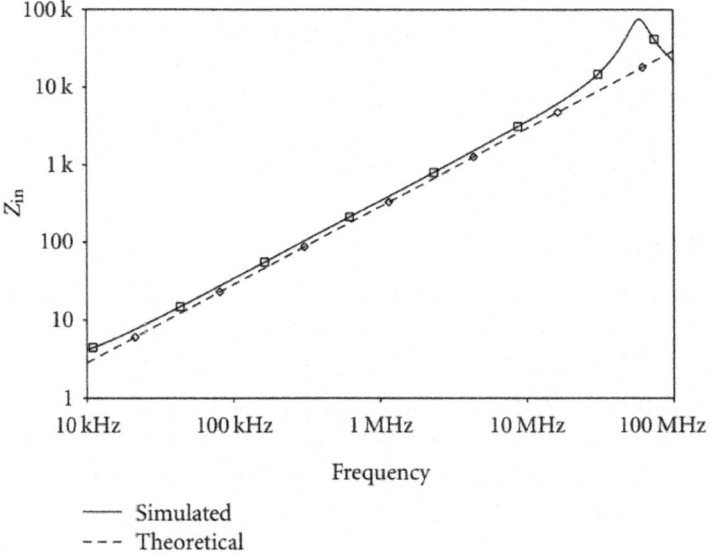

— Simulated
--- Theoretical

Figure 17: Frequency response of grounded inductor.

Voltage Amplifier, Current Amplifier, and Grounded Inductor

The proposed DDCCTA may also be configured for voltage and current amplifiers and grounded inductor simulator as shown in Figures 12, 13, and 14, respectively. The transfer functions may be expressed as follows:

(i)   voltage amplifier:

$$\frac{V_{out}}{V_{in}} = \frac{1}{Rg_m},$$

(21)

(ii)   current amplifier:

$$\frac{I_{out}}{I_{in}} = Rg_m,$$

(22)

(iii)   grounded inductor simulator:

$$Z_{in} = \frac{V_{in}}{I_{in}} = \frac{sCR}{g_m}.$$

(23)

It may be noted that the gain of amplifiers and inductance can be adjusted by g_m, that is, by varying bias current of DDCCTA.

To verify the proposed amplifiers, the simulations have been carried out for $R = 0.5\ k\Omega$ ($2\ k\Omega$) for voltage (current) amplifier and bias current of $25\ \mu A$ to $400\ \mu A$. The results are depicted in Figures 15 and 16. The inductor simulator is also validated through simulation with $R = 0.5\ K\Omega$, $C = 100\ pF$, and bias current of $90\ \mu A$. Figure 17 shows the simulated and theoretical results and there is close agreement between the two.

CONCLUSION

A new analog building block, namely, DDCCTA, is presented and some of its properties are discussed. It is found that the proposed DDCCTA is useful up to about 201 MHz. As applications of the proposed DDCCTA, VM multifunction filter, CM universal filter, quadrature oscillator, voltage and current amplifiers, and grounded inductor simulator topology are presented. The resistor being grounded may easily be implemented as a variable one using only two MOS [17]. The simulation results verify the theory.

REFERENCES

1.   G. Ferri and N. C. Guerrini, Low-Voltage Low-Power CMOS Current Conveyors, Kluwer Academic, London, UK, 2003.

2.   R. L. Geiger and E. Sanchez-Sinencio, "Active filter design using operational transconductance amplifiers: a tutorial," IEEE Circuits and Devices Magazine, vol. 1, no. 2, pp. 20–23, 1985.

3. A. S. Sedra and K. C. Smith, "A second generation current conveyor its application," IEEE Transactions on Circuit Theory, vol. 17, no. 1, pp. 132–134, 1970.

4. A. Fabre, O. Saaid, F. Wiest, and C. Boucheron, "High frequency applications based on a new current controlled conveyor," IEEE Transactions on Circuit and Systems I, vol. 43, pp. 82–91, 1996.

5. I. A. Awad and A. M. Soliman, "Inverting second generation current conveyors: the missing building blocks, CMOS realizations and applications," International Journal of Electronics, vol. 86, no. 4, pp. 413–432, 1999.

6. H. O. Elwan and A. M. Soliman, "Novel CMOS differential voltage current conveyor and its applications," IEE Proceedings—Circuits Devices Systems, vol. 144, pp. 195–200.

7. W. Chiu, S. I. Liu, H. W. Tsao, and J. J. Chen, "CMOS differential difference current conveyor and their applications," IEE Proceedings—Circuits Devices Systems, vol. 143, pp. 91–96, 1996.

8. C. Toumazou and A. Payne, "Current feedback opamp: a blessing in disguise?" IEEE Circuits and Devices Magazine, vol. 10, pp. 43–47, 1994.

9. R. Prokop and V. Musil, "CCTA-a new modern circuit block and its internal realization," in Proceedings of the International Conference on Electronic Devices and Systems, (IMAPSCZ '05), pp. 89–93, Brno, Czech Republic, 2005.

10. W. Jaikla, P. Silapan, C. Chanapromma, and M. Siripruchyanun, "Practical implementation of CCTA based on commercial CCII and OTA," in Proceedings of the International Symposium on Intelligent Signal Processing and Communication Systems, (ISPACS '08), pp. 1–4, Bangkok, Thailand, February 2009.

11. M. Siripruchyanun and W. Jaikla, "Current controlled current conveyor transconductance amplifier (CCCCTA): a building block for analog signal processing," Electrical Engineering, vol. 90, no. 6, pp. 443–453, 2008.

12. D. Biolek, "CDTA—building block for current-mode analog signal processing," in Proceedings of the European Conference on Circuit Theory and Design, (ECCTD '03), pp. 397–400, Krak ow, Poland, September 2003.

13. M. Siripruchyanun and W. Jaikla, "CMOS current-controlled current differencing transconductance amplifier and applications to analog signal

processing," International Journal of Electronics and Communications, vol. 62, no. 4, pp. 277–287, 2008.

14. A. Jantakun, N. Pisutthipong, and M. Siripruchyanun, "A synthesis of temperature insensitive/electronically controllable floating simulators based on DV-CCTAs," in Proceedings of the 6th International Conference on Electrical Engineering/Electronics, Computer, Telecommunications and Information Technology, (ECTI-CON '09), pp. 560–563, Pattaya, Chonburi, Thailand, May 2009.

15. W. Jaikla, M. Siripruchyanun, and A. Lahiri, "Resistorless dual-mode quadrature sinusoidal oscillator using a single active building block," Microelectronics Journal, vol. 42, no. 1, pp. 135–140, 2010.

16. H. P. Chen, "Versatile universal voltage-mode filter employing DDCCs," International Journal of Electronics and Communications, vol. 63, no. 1, pp. 78–82, 2009.

17. H. P. Chen, "High-input impedance voltage-mode multifunction filter with four grounded components and only two plustype DDCCs," Active and Passive Electronic Components, vol. 2010, Article ID 362516, 5 pages, 2010.

18. N. Herencsar, K. Vrba, J. Koton, and I. Lattenberg, "The conception of differential-input buffered and transconductance amplifier (DBTA) and its application," IEICE Electronics Express, vol. 6, no. 6, pp. 329–334, 2009.

19. N. Herencsar, J. Koton, K. Vrba, and I. Lattenberg, "New ´ voltage-mode universal filter and sinusoidal oscillator using only single DBTA," International Journal of Electronics, vol. 97, no. 4, pp. 365–379, 2010.

20. M. Siripruchyanun and W. Jaikla, "Electronically controllable current-mode universal biquad filter using single DOCCCDTA," Circuits, Systems, and Signal Processing, vol. 27, no. 1, pp. 113–122, 2008.

21. S. Mangkalakeeree, D. Duangmalai, and M. Siripruchyanun, "Current-mode KHN filter using single CCCCTA," in Proceedings of the 7th PSU Engineering Conference, pp. 306–309, HatYai, Thailand, May 2009.

22. S. I. Liu, "High input impedance filters with low component spread using current-feedback amplifiers," Electronics Letters, vol. 31, no. 13, pp. 1042–1043, 1995.

Chapter 8

POWER AMPLIFIERS FOR NEXT GENERATION WIRELESS PLATFORMS

Kevin Tom[1], Vandana Basoo[1], Mike Faulkner[1], Thomas Lejon[2]

[1]Centre for Telecommunications and Microelectronics Victoria University, Melbourne-3011, Australia

[2]Ericsson AB, Business Unit Access, SE 16480, Sweden

ABSTRACT

Class-E amplifier has the potential to deliver high efficiency required for the next generation wireless systems. In this journal, we discuss a novel load pull analysis technique to characterize the efficiency performance of Class-E amplifier in an outphasing power combining scheme. Class-E amplifier is not an ideal current or voltage source as is required for the traditional analysis of outphasing structures. It requires a phase modulated input signal and has a non-linear transfer characteristic which is a function of load impedance. Here we define an operating load locus based on the load pull analysis which can be used to predict the non-linear transfer function, efficiency, output power, input drive phase and many other factors associated with the outphasing class-E amplifier. This scheme could also be used to characterize any amplifier class in an outphasing structure. Finally modulation performance of Class-E amplifier using PWM technique is also presented.

INTRODUCTION

Wireless basestations will need to adapt to handle the new requirements of the next generation of wireless system (4G). Wider bandwidths, higher bit rates, multiple antenna structures and new bandwidth efficient orthogonal frequency division multiplexing (OFDM) modulation schemes are required [1-5]. Transceivers will need higher transmission powers, greater fidelity (reduced EVM – error vector magnitude) and greater integration. However they must fit into the same volume and require the same or less basestation air-conditioning load (or battery life for terminals). Heat dissipation must therefore be reduced, and this requires an increase in efficiency. The RF power amplifier (PA) dominates the efficiency performance.

Power amplifiers will play an important role in these next generation systems. This critical component dictates the size, cost and performance of the overall system. The main problem in power amplifier design is the efficiency. By improving the efficiency of a power amplifier less heat is dissipated in the device, this has many benefits. By reducing heat dissipation lower rated devices can be used for amplification and smaller heat sinks designed, this reduces the size of the overall system and increases its reliability, since heat is a major contributor of device failure. These size and reliability benefits, add up producing a cost effective solution.

There are two general types of PAs-linear PAs and switch mode PAs. Linear PAs like class-A and class-B output scaled versions of the input at the expense of efficiency. In contrast switch mode PAs offer the potential for better efficiency but are generally nonlinear. Recent advances in linearization techniques (Feed Forward [6], Pre-Distortion [7]) can overcome the latter deficiency.

Switch mode PAs run on one general principle of minimizing the power dissipated in the transistor. The ideal case would be if the transistor behaved as a switch. When the switch is open, no power is dissipated since there is no current flowing. When the switch is closed, no power is dissipated since there is no voltage across the switch. Since transistors are not perfect switches, the power can be minimized by making sure current and voltage are never high simultaneously. The efficiency of switch PAs varies with the output signal level which reduces their performance with modulated signals.

A tradeoff between efficiency and linearity always exists in PA design. Conducting class PAs such as class-A and class AB offer good linearity but are inefficient with envelope varying signals. On the other hand switching class PAs such as class-D, E and F have excellent efficiency but are very non linear [6]. A typical feed-forward class AB power amplifier has efficiency in the range of 10% with modulated signals [7]. Alternatively a switch mode class-E amplifier attains drain efficiency of 45% when operated with 8dB peak to average Rayleigh enveloped signal [8]. But they are highly non linear forcing the need for compensation in the modulator circuits. This journal looks at techniques to obtain the non-linear modulation characteristics which are optimized for power output and efficiency.

Linear amplification using non linear components (LINC) also known as outphasing [9-10] is one technique for obtaining linear operation from switched outputs, shown in Fig 1.

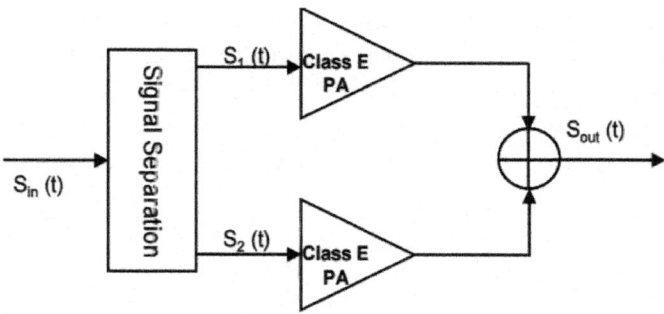

Figure 1: Outphasing Principle.

Here an amplitude and phase modulated signal $S_{in}(t) = |S(t)|exp(j\varphi(\tau))$ is decomposed into two constant amplitude signals ($S_1(t)$, $S_2(t)$) that can then be amplified using two highly efficient, non linear PAs, without adding any further distortion. The PA outputs are then summed to yield the signal for transmission $S_{out}(t)$. V_0 denotes the amplitude of each branch signal and θ the input drive phase that determines the amplitude, $|S(t)|$.

$$S_{in}(t)=S(t)cos[\omega t+\phi(t)]$$ (1)

$$S_1(t)=V_0cos[\omega t+\phi(t)+\theta(t)]$$ (2)

$$S_2(t)=V_0cos[\omega t+\phi(t)-\theta(t)]$$ (3)

$$S_{out}(t)=S_1(t)+S_2(t)=2V_0A(t)cos[\omega t+\phi(t)]$$ (4)

This scheme enables linear amplification of the original signal with system efficiency comparable to that of the used PAs. High efficiency of outphasing system is possible only if the PAs behave as ideal voltage sources. In such cases the efficiency of PAs is independent of the load impedance. Traditional LINC analysis using overdriven class-B, class-C or class-D amplifiers assume a voltage source [11, 12] for both amplifiers as this approximation is true. When it comes to the class-E amplifier it's not an ideal voltage or current source. Its output characteristic is a function of load impedance [13]. Therefore the traditional LINC analysis does not apply for class-E amplifiers instead we propose a load pull analysis method.

In practice saturated class B/C PAs cannot behaves as perfect voltage source when the load impedance goes reactive [12]. So this load pull analysis method is also applicable to any amplifier class used in an outphasing structure.

In this journal a load pull analysis of class-E amplifier is performed to chararcterise its LINC performance. This is possible as LINC is a load modulation. Load modulation is a technique in which amplitude modulated signals are produced by the dynamic variation of the load impedance of the power amplifier. Section II discusses class-E amplifier operation. Section III examines outphasing power combining schemes. Section IV considers the load pull analysis of class-E amplifier using LINC power combining technique. Section V discusses pulse width modulation performance of the Class-E amplifier.

CLASS-E AMPLIFIER

The basic topology of class-E amplifier is shown in Fig.2. The circuit includes a transistor operated as a switch, a shunt capacitor C_1 which includes intrinsic transistor output capacitance (C_{ds}), RF choke L_1, a series tuned output circuit $L_0 C_0$ and the load resistor, R.

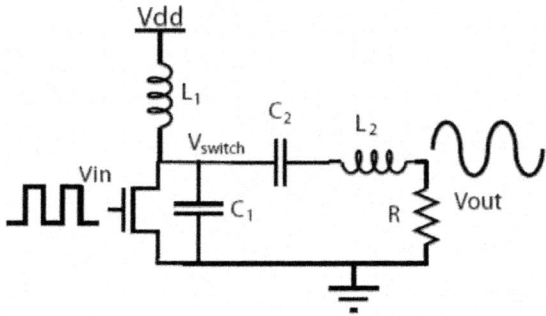

Figure 2: Class-E Amplifier.

The class-E amplifier is a switch based power amplifier that achieves high efficiency at constant power output [13-17]. Theoretically it can achieve 100% efficiency. High efficiency can be obtained by keeping the voltage-current overlap across the transistor small. This is done using a technique called zero voltage switching (ZVS). This technique ensures that the voltage across the switch is near zero when the device turns on. This is achieved through a matching network consisting of a shunt capacitor and a series resonant tank. This matching network also sets the slope of the switch voltage to zero at turn on there by reducing the stress on the device. This is called zero derivative switching (ZDS). At the turn-off transition little power is lost as well. The shunt capacitor keeps the voltage low while the current shuts off. Most of the power loss occurs when the switch is on. The current through the device is large enough that with a small on resistance, the power loss is significant. A typical voltage and current waveform across the switch is shown in Fig-3. The switch voltage reaches zero with zero slope when the device turns on. (Time 77nsec in Figure-2).

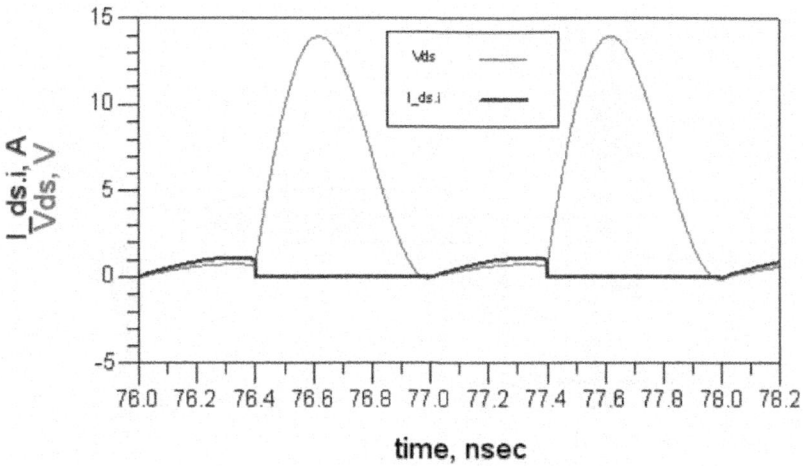

Figure 3: Current/Voltage Waveform across transistor.

The design of class E power amplifier can be done analytically if the RF output is assumed sinusoidal (i.e. infinite loaded quality factor, Q_L). When the loaded Q is not finite the solution cannot be derived analytically. This causes error in output power calculation [13]. This error gets larger as the loaded Q becomes smaller. To account for the output power error from the loaded Q, design equations that take into account loaded Q are used in this design [13].

$$P = \frac{Vcc^2}{R} 0.576801 \left(1.0000086 - \frac{0.014395}{QL} - \frac{0.577501}{Q^2L} + \frac{0.205967}{Q^3L} \right)$$

(5)

$$C_1 = \frac{1}{2\pi fR \left(\frac{\pi^2}{4} + 1 \right) \frac{\pi}{2}} \left(0.99866 + \frac{0.91424}{Q_L} - \frac{1.03175}{Q^2_L} \right) + \frac{0.6}{(2\pi f)^2 L_1}$$

(6)

$$C_2 = \frac{1}{2\pi fR} \left(\frac{1}{Q_L - 0.104823} \right) \left(1.00121 + \frac{1.01468}{Q_L - 1.7879} \right) - \frac{0.2}{(2\pi f)^2 L_1}$$

(7)

$$L_2 = \frac{Q_L R}{2\pi f}$$

(8)

The series resonant circuit $L_0 C_0$, is usually not tuned to the operating frequency, f, having at this frequency a net series reactance $X_2 = \pi f L\Delta$ (Fig.-4) produced by the difference in the reactances of the inductor and capacitor of the series tuned circuit given by Eq. (9).

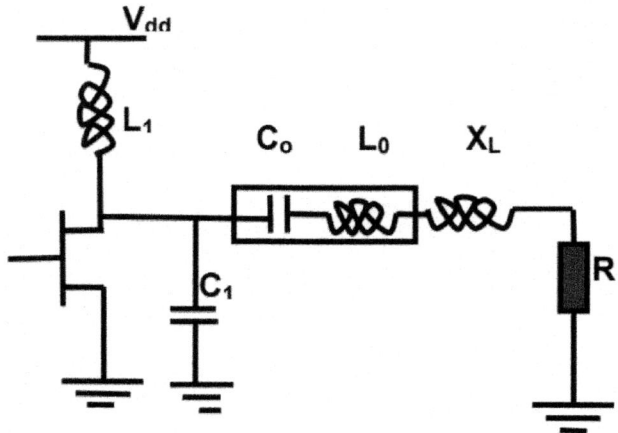

Figure 4: Class-E amplifier with residual load impedance.

$$X_L = 2\pi f L_0 - \frac{1}{2\pi f C_0}$$

(9)

In this analysis $L_0 C_0$ is considered as a resonant circuit tuned to the operating frequency f, in series with a net inductance XL. This design uses Motorola (MRF9745T1) high frequency LDMOS FET. This model has an output capacitance of 11pF (C_1), drain source resistance (R_{ds}) of 1Ω, switch breakdown voltage of 35V and supply voltage of 5.8V. Selecting frequency (f) as 1GHz and Q to be 10 the resonant tank components C_0, and L_0 can be determined as given in [3] (Table 1). R is taken as 3.5Ω to get an output power of 2.2W. RF choke (RFC) L_1 is selected to be 20nH, which keeps the ac ripple less than 5%. Simulations are done using Agilent® ADS.

Table I: Class-E Component Values

Component Name	Value
L_0	6.65nH
X_L	0.826nH
C_1	11pF
C_0	3.81pF
R	3.5Ω
L_1	20nH

OUTPHASING POWER COMBINING SCHEMES

In order to exploit the inherent efficiency benefit of the ouphasing system, a suitable power combining scheme needs to be analysed. A LINC power combiner using class-E amplifier is shown in Fig 5.

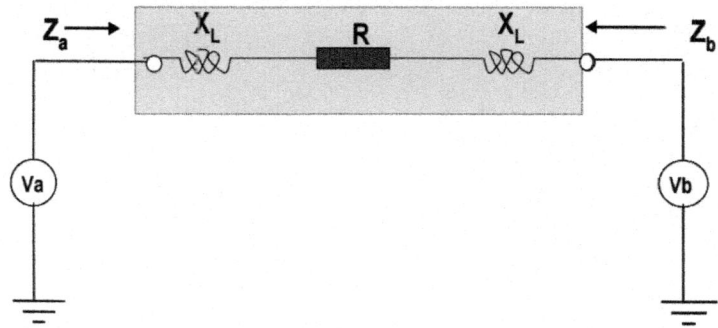

Figure 5: LINC power combiner using class-E.

Here the class-E amplifiers are connected differentially to a common-mode load R. The residual load reactance X_L of each amplifier is considered as part of the power combiner network. This combiner network is linear. Za and Zb correspond to the effective load impedance seen by each amplifier while Va and Vb correspond to the output voltage of each amplifier for a given input drive phase (Ø). Both amplifiers are driven by square pulses having the same amplitude and frequency, but their phase is varied. The phase of Va and Vb is changed by varying the input drive phase (Ø). When Va and Vb are in phase the amplifiers see infinite impedance and when they are 180° out of phase they experience an effective load impedance of $R+jX_L$ resulting in class-E operation of both amplifiers. There exists a unique relationship between $[Z_a, V_a]$ and $[Z_b, V_b]$ in terms of the power combiner circuit elements R and X_L. This is obtained by solving the combiner network current equations and the solution is given in Eq. (10) and (11).

$$V_b = \frac{[Z_a - R - j2X_L]V_a}{Z_a} \tag{10}$$

$$Z_b = -[Z_a - R - j2X_L] \tag{11}$$

As Za is a function of the input drive phase (Ø), it can be seen that as Ø changes the effective load impedance seen by each amplifier varies. This determines the efficiency performance of each amplifier in the LINC structure.

CLASS-E LINC LOAD PULL ANALYSIS

To get the operational load locus of the LINC combiner a load pull analysis of a class-E amplifier is performed to obtain a plot of V_a and Z_a $(R+jX_L)$ Fig. 6.

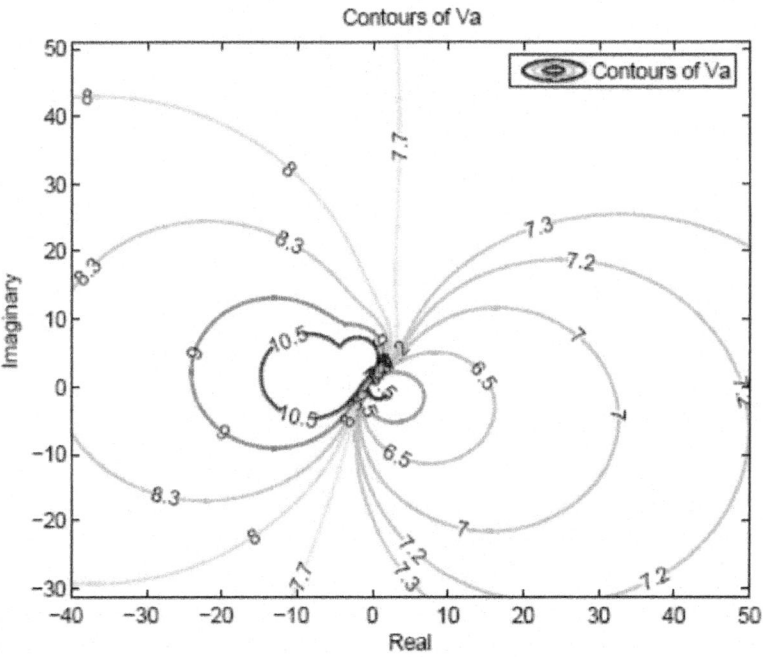

Figure 6: Load pull plot of single amplifier. Contours of Va vs. output load impedance (complex).

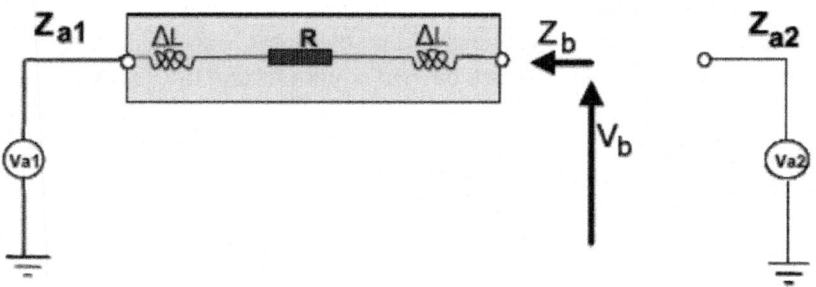

Figure 7: LINC Power Combiner Analysis.

From Fig. 7 it can be noticed that for each value of V_{a1} and Z_{a1} there exists a unique value for V_b and Z_b seen from the other side of the combiner network, which is the load pull characteristic of amplifier B [Vb, Z_b]. [V_b, Z_b] must be a valid combination of second amplifier [V_{a2}], which implies [V_b, Z_b] is only valid when it equals [V_{a2}, Z_{a2}]. Equilibrium is reached only when load pull seen at amplifier B equals load pull at amplifier 2. Superimposing both plots shows the impedance where [V_b, Z_b] equals [V_{a2}, Z_{a2}], from which we obtain the operational load locus (Fig. 8) of the class-E amplifiers in the LINC combiner structure. This load locus is dependent on the operational frequency and changes with frequency of operation.

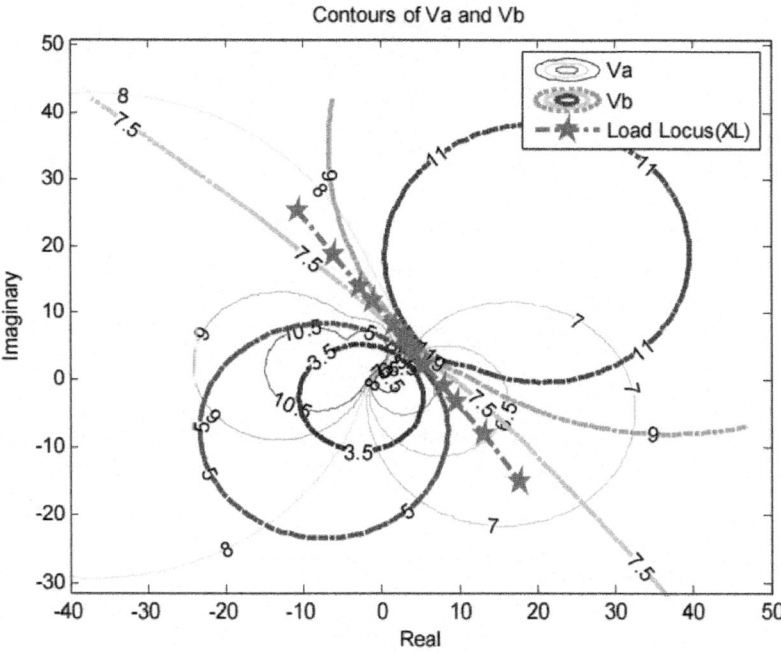

Figure 8: Class-E LINC load locus. Both amplifiers must operate on this line. The phase of the input drive signal defines the actual positions.

In Fig. 9 two new load lines are obtained by taking alternate values of X_L in the combining network. These are superimposed on the output power contour of a single class-E amplifier. A_1 and B_1 shows the operating points of both amplifiers in the LINC structure for a specific input drive phase. When one amplifier is operating at A_1 the amplifier at the other side of LINC will be at position B_1. So the total output power is obtained by adding the individual powers of each amplifier. Sometimes an amplifier sinks power depending on the drive phase.

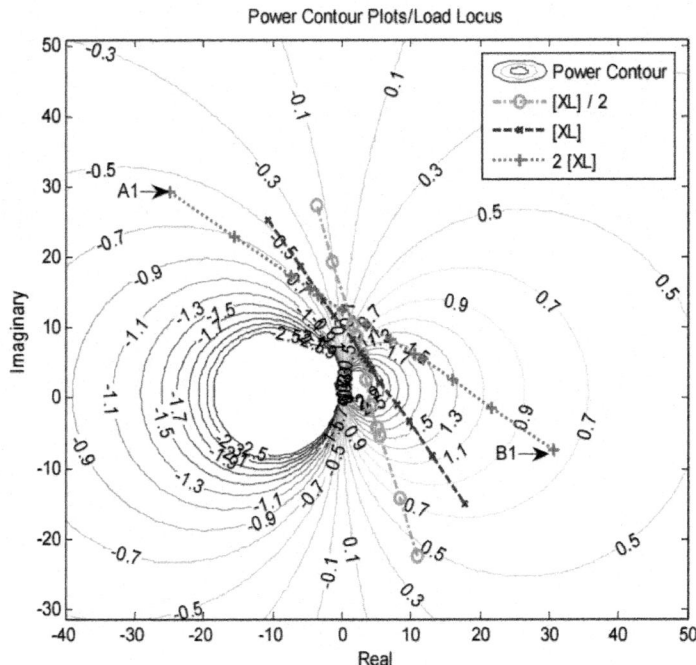

Figure 9: Amplifier output power contours as a function of load impedance. Output power is modulated by moving the amplifier operating point up and down the load locus.

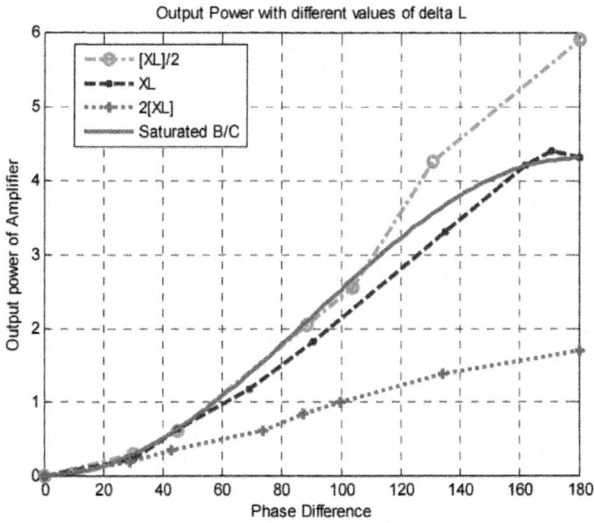

Figure 10: Output power of class-E LINC.

Fig. 10 shows the plot of output power of Class-E LINC for different drive phases and XL values. Reducing XL increases output power. Output power is approximately linear with drive phase. Fig. 8 also shows the traditional cosine shape for an ideal saturated class B/C voltage source LINC system.

Fig. 11 shows the power loss across the class-E LINC for different X_L values and drive phases. Losses in class-E amplifier are a combination of conduction and switching losses. In the optimal operation of class-E amplifier the only loss mechanism is conduction loss, which occurs in the LINC combiner at an input drive phase of 180° (blue curve) in Fig. 11. For all other drive phases the loss in each amplifier is a combination of both losses and clearly dominated by switching losses as the amplifier moves away from zero-voltage switching (ZVS). Note the higher losses in $X_L/2$ curve.

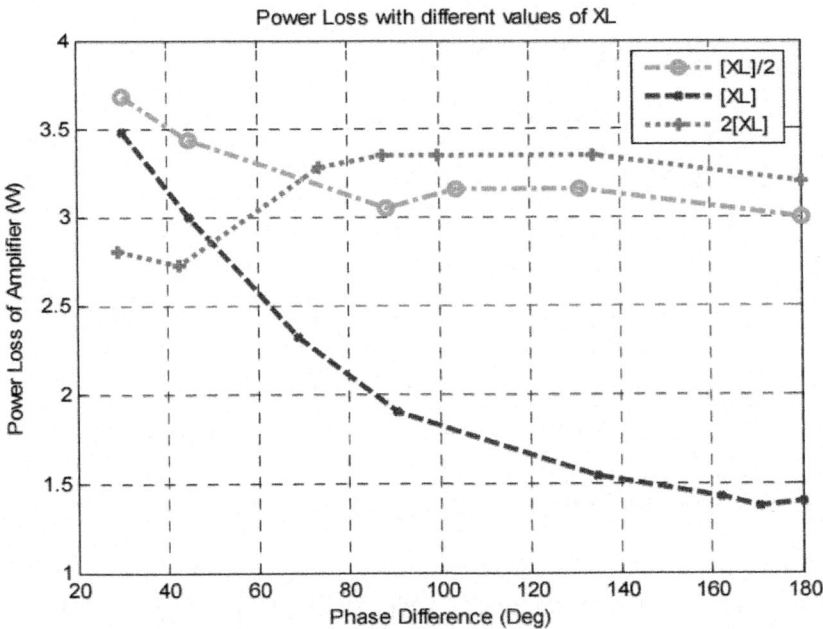

Figure 11: Total losses[conduction + switching] of class-E LINC.

In Fig. 12 output power is plotted against efficiency. Peak efficiency of 81% is obtained for the LINC combiner using $X_L/2$ but the power is restricted to 1.8W. The plot gives a good indication of the performance of the Class -E LINC structure with various values of residual inductance X_L. A suitable XL could be selected based on the power probability distribution function of the modulation. Ideally, a structure with high efficiency and high output power is preferred, which indicates X_L is the best choice.

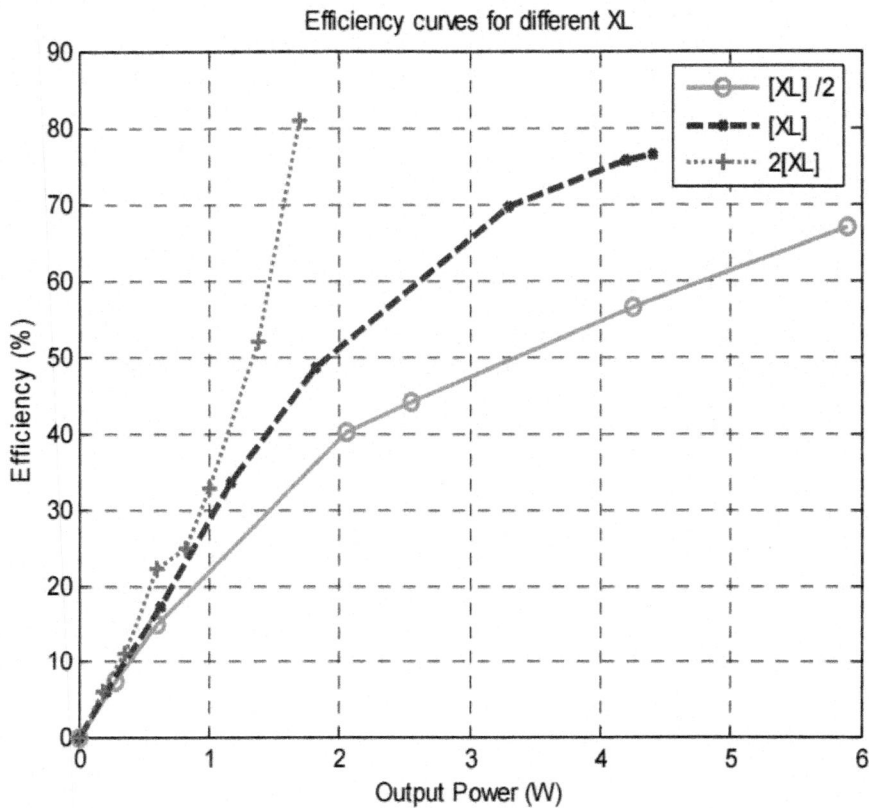

Figure 12: Efficiency vs. output power of class-E LINC for different XL values.

PULSE WIDTH MODULATED CLASS-E POWER AMPLI-FIER

Class-E amplifier achieves highest efficiency and constant output power when operated with a square wave having 50% duty cycle. By applying a pulse width modulated (PWM) input the output power can be varied. Simulating the design in ADS for different duty cycles results in the following waveforms, Fig. 13-14. Output power and efficiency at various duty cycles is plotted in Fig. 15.

Output power is low for duty cycles less than 20% and more than 80%. Efficiency is high for duty cycles around 50%. At low duty cycles the RF choke has less time to store energy and at high duty cycles the stored energy is released not at the fundamental frequency. Tweaking the components C_1, C_2, and L_2 the power and efficiency curves can be shifted towards lower or higher duty cycles.

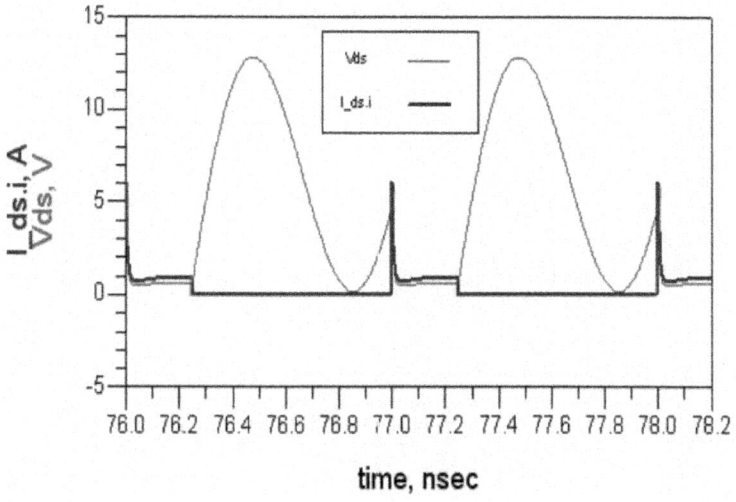

Figure 13: Switch waveform for 25% Duty Cycle.

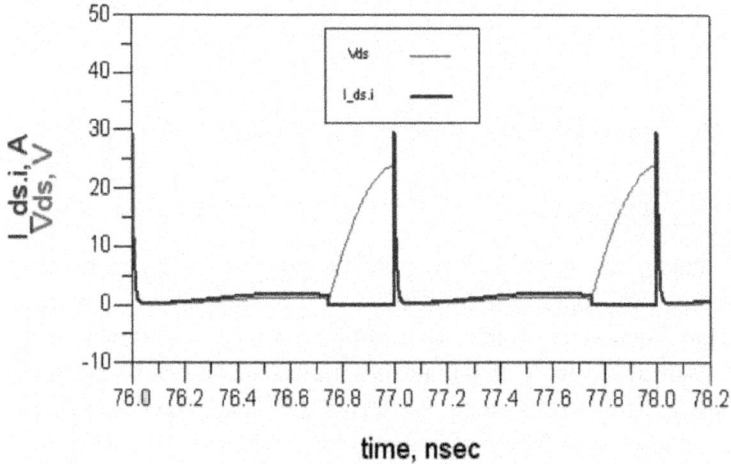

Figure 14: Switch waveform for 75% Duty Cycle.

The class E power amplifier is designed for maximum efficiency when driven by a square wave with a 50% duty cycle. When the power amplifier is driven by a pulse width modulated square wave, the efficiency is expected to drop since ZVS is violated. Since the figure of merit is system efficiency, tweaking the system may result in better overall efficiency. Thus, tuning the matching network is explored in [19]. Perturbations are done on three components of the matching network – the shunt capacitor (C_1), resonant capacitor (C_2) and resonant inductor (L_2).

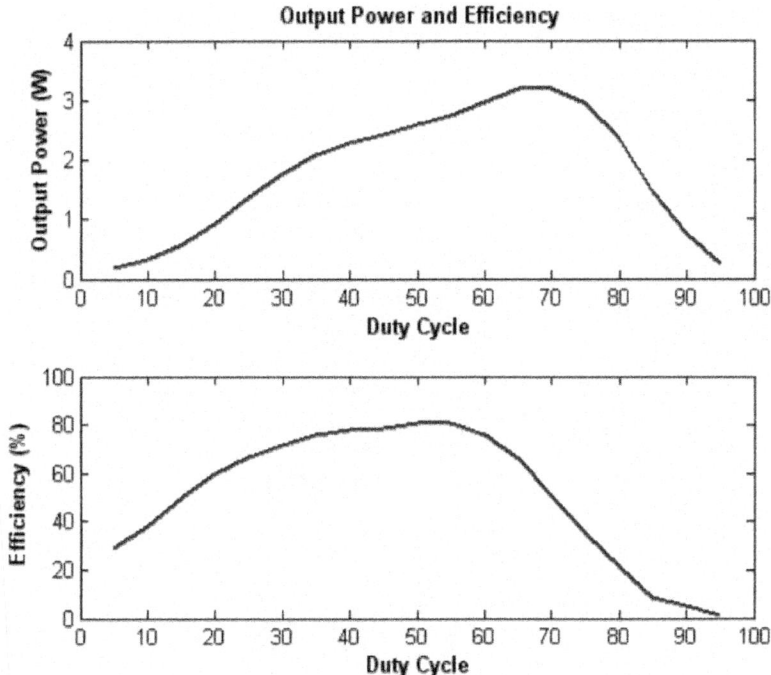

Figure 15: Output Power and Efficiency Variation with Duty Cycle.

CONCLUSIONS

In this paper we discussed a suitable scheme to characterize the LINC performance of class-E amplifier. Class-E amplifier is neither an ideal voltage nor a current source, so traditional LINC analysis does not apply to class-E amplifiers. Instead a load pull analysis is proposed to get its transfer characteristics and its efficiency performance. The Class-E amplifier is capable of achieving an efficiency of around 81% when operated in an outphasing scheme. In this mode of operation switching losses dominate and account for more than 60% of the total losses. This load-pull analysis technique can be utilized to characterize any amplifier class in an outphasing power combining scheme. Further pulse width modulated performance of class-E amplifier is also discussed. The amplifier achieves an efficiency of 80% when operated in PWM mode.

REFERENCES

1.   H. Yang, "A road to future broadband wireless access: MIMO-OFDM based air interface," IEEE comm. Mag., vol 43, pp 53-60, Jan. 2005.

2. A. Diet, C. Berland, M. Villegas and G. Baudoin, "EER architecture specifications for OFDM Transmitter using a class E amplifier", IEEE Microwave and Wireless Components Letters, vol. 14 no. 8, pp. 389-391, August 2004.

3. C. Berland, I. Hibon, J.F. Bercher, M. Villegas, D. Belot, D. Pache and V. le Goascoz, "A Transmitter Architecture for Nonconstant Envelope Modulation", IEEE Transactions on Circuits and Systems-II: Express Briefs, Vol. 53 no. 1, pp. 13- 17, January 2006.

4. B.Berglund, J.Johansson, T.Lejon, Ericsson Radio Review,Available: www.ericsson.com/ericsson/corpinfo/publications/r eview/2006 03/01. shtml

5. F.H. Raab, P. Asbeck, S.Cripps, P.B. Kennington, Z. B. Popovic, N. Pothecary. J. F. Sevic and N.O. Sokal, " Power Amplifiers and transmitters for RF and microwave," IEEE Trans. Microw. Theory Tech., vol.50, no.3, pp. 326-334, Mar.2002.

6. P.B. Kennington, "High Linearity RF Amplifier Design," Artech House, 2000.

7. M. Faulkner, M. Mattsson, W. Yates. Adaptive Linearisation using Pre-Distortion. VTC 1991.

8. K. Tom, M. Fauklner, T. Lejon, "Performance analysis of Pulse Width Modulated RF Class-E Power Amplifier," IEEE-VTC-S, pp 56-60, May 2006.

9. H. Chireix, "High Power Outphasing Modulation," Proc. IRE, vol. 23, pp. 1370-1392, Nov.1935.

10. D.C. Cox, "Linear amplification with nonlinear components," IEEE Trans. Commun., vol. COM- 23, pp. 1942-1945, Dec. 1974.

11. Ilkka Hakala, Leila Gharavi, Risto Kaunisto, "Chireix Power Combining with Saturated Class-B Power Amplifiers," 12th GAAS® Symposium, Amsterdam August 2004.

12. Ilkka Hakala, David K. Choi, "A 2.14GHz Chireix Outphasing Transmitter," IEEE transactions on Microwave Theory and Techniques vol. 53, No. 6 June 2005.

13. N. Sokal, "Class-E High Efficiency Power Amplifiers, from HF to Microwave," Microwave Symposium Digest June 1998.

14. M. Kazimierczuk and K. Puczko, "Exact Analysis of Class-E Amplifier at any Q and switch duty cycle," IEEE transactions on Circuits and Systems, February 1987.

15. F. H. Raab, "Efficiency of outphasing RF poweramplifier systems," IEEE Trans. Commun. Vol. COM-33, pp. 1094-1099, Oct. 1985.

16. F. H. Raab, "Idealized Operation of the Class E Tuned Power Amplifier," IEEE transactions on Circuits and Systems, vol. 24, No. 12, Dec 1977.

17. F. H. Raab, "Effects of Circuit Variations on Class E Tuned Power Amplifier," IEEE journal JSSC , vol. Sc-13, No. 2, April 1978.

18. "MRF9745T1 datasheet," Motorola, Available: www.chipdocs.com/pndecoder/datasheets/MOT/M RF9745T1.html

19. K. Tom, M. Faulkner, "Performance analysis of RF class-E power amplifier with varying duty cycle", AtCRC Conference, pp 26-30, Nov-2005.

CITATION

CHAPTER 1

Anamarija Juhas and Ladislav A. Novak, "Conflict Set and Waveform Modelling for Power Amplifier Design,"Mathematical Problems in Engineering, vol. 2015, Article ID 585962, 29 pages, 2015. doi:10.1155/2015/585962.

CHAPTER 2

Anthony N. Laskovski and Mehmet R. Yuce (2011). Power Amplifiers for Electronic Bio-Implants, Biomedical Engineering, Trends in Electronics, Communications and Software, Mr Anthony Laskovski (Ed.), ISBN: 978-953-307-475-7, InTech, DOI: 10.5772/13557.

CHAPTER 3

Inderpreet Kaur and Neena Gupta (2012). Hybrid Fiber Amplifier, Optical Communications Systems, Dr. Narottam Das (Ed.), ISBN: 978-953-51-0170-3, InTech, DOI: 10.5772/33240.

CHAPTER 4

Paolo Colantonio, Franco Giannini, Rocco Giofre and Luca Piazzon (2010). The Doherty Power Amplifier, Advanced Microwave Circuits and Systems, Vitaliy Zhurbenko (Ed.), ISBN: 978-953-307-087-2, InTech, DOI: 10.5772/8431.

CHAPTER 5

Essra E. Al-Bayati, R. S. Fyath, Design and Performance Investigation of a New Distributed Amplifier Architecture for 40 and 100 Gb/s Optical Receivers, ISSN 2277-3061.

CHAPTER 6

Roman SOTNER, Jan JERABEK, Norbert HERENCSAR, Voltage Differencing Buffered/Inverted Amplifiers and Their Applications for Signal Generation, http://www.radioeng.cz/fulltexts/2013/13_02_0490_0504.pdf.

CHAPTER 7

Neeta Pandey and Sajal K. Paul, "Differential Difference Current Conveyor Transconductance Amplifier: A New Analog Building Block for Signal Processing," Journal of Electrical and Computer Engineering, vol. 2011, Article ID 361384, 10 pages, 2011. doi:10.1155/2011/361384.

CHAPTER 8

Kevin Tom, Vandana Basoo, Mike Faulkner, Thomas Lejon, Power Amplifiers for Next Generation Wireless Platforms, ISSN: 1449-2679.

INDEX